Lisa Lee

About the Author

Michael Largo is the author of *Final Exits: The Illustrated Encyclopedia of How We Die*; *Portable Obituary: How the Famous, Rich, and Powerful Really Died*; *Genius and Heroin: The Illustrated Catalog of Creativity, Obsession, and Reckless Abandon through the Ages*; and three novels. He currently lives in the United States.

God's Lunatics

Also by Michael Largo

Nonfiction

Genius and Heroin

The Portable Obituary

Final Exits

Fiction

Southern Comfort

Lies Within

Welcome to Miami

God's Lunatics

Lost Souls, False Prophets, Martyred Saints, Murderous Cults, Demonic Nuns, and Other Victims of Man's Eternal Search for the Divine

Michael Largo

HARPER

NEW YORK • LONDON • TORONTO • SYDNEY

HARPER

 For information address HarperCollins Publishers, 10 East 53rd Street, New York, NY 10022.

HarperCollins books may be purchased for educational, business, or sales promotional use. For information please write: Special Markets Department, HarperCollins Publishers, 10 East 53rd Street, New York, NY 10022.

FIRST EDITION

Designed by Justin Dodd

Library of Congress Cataloging-in-Publication Data is available upon request.

ISBN 978-0-06-173284-3

10 11 12 13 14 OV/RRD 10 9 8 7 6 5 4 3 2 1

For
Lynn L. Riggle

When you knock, ask to see God—none of the servants.

—Henry David Thoreau, from *Letters to Various Persons* (1879)

CONTENTS

INTRODUCTION / xiii

GOD'S LUNATICS A TO Z / 1

SOURCES / 537

ACKNOWLEDGMENTS / 557

INTRODUCTION

As It Was in the Beginning . . .

What happens to us when we die? Is this all there is to life? What's the point of it all? The mystery of death—and the shadow mortality casts over life—induced humanity to invent religion for itself. Shocked by the unacceptably finite nature of life, we needed a story—and ultimately a theology—to make sense of it. Can it be possible that we are born, live for an indiscriminate amount of time, and then die for no reason? Religion has been our way of explaining life's unanswerable mysteries. Scientists agree that 100,000-year-old human remains discovered in a cave near Nazareth, Israel, were buried ritualistically, with bones painted red, magic seashells placed about, and skulls positioned to face north. This indicates that our remotest ancestors had a belief, at least, in the afterlife. It marked a point in ancient history when humans began to invent a god of their imagining to explicate their fate. The world seemed as scary then as it does now without believing it had some sort of direction from a divine source. From these earliest times, humankind has dreamed up explanatory narratives, symbolic props and rituals, and complex theological frameworks to lend life objective meaning.

God's Lunatics examines the stories and beliefs we have told ourselves about God through the ages. It chronicles the lives of celebrated mystics, martyrs, wizards, shamans, participants in famous demonic possessions, cult leaders, the founders of working experiments in utopias, and originators of New Age movements who dedicated their lives to solving the secrets of the universe. It also shows how insane much of this behavior has been.

Each generation has readily given audience to those who attested to be prophets, offering us their guidance by proclaiming a divine connection. Often the messages of these celestial interpreters consumed the lives and minds of vast segments of the population, even if the roads they commanded as a means to find certain truths were widely varied, ranging from the ridiculous or horrific, to the sublime. Whether it was through self-sacrifice (or the sacrificial offerings of others), by praying, dancing, starvation, turning oneself into stone, or staring into the sun, we have certainly not lacked imagination in our quest to understand the Divine.

Today, if a person claims to *actually* hear the voice of an angel, they are often diagnosed with a mental disorder—and quickly medicated. But it was not that way in previous centuries when such pronouncements could very well gain a following. A thousand years ago, such persons were likely hailed as prophets or saints. In the olden days, there were big-budget miracles like parting seas, bread raining from heaven, and resurrecting the dead. Now, we look for godly signs in grilled sandwiches, and such, though we are no less eager to seek tangible proof that we are not alone.

Historians cite religion as an important factor in the formation of societies. It helped civilizations to cohere by displacing brute force with a divine club, which convinced followers to abide by laws based on their own invented divinity's wishes. Many of these ancient rites still permeate our culture, effecting laws and governments, regardless of their origins. Today, there are more than three hundred thousand places of worship in the United States alone; people still want to know the answers to the big questions: Where did we come from? How should we live? Where do we go when we die? It is comforting to believe in something greater.

Nevertheless, religions base their entire foundations on ideas and concepts that cannot be proven—and because of this—anything can be offered as truth. The craziest notions became dogma, demanding "faith" to accept it. The tendency of religions to foster delusional and irrational behavior, in fact, has caused mass fatalities. How many people died because of religion? I estimate that several hundred million perished due to it. (And that's not including the roughly three million personally wiped out by God during the time of Noah and the Great Flood.)

By simply counting deaths due to religious warfare and backward religious doctrines—which have not infrequently prevented advancements to stave suffering and mortality—religion has a questionable record of accomplishment. Conversely, how many has religion helped and comforted? I could not guess, even if theologians argue that faith, in theory, has saved billions of souls.

I have been researching death for many years. While tracking the endless tide of human lives that have come and gone—usually leaving no greater mark of their existence than a mere digit added to the tally—I admit there were times I wondered, *What purpose does this all have?* I was educated by Jesuits, though since the 1970s—especially after reading the work of anthropologists like Michael J. Harner—I examined the explosion of new spiritual ideas that seemed to flourish during that era. I was particularly interested in beliefs regarding death and the afterlife. I have included in the sources section of this book a sampling of the comparative religion volumes and holy texts I collected.

My field experience began as a Catholic altar boy who memorized Latin. I then practiced a number of Eastern theologies, from Zen to transcendental meditation (TM). I went on Christian spiritual weekend retreats, as well as Eastern ones; attended New Age lectures, including talks on theories about angelic time travelers to seminars on the esoteric value of crystals. I had my palms read, did a past-life regression, attended a Santeria session in Miami, and visited many holy shrines in Rome. Also, I attended services held by Episcopalians, Methodists, Baptists, and Calvinists. I went to Pentecostal meetings, as well as sat among the crowds of a Mega Church. In addition, I ate matzo ball soup with Jewish friends in their *sukkah* in Brooklyn, and once did mushrooms with some women who were into Druids and Earth religions in upstate New York. In the late 1970s, I sat in on a Rasta ceremony in Negril, Jamaica, and partook of "lamb's bread," and met a "medicine man" in the mountains of Baja, California, a hippie-type Mexican reputed for his "mind-expanding tea." More recent, I went to Port-au-Prince, Haiti, and heard Voodoo drums, though due to security concerns I could not get close enough to witness the ceremony.

The above was not as Indiana-Jonesing as it might seem, since it took place over more than twenty-five years. However, during the last year, I hunkered down and committed to the ascetic route, working among ancient texts, reading doctrines and dogmas put forth by the religions themselves, immersed in what the early monks called "*lectio divina*" or "divine reading." Instead of arguing for or against god, I thought it better to look at the hierographies or the lives of holy iconic figures as cited by a religion's own texts, and I summarized high and low points, the way one might read a consumer report before purchasing a product. I approached no religious ideology with preconditioned condemnation, or near-sighted praise, which permeates the boundless literature on the subject. I remained awestruck by those who strove to pierce the "doors of perception," hunting for ecstasy, redemption, salvation, or enlightenment, and by the utterly odd details of history's large cast of fanatical saints and prophets. I wanted to compile a reference for everyone, regardless of religious persuasion, or lack thereof: It is better to check out the details of a religion, including the one you may practice, or the one you might join, before "buying into it." I also tried not to disrespect any faith (though do unapologetically call "cults," cults).

On a recent trip to New York, I asked a Moslem cabdriver why he faces Mecca to pray, and he knew nothing about the stone from out of space located there. I interviewed a sidewalk preacher in San Francisco, as he was perched on a crate, giving a portable speaker-blaring sermon about an impending apocalypse, if he knew when the Bible was first written. He did not. I asked a woman, always dressed in black (who famously attended the seven o'clock mass every day for uninterrupted twenty years at St. Teresa's Church, Staten Island), if she knew how many times a day she was allowed to receive the Eucharist. She was unable to recall the specific Church rule or its relationship to sundown. I quizzed my dog's vet, at the clinic with a waiting room filled with Evangelical literature, if she thought my pet might someday go to heaven. Not figuratively, but literally: she only stared back at me. After reading *God's Lunatics*, you'll know these answers. It is also my hope that we understand that all of our religions have in them many things that are out of kilter.

To argue which religion is right, as you will find out from this book, is futile, counterproductive, and even lethal. Society still swallows religious dogma hook, line, pew, and steeple, while condemning, committing violence, and even killing in its name. It remains noble to be open-minded and seek a sense of sacred or a higher purpose in life, if nothing else than for the betterment of humanity. Yet religion has always been a beacon for fanatical people; this is certain. Many continue to accept soul-saving fantasies and sheer ludicrousness to become another one of God's lunatics. It is not unreasonable to imagine our own "end of days," or at least a significant transformation of our culture, instigated by some form of religious insanity.

A

ABRACADABRA
ABRACADABR
ABRACADAB
ABRACADA
ABRACAD
ABRACA
ABRAC
ABRA
ABR
AB
A

ABRACADABRA

MYSTICAL CHANT

The most famous magical incantation, "Abracadabra," was originally uttered with serious intent by Roman physicians, and its mysterious origins perhaps go back as far as ancient Egypt. The first documented reference was made by a second-century Roman doctor named Sammonicus Serenus, who was the personal health adviser of Emperor Caracalla, a despot remembered for a short and savage temper. Sammonicus, after study in Egypt, bravely recommended an amulet to heal the emperor, with the letters spelling *Abracadabra* arranged to form an inverted pyramid. It was to be worn around the neck to protect from sickness, especially military fever. The doctor wrote the exact prescription as a lyrical poem, as was traditional. A scroll attributed to the doctor was later discovered by medieval scribes who once again put the phrase's mystical connotation back to use. Linguists believe it's a corruption of the Aramaic phrase *avda kedavra*, meaning loosely, "So I say, it shall be done."

Even if its overuse by stage magicians, especially during the vaudeville era, has in modern times reduced the word to cartoon gibberish, others think its reversed-pyramid arrangement leads it to an earlier and a more scholarly genesis: the letters on the amulet were actually encrypted, and only read correctly by knowing healers, dating to Egyptian priest-physicians. This esteemed and selective group was known for deceptively

Military fever was any flu-like disease, though likely malaria. Roman martial strategists feared it more than they did foreign enemies. Romans believed gods, spirits, and the alignment of stars were the primary cause of illness. For diagnosis, a sick patient's exhaled breath was examined by placing a mirror under the nostrils. The infectious cloud was placed against the grid of the abracadabra amulet to determine the treatment for a cure.

concealing their secrets with codes that they believed had been revealed to them by the gods.

In 212 A.D., Dr. Sammonicus accepted an invitation to a banquet he thought was being held in his honor, and fell victim to Emperor Caracalla's revengeful scheme. Apparently dissatisfied with the enchanted incantation, the ruler had the old physician murdered upon arrival, his chest pierced by multiple swords quicker than the good doctor could say "Shazam." (Shazam, a twentieth-century invention of Fawcett Comics, was a superhero whose name was an acronym for Solomon, Hercules, Atlas, Zeus, Achilles, and Mercury.)

HOCUS POCUS

Now a byword for any silly magic spell, hocus pocus was originally the name of a seventeenth-century traveling magician. "The Most Excellent Hocus Pocus" was so good at trickery—some say sorcery—that he became the personal favorite of King James of England. It seemed he had a particular knack for making things disappear, even if it seemed he never learned how to make them come back. The jester-cum-wizard fell out of favor when the king began to take interest in more serious entertainment, namely a new playwright named William Shakespeare and his theater troupe. A book written by Hocus Pocus, Jr., appeared in 1634, but the fate of the senior Pocus is unknown, as records of his existence have simply vanished. Ever since, magicians on the street corner or stage have added a "hocus-pocus" to their mostly ad-lib incantations, hoping to conjure the spirit of the original trickster.

ABSALOM

BIBLICAL EGOTIST

Absalom was King David's third son. Possessing a tabloid idol's heart-throbbing good looks, he prided himself on his long, beautiful hair and preferred the finest in fashions. Absalom was a hothead, though, who aspired to be more than a mere biblical playboy. He made his first serious appearance in scriptures for scheming to murder his half brother Amnon, who was rumored to have had sex with his younger sister, Tamar. For two years Absalom concealed the hatred for his elder brother—and heir apparent to David's throne—until he finally

convinced Amnon to attend one of Absalom's infamously wild all-night sheep-shearing parties. When Amnon was sufficiently drunk he ordered servants to stab his elder brother to death. Four years later, Absalom, now next in line, became impatient to take over the reins and organized a successful revolt. Initially the people of Israel and Judah supported his takeover, with many swayed by his charismatic looks and promises of good times for all, and forced David to flee across the Jordan River. A court servant still loyal to David persuaded Absalom not to pursue his father—which would have been the end of David—but rather to wait for the final blow after Absalom assembled his own, stronger army.

In the waning years of the eleventh century B.C., the father's and son's forces finally clashed at a place called the Wood of Ephraim. Absalom's long locks were his undoing. While in retreat and separated from his soldiers, Absalom was swooped off his galloping steed when his hair became tangled in a low-hanging oak branch. David had ordered his men to treat his young son gently if captured. However, David's senior general, Joab, rushed to the site where Absalom dangled. He grabbed three spears and thrust them into Absalom's heart. David was distraught, despite the betrayal of his son. Years later, on his deathbed, David's last words were directed to the soon-to-be king, Solomon, whom the dying king ordered to murder General Joab (David's nephew) in retaliation for killing his handsome son. (*See also*: Divine Hair)

Pride goeth before destruction and a haughty spirit before a fall.
—Proverbs

The Bible lists seven deadly sins and considers hubris or vanity as the most evil, reasoning that it is the cause of other sins: self-centeredness leads to transgressions of lust, gluttony, greed, wrath, sloth, and envy. Absalom didn't need to go into battle but liked how he looked in armament, saddled upon a horse. At a gallop he knew his blowing mane made quite a spectacle. Disregarding military advice concerning a number of issues in the campaign against his father, the most serious was Absalom's refusal to wear a helmet, or tie his hair into a ponytail.

ADAMITES

CHURCH OF NUDISTS

A second-century North African sect believed salvation was to be found by living as Adam and Eve did before their eviction from the Garden of Eden. Calling their church Paradise, they abolished marriage and advocated "holy nudism." Sin, it was reasoned, was the true catalyst for donning clothes and ultimately for matrimony. Adam and Eve—when they possessed pure, primal innocence—needed neither. Adamites shared property and refused to acknowledge governments or laws, since they asserted none of these existed in the Garden of Eden. Nudity was proof they had achieved a state of grace, and were freed of all human-devised moral constraints. To them "possessing" a mate was the primary cause for centuries of grief, so instead they had sex the way we might now shake hands upon greeting. Followers of these communal "Paradises" that sprung up never lasted long and were eventually condemned as heretics. They were usually disbanded by force and had to continue in strict secrecy.

In the 1400s the Adamites received more negative publicity when a Catholic priest, Peter Kanis, formed a group with about two hundred followers. Again, nudity, free love, and communal property were advocated as a way to return to idealist humanity as it was before the invention of sin. They found reprieve from criticism when they moved to an island on the Nerarka River in Bohemia, now the Czech Republic. However, when they began to have financial problems, some blamed their lack of industry on their excessive sexual activity. Subsequently, the Bohemian Adamites began to make night raids into surrounding towns to pillage and rob, though donning cloaks and masks for that. The island was eventually attacked; Kanis was captured, and along with seventy followers they were all burned at the stake.

The Adamites developed offshoots, such as the Netherlands branch, Brethren and Sisters of the Free Spirit, which counted more than half a million members in Germany during the late 1700s. (*See also*: Beguines) However, by 1849 the last true Adamites were entirely wiped out by coordinated military action throughout Europe.

WHO TOLD THEE THAT THOU WAST NAKED?

A modern revival in "religious nudism" seems imminent. David Blood, the project manager in charge of building a nudist-Christian theme park near Tampa, Florida, recently noted, "The Bible very clearly states that when Adam and Eve were in right with God, they were naked. When people are in right with God, they do not have to fear nudity."

The first American naturist Christian church was founded in the 1920s by Reverend Ilsley Boone, known familiarly as "Uncle Danny." He was the pastor of the New Jersey Church of Ponds, affiliated with the Dutch Reformed Protestants, and is credited with fighting the U.S. Postal Service to allow nude images to be sent through the mail. (Hugh Hefner cited the Reverend Boone ruling as the legal footing to start *Playboy*.) Uncle Danny eventually went secular and helped form the American Association for Nude Recreation, which claimed more than sixty thousand members in 2009. The Nudist Calvary Baptist Church in Texas states on their website: "Every Sunday we give thanks to Him with all of our body and soul. And it includes the sacrifice of our garments, for it was when Eve ate from the Tree of Life and gathered fig leaves that mankind fell into sin."

ALBERT9
DVRER
NORICVS
FACIEBAT
1504

ST. AGATHA

BREASTLESS VIRGIN MARTYR

A model of chastity, St. Agatha is often depicted carrying her own breasts on a silver platter. Born in the mid-second century to a wealthy Sicilian family, she became a Christian and refused numerous offers for marriage. She was apparently so beautiful that one jilted Roman official, Quintian, was certain he could persuade her to change her mind. He ordered Agatha's arrest and brought her before a judge—himself. When she argued that she couldn't marry because it would require having sex with a husband, and that her life was dedicated to God, Quintian ordered her to spend a month in a brothel; Quintian presumably believed exposure to the wild side might loosen her morals. Forced to fight off advances, Agatha was relieved when Quintian reversed course and simply sent her off to prison for physical torture. During the procedures, which were directed to inflict pain on her sexual organs, she was visited by the ghost of St. Peter, who she claimed gave her the "patience to suffer." Having lost his patience, Quintian thought that if she wanted to offer her body up to God, he'd help her, and ordered Agatha's breasts presented to him on a dish.

In a strange tribute, each year on her feast day, Sicilian bakeries make a delicacy, *capezzoli di St. Agatha*, breast-shaped pastries, with a cherry on top.

ST. CATHERINE OF ALEXANDRIA

Daughter of a fourth-century governor in Alexandria, Egypt, Catherine was supposedly even more beautiful than Agatha and likewise refused marriage. When she converted to Christianity in her teen years, Catherine claimed the Holy Mother herself came down and transported her to heaven, where she was married to Jesus. Apparently the honeymoon was short, for she was returned immediately back to earth, where her claims of divine marriage got her sent to the torture chamber, and eventually sentenced to death by wheel. This device was a public execution technique, in which the victim was tied naked to spokes and bludgeoned to death. But when Catherine touched the wheel intended for her demise, it crumbled to sawdust. She was beheaded instead, and earned the name of Catherine of the Wheel. Angels allegedly gathered up her head and body and transported it to Mount Sinai, where it's claimed her relics are kept in a monastery that still operates, and which is named in her honor.

AHAB AND JEZEBEL

THE ORIGINAL TROUBLED MARRIAGE

Ahab ruled as king of Israel for twenty years in the ninth-century B.C. The Bible called him "more evil than all the kings before him." Even though Ahab seemed a shrewd commander, mustering massive armies to protect his people, his marriage to Jezebel, a Phoenician princess and non-Jew, raised concern. His new wife venerated the Phoenician deity Melqart (sometimes called Bael), a figure similar to Hercules in Greek

mythology, and who is depicted as a bearded figure holding both an ax, the symbol for death, and a staff, signifying life. Worshipers of this deity were cultlike in behavior, as noted by ancient writer Heliodorus: "Now they leap spiritedly into the air, now they bend their knees to the ground and revolve on them like persons possessed." Jezebel's sway was considerable: Ahab built lavish temples to her god, and some suspected he abandoned the God of Abraham.

Today the term *Jezebel* signifies a woman who controls her husband and uses sex as her most potent tool. In reality, their marriage was one of convenience, for purposes of alliance, and she simply refused to

disown her own culture. She was sharp-tongued and clever enough to retain power through her sons even after Ahab died.

In the classic novel *Moby-Dick*, Herman Melville chose the name Ahab for his own tyrant—the *Pequod*'s captain, who is fixated on revenge, blinded by all else except for a great white whale that bit off his leg. Historically, King Ahab was credited with building a palace made entirely of elephant ivory, and his kingdom remained relatively prosperous during his reign, despite the moniker as "most evil." To further draw parallels to the biblical king, Melville's whale was prized for its ivory, and Captain Ahab used a prosthetic leg made from whalebone. The biblical names Melville borrowed for his story explored themes of obsession, retribution, and religion; Melville's Ahab died when dragged into the sea by a harpoon's rope wrapped around his leg. Jeze-

After ignoring Jezebel's advice, Ahab had joined a battle disguised as an ordinary foot soldier and was killed by a stray arrow.

bel's fate was equally as brutal: When her end came she put on makeup and fixed her hair, and sat at the palace window provoking her enemies to get her if they dared. The succeeding ruler, Jehu, ordered Jezebel's eunuchs to toss her out the window. The Bible seems to make a special point of detailing her demise: "They threw her down; and her blood spattered on the wall and on the horses, and they trampled her."

Heaven have mercy on us all—Presbyterians and Pagans alike—for we are all somehow dreadfully cracked about the head, and sadly need mending
—Herman Melville

AFTERLIFE, THE

WHAT'S NEXT?

More than 85 percent of Americans believe in an afterlife. Most conceive of it as a pleasant place where a person's soul goes to after physical death. The landscape and details of heaven remain open for debate, and have changed throughout the ages.

- The ancient Egyptians believed the soul was ferried across the Nile River to the Kingdom of the Dead and put to work in a field, among other obstacles, such that the rich always added a few living servants in their tomb to labor for them on the other side. After passing an afterlife quiz, the soul might make it to where the gods resided and live in a more luxurious version of Egypt. (*See also*: Book of the Dead)

- In Greek mythology, heaven was the Elysian Fields, a landscape of waving grasses beside a stream, and reserved for the virtuous and heroic.
- Valhalla was the Viking heaven, which required heroic deeds and a warrior's death for entry, and was imagined as a great beer hall.
- Zoroastrianism, originating around 1,000 B.C., viewed the soul as put through a long and horrific purification process, like a spiritual blender, lasting a terribly long time until the world's end.
- In Judaism, the Talmud describes how souls are judged, and if not completely incapable of rehab, are allowed to work out defects and eventually enter "the World to Come."
- Christians count on what Jesus said—that good souls become like angels in heaven.
- Islamic religions believe the soul goes to Paradise, or Jannah, though it cannot get there until final judgment day, and so remains waiting at varying levels of comfort or torment depending on the person's deeds. For Muslims as well as Christians, heaven will be a place where all earthly needs, wants, and wishes are fulfilled. (*See also*: Soul's Weight)

HOW MANY PEOPLE ARE IN HEAVEN?

According to the Bible, not many people will get to heaven. And the rules and laws of most other religions contain so many loopholes and specific "purification" requirements that make damnation more likely. A number of only 144,000—as mentioned twice in the book of Revelation—will ultimately reside in heaven, and that is considered by biblical literalists as the maximum occupancy of eternal paradise. (*See also:* Jehovah's Witnesses) Officially, the Roman Catholic Church has only pinned down about ten thousand individuals who are in the presence of God, having been let into heaven—those are the canonized saints.

CAN MY PET GET IN?

As far as your dead pets and their chances of an afterlife, the Bible is unclear, though it seems they simply turn to dust. In Ecclesiastes, animals are "from the dust, and all turn to dust again." However, 47 percent of people surveyed believe they will be reunited with their favorite animals in the afterlife. Even a slipper-shredding dog will wag its tail past the Pearly Gates, since animals never "fall," as mankind has done, and so they require no redemption. In theory they are automatically saved.

A SAINTLY DOG

For more than six hundred years, a greyhound dog by the name of Guinefort has been venerated in France as a saint. St. Guinefort is prayed to for protection from crib death. The dog had guarded his master's newborn from getting bitten by a venomous snake. However, when Guinefort's owner first saw blood on the infant, he thought the dog had done it, and so killed the greyhound instantly. When he discovered his error, he buried Guinefort in a barren field, which by the next spring had a grove of fruit trees in full bloom. Although Guinefort has been denied official canonization because he is not a human, many believed this miracle, and others attributed to the canine, indicate that there is at least one dog in heaven.

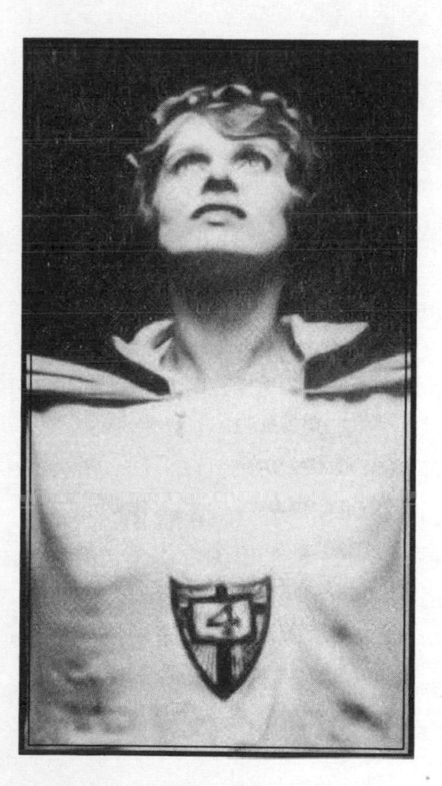

SISTER AIMEE

FIRST MULTIMEDIA PREACHER

Young Aimee Kennedy played a make-believe version of Salvation Army as a girl, preaching to congregations she made from her doll collection. Her mother was a Salvationist, and by the age of eighteen Aimee had met a preacher at a religious tent revival. They soon married and traveled to China to spread the gospel. When her first husband died of dysentery, she returned and married an accountant, Harold McPherson. After giving birth to a child, Aimee suffered what we now call postpartum depression. Bedridden for a number of years, she asked God's help and made a promise to spread the word if healed.

Once on her feet, she left home in 1915 and headed up and down the East Coast in a "Full Gospel Car," an automobile painted with religious slogans, and preaching from a megaphone. She began to muster rousing tent revival meetings of her own, at which she frequently practiced speaking in tongues and faith healing. When her Pentecostal road trip reached San Diego, Aimee McPherson drew a crowd of more than thirty thousand that required the National Guard to retain order. In 1923, then known as "Sister Aimee," Los Angeles became her permanent base of operation, where she built a 5,200-seat church complex-cum-theater, dubbed the "Angelus Temple."

Sister Aimee became one of the first to use radio and newsreels to reach even wider audiences. Her charisma was such that during the 1920s, when women were often marginalized, McPherson became one of the most influential national personalities, giving her opinions on social as well as religious issues. For example, she was adamant that public schools should teach evolution, stating, "[Darwin's theory] is the greatest triumph of satanic intelligence in 5,931 years of devilish warfare, against the Hosts of Heaven." Her campaign on the issue ultimately influenced the outcome of the infamous Scopes Monkey Trial, which banned evolution from being taught in some schools. She broke affiliation with other Christian groups and started her own denomination, the International Church of the Foursquare Gospel. She discovered that it was possible to attract even more converts when her Pentecostal style added techniques and symbols from New Age mystical groups flourishing then in Southern California.

However, in 1926 Aimee went missing, with newspapers declaring she had drowned. When she reappeared in Mexico five days later, insisting she had been kidnapped, her credibility was questioned. During a trial, when charged with filing a false police report of her abduction, it was revealed Aimee had apparently gone on a long weekend tryst with a married engineer who worked at her Angelus Temple radio station. Pressed about her kidnapping testimony she said, "It's my story and I'm sticking to it." The episode led to her downfall as a religious icon, and the media painted Sister Aimee to be a spiritual huckster. Nevertheless, she kept the church afloat, and actually amassed a fortune, even if she suffered occasional nervous breakdowns and later died of a barbiturate overdose in 1944 in a hotel room while on tour preaching her popular "Story of My Life" sermon. Her son, who found her dead, took control of the International Church of the Foursquare Gospel, and as of 2008, according to their count, it claimed two million members.

THE BUSINESS OF MEGACHURCHES

Sister Aimee's techniques used to recruit a massive congregation remain standard study for "pastorpreneurs," or ministers who set out to create a profitable megachurch. By definition, a megachurch needs more than two thousand attendees per service, which in itself attracts attention, garnering even more followers. Pastors employ the marketing principles of capitalism and brand their message, eventually franchising it to other churches that pay to televise sermons on giant screens. More than five million people go every Sunday to a theater-style church to watch broadcasts of a ceremony. California, Texas, Florida, and Georgia have the largest number of megachurches, with more than 1,200 in operation nationwide. Although fundamentalist Christianity, or "falling into the arms of Jesus," is the primary message, church-sponsored social activities such as weight-loss groups and dating services are part of the "soulcraft" and the marketing plan of the pastorpreneurs to entice more "customers."

AKASHIC RECORD

UNIVERSAL SPIRITUAL DATABASE

Literally everything, from the Big Bang to the growl of a dinosaur, to a leaf falling in a forest, to a caveman's war grunt—the entirety of existence, including all your thoughts and actions, from the womb till now—has been dutifully recorded by a cosmic force. In Hinduism this infinitely large cache is referred to as the Akashic Record. In Judaism and Christianity, it is similar to the Book of Life, where the detailed doings of the wicked and the righteous are consulted at Judgment Day. Spiritualists call it the Hall of Records (a secret library buried under the

Great Sphinx of Giza), and to New Age religions the Akasha is often referred to as Universal Consciousness. Akasha was first conceived by Hindu sages, who claimed the record was ten thousand years old. It's believed that all knowledge, including past, present, and the probability of future events, is contained in this celestial database. According to Akasha devotees, the record is still constantly being encrypted at seven times the speed of light. Although this tome is difficult to access, it's experienced in some degree in the form of an intuition, or when having a "gut feeling," or intrinsic knowledge about some circumstance. These little knowings are, theoretically, always available to the seeker. However, obtaining the password to use this divine search engine is rarely achieved by means of one's ordinary senses. The Hindus consider this knowledge-crammed force to be the underlying symmetry and cohesiveness present in everything, from the galaxies to the design of a flower.

BIG BANG AND AKASHA

In 1931, Catholic priest Georges Lemaître was the first to propose how the universe was created from a "primeval atom," or a "Cosmic Egg [that exploded] at the moment of the creation." His work has since been generally accepted by scientists and is known as the Big Bang Theory; it describes a universe in constant expansion, emitting from a singular, though unknown source. Scientists had believed that before this point of origin, about 12 billion years ago, time and space did not exist. Lemaître also postulated that the universe will expand in forward time, until contracting back again into the primordial atom. Akasha explains a similar theory, though universal expansion is due to the increase of information constantly added to the cosmic record.

The early Hindu sages wrote codes needed to retrieve the Akasha on palm leaves, or *nādi* leaves, which were the size of small slats, like a Venetian blind, though thoroughly inscribed with miniature letters and symbols. When the British colonized India, many ancient leaves were discovered in libraries, the oldest being more than two thousand years old.

The skilled reading of the *nādi* leaves can guide one in finding the record of one's particular soul. The more proficient readers can deci-

pher fastidious details about one's past and future, such as, perhaps, the name of the child that you will have in ten years' time, or the exact age at which you will marry. Most ethical *nādi* interpreters will not tell the precise moment of your death, even if they know it.

THE FUTURE IS WRITTEN ON YOUR HAND

Some think that the Creator stamped a personalized access code to universal knowledge on each person's hand before birth, and thus fingerprints and lines on palms are unique. Some form of palm-line interpretation was believed to be practiced as early as the Stone Age. Prehistoric handprints found on cave walls were perhaps more than signatures and instead held mystical significance, the story of their lives as told in the crisscrossing lines in the palms. Palmistry, as practiced today, can be traced to gypsies, with hereditary origins linking them to India at the time *nādi* records were written. To palmists, an examination of the size, shape, proportion of fingers, and the differences in line formation on the palms of the right and left hand reveals the soul's personal history and offers predictions of future events. Historical figures who believed or dabbled in palmistry include Alexander the Great, Napoleon, Mark Twain, Jane Austen, Sir Arthur Conan Doyle, Jules Verne, Oscar Wilde, Sarah Bernhardt, Mary Pickford, Marilyn Monroe, Grover Cleveland, Thomas Edison, and Eleanor Roosevelt. At first skeptical, Twain said after a reading: "[The palm reader] exposed my character to me with humiliating accuracy. I ought not to confess this accuracy, still I am moved to do so."

ST. ALBANS

HEADLESS ENGLISH MARTYR

St. Albans is now a suburban city twenty miles north of London. It was named in honor of Britain's first Christian martyr. Albans was just an ordinary pagan farmer living in the town originally called Verlamchester during the Roman occupation of Great Britain, sometime during the late second or early third century. When the Roman soldiers went on a sweep to collect and execute all those with these religious leanings, the good-natured Albans gave the local priest a place to hide. After a while Albans liked the preacher's ideas and converted to Christianity. When the Romans persisted, insisting there was a Christian priest rumored to be about, Albans opened the door to his cottage and walked out dressed in the priest's robes. He was immediately dragged to the magistrate, who knew Albans was not a priest but ordered his execution anyway after hearing Albans's claim that he had been baptized to the new faith. On the way to a hilltop overlooking the city, where offerings were made to the Roman gods, a bridge was packed with people heading to market. Albans knew the soldiers would clear a path through the citizenry with their swords, so instead he led them to the bank of the river. According to legend, with a raised hand like Moses, Albans miraculously halted the flow of water. He then went with his Roman captors, leading them across the momentary dry gap to his death sentence. When the executioner heard of this, he refused to kill Albans and immediately converted to Christianity. After some delay, a replacement volunteered to serve as slayer, killing Albans and the first executioner—making this fresh convert the second of England's Christian martyrs. However, just as the stand-in executioner stood over his handiwork of the newly severed heads, he suddenly dropped his sword and clutched his face in dire pain. The large crowd gasped when the man wailed as if possessed. An instant later, his eyeballs popped out of their sockets. Both rolled until at rest, staring up at the two frozen-looking saintly faces before them. St. Albans is often depicted in art carrying his head in his hands.

MARTYR FACTORY
ENGLAND, 1535–1681

Sixteenth- and seventeenth-century England swelled the ranks of martyrs and saints for both Catholics and Protestants. The bloody times began in earnest with King Henry VIII; according to eyewitness chronicler Raphael Holinshed, the impatient monarch ordered the execution of at least seventy-two thousand people for either political or religious offenses. After the pope refused to grant a divorce, Henry broke from the Catholic Church and made the newly formed Protestant Church of England the official religion. Practicing Catholicism was deemed an act of treason. Followers of the pope were dragged through the streets, and then hanged until near dead. The condemned then watched as they had their own intestines and genitalia removed and burned before their eyes. Then each limb was severed, followed finally by beheading. In addition, the body parts were displayed about town for all to see. In 1535, when Henry ordered his staunch Catholic adviser Sir Thomas More put to death, he thought it merciful to have More only beheaded, of which More said, "God forbid, the king should use any more such mercy on my friends." The king did; the Catholic Church credits Henry VIII with making thirty-eight Catholic saints, while many other victims were deemed as "Venerable," or simply martyrs. His daughter, Queen Elizabeth I, added twenty-five saints to the Catholic saint rolls, though she usually preferred chaining the accused in the Tower. Typical of Elizabethan martyrs, records concerning one obscure figure, Dr. Nicholas Harpsfield, a distributor of Catholic books, indicated how he died a martyr in 1580 of "hunger, cold and stench." In Elizabeth's defense, however, her earnestness in prosecuting Catholics was in retaliation to a papal bull that urged all faithful Catholics in England to assassinate the queen. (*See also:* Bloody Mary)

AMEN

THE MOST MISUNDERSTOOD WORD

In historic Hebrew writings *amen* was used as an adjective, meaning "reliable." When people spoke and wanted it made known that they emphatically meant what they said, they added their own *amen* to the sentence. For example, a marketplace butcher might hold up a dead chicken to entice passing customers and say, "This meat is fresh, amen," meaning that he certifiably attests that the carcass is truthfully not rancid. There are not too many *amens* found in the Old Testament, but after Jesus reportedly added *amen* to many of his sentences, it became one of the most repeated of all Christian words used in writings and services. Most modern churchgoers think it means "I second that," or "I agree with that." Many children learning to recite prayers think it is a religious version of "The End." (*See also*: Holy Laughter)

"ALHAMDULILLAH"

In Islam the most commonly used religious phrase is *Alhamdulillah,* meaning "praise to Allah," and which is applicable in a variety of instances. For example, after a road trip and one is glad to be home, an Alhamdulillah is in order. At the end of a satisfying meal, it could be used again or could be uttered for more substantial situations such as receiving a sizable tax refund, or even at the birth of a child. However, in modern times, especially in the Middle East, saying this, or *Allahu Akbar* (God is great) while smiling inappropriately is often the only clue a suicide bomber may give before detonation.

ANGELS

GOD'S POSTAL SERVICE

In many religions, angels are ethereal beings, generally considered messengers of gods that can take on human shape. They wear great birdlike wings in order to travel between heaven and earth. Early scholars trying to describe the unseen workings of nature, or the serendipity of events, used angels as the unexplained forces behind certain outcomes.

The most famous angels are archangels Michael, Gabriel, and Raphael. Michael is a warrior angel and credited in the book of Revelation with defeating the once-good angel Satan, sending him to hell. In the book of Daniel, Michael is portrayed as a protector of Israel. Gabriel, on the other hand, lives in the light of the moon and is the angel frequently encountered by individuals confronting death, although he was also the messenger who informed Mary that the Holy Spirit would place baby Jesus in her womb. Gabriel is said to have the doomsday trumpet, with which he will signal that Judgment Day has arrived. Raphael is considered a healer, facilitating cures when all is thought hopeless.

BABY-SIZED DIVINE BEINGS

While archangels are often depicted as full-grown, cherubs are much cuddlier and no taller than a toddler. However, according to the book of Genesis, Satan was originally a cherub, and thus a type of angel easily provoked into battle. When God threw Adam and Eve out of the Garden of Eden, two of the baby-sized cherubs guarded the Tree of Life, waving flaming swords and just itching for a fight.

The early Christian church tried to place angels in a chain of command and thus adopted many of the ideas that Dionysius the Areopagite wrote in his fifth-century book *The Celestial Hierarchy.* Accordingly, there are nine different types of celestial beings: Seraphs, Cherubs, Thrones, Dominions, Virtues, Powers, Principalities, Archangels, and Angels. Seraphs are the highest and closest to God, seated at the foot of his throne. As described in the book of Revelation they perpetually praise God, repeating: "Holy, holy, holy is the Lord God Almighty, who was, and is, and is to come." Cherubs are the second highest of the heavenly beings. Next in line are Ophanims, and like the Seraphs, they surround God's throne, although they are silent and supposedly never sleep. Thrones in the heavenly order do not take on human form, but are in charge with transporting God's throne. They have been likened to giant wheels covered with a million eyes. However, according to St. Thomas Aquinas, Thrones are likewise responsible for turning the wheels of human justice.

Dominions, Virtues, and Powers reside in the second sphere of heaven. Dominions are the managers, relegating and assigning tasks to lower angels, while Virtues supervise planetary orbits and such. The Powers are part of heaven's think tank and come up with new ideas and philosophies. The final level of heaven is where Principalities, Archangels, and ordinary Angels reside. Principalities work with certain like-thinking human groups and are said to favor science and the arts. Archangels do the hands-on tasks as assigned by the Dominions, though they are usually only sent for big assignments of historical proportions. And Angels, the ones most interacting with humans, deal with the details of mankind's everyday living.

THE PRESENCE OF ANGELS

On February 24, 1989, a forty-foot section ripped off the top of a United Airlines Boeing 747 flying one hundred miles south of Hawaii. Nine passengers were instantly sucked out and disappeared into the heavens. The pilot and the other 328 passengers in this aerodynamically improbable aircraft attested to seeing a giant hand holding up the wings, ultimately bringing the craft to a safe landing.

ANGLICAN CHURCH

CHILD OF DIVORCE

Anglicanism and the Episcopalian Church trace origins directly to English King Henry VIII in 1534, when Pope Clement VII did not grant him a divorce from Catherine of Aragon. If the pope had granted it, it's likely the Church of England would never have been formed. In the Italian pope's defense, he was awash in political turmoil—he had himself been imprisoned and Rome had been besieged by the brother of the wife whom Henry wanted to divorce.

Even though Anglicanism denied papal authority, it retained nearly all Catholic theology, including sacraments. The differences include allowing clergy to have families, women and gays as priests, blessings of same-sex marriage, and viewing online services that count as much as attending church. In addition, heaven and hell is not a place, but rather a state of mind. There are about 77 million people counted as members of Anglican religions.

SWITCHING SIDES

In 2009, the widely popular Catholic priest Father Cutie was spied ogling a bikini-clad female on Miami Beach. When the mystery woman turned out to be his longtime girlfriend, the Catholic archdiocese yanked him from his church position—despite the protest of hundreds of adoring women parishioners. Father Cutie abruptly ended the controversy by becoming an Episcopalian priest, and marrying his love interest with the blessing of the Church of England.

ST. ANTHONY

FINDER OF LOST OBJECTS

Not to be mistaken with St. Anthony, the patron saint of grave diggers (a hermit in Egypt who died in 356 at the age of 105), the most famous St. Anthony was a Portuguese Franciscan friar who spent most of his time in Italy during the twelfth century. When a religious speaker failed to show for an ordination ceremony in 1220, Anthony was taken from his kitchen duties and told to give a lecture with whatever words the Holy Spirit put in his mouth. He subsequently blew everyone away with his eloquence—no one knew Anthony was fluent in half a dozen languages. It was said that when he spoke, even fish came to the surface to hear him. From then on, he was sent out to preach.

Due to another freak occurrence, St. Anthony came to be petitioned more than any other saint to help find missing objects: everything from car keys, to wallets, to runaway spouses. Since Anthony was such a great orator, he was given a few reference books, and there was one in particular in which he jotted down important marginal notes. It seems

that one of his students, deciding monastery life not for him, abruptly left the order, pilfering St. Anthony's favorite book in the process. St. Anthony had no idea it was stolen and thought he had simply misplaced it. He prayed for the book to be found, and within a day the absconding novice, seemingly possessed, came banging on the monastery door. He tossed back the sacred contraband, saying a devil had been chasing him with an ax since he had committed his misdeed. From then on St. Anthony has been asked to help find things through many formal prayers.

However, the more forgetful employing his aid often have used a more familiar version: "Tony, Tony, turn around. Something's lost and must be found."

St. Anthony, the grave digger saint, was such a successful hermit that people forgot he was living in a cave. He had once been quite a lady's man in Egypt, and became obsessed with sex. So to quell this temptation he stayed alone in a cave for more than sixty years. Right before he died, though, he came into town, and asked people to give him a proper burial. Hence he became the patron of cemetery workers.

RELIGIOUS PLAGIARISM

In 561 A.D., St. Columba was accused of copying the writings of his teacher St. Finnian and calling them his own, and so the matter was brought before Irish king Dermott. After St. Columba lost the case, he started a feud among priests that eventually led to the Battle of Cúl Dreimhne, where three thousand were killed. When threatened with excommunication for his hand in the carnage, St. Columba convinced the church hierarchy that if allowed to work as a missionary, he would save twice as many souls as he had caused to die. St. Columba is credited with "writing" three hundred other books and for converting the Picts of northern Scotland to Christianity.

ANTICHRIST

BIZARRO JESUS

Being called an Antichrist is much worse than being dubbed a mere false prophet. According to the Bible, and in the epistles of John in particular, we are warned of the danger of an Antichrist sure to show at the end of the world, though the definition was later broadened to deem as an Antichrist anyone "who denieth the Father and the Son." Generally carrying demonic implications, the Antichrist possesses deceptively charismatic attributes that will place him in a position of great power. The clout the Antichrist will wield is predicted to rival that of Jesus.

In the late sixteenth century, Matthias Flacius, a Lutheran scholar, wrote a twelve-volume history explaining in detail how certain popes were Antichrists in disguise. Catholics, meanwhile, were certain that

the son of Martin Luther, an ex-priest who married an ex-nun, had the possibility of bearing an Antichrist. As recently as 1999, evangelist Jerry Falwell announced: "The Antichrist was probably alive on Earth, and certainly a Jewish male," which he later said was not an anti-Semitic statement but instead merely an interpretation of scriptures. And even in the 2008 U.S. presidential race, charismatic politicians inspired rumors spread on Internet bulletin boards, alerting those waiting for the Antichrist to look for signs. Those wishing for the Second Coming in a way *hope* that an Antichrist will arrive, since the Bible relates that once the figure appears, Jesus will return to slay him.

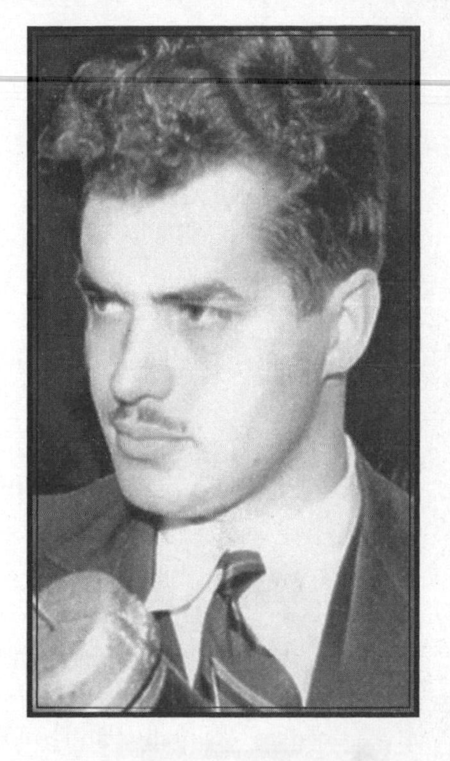

On the other hand, some people claimed fame by declaring themselves as such. In the 1940s a brilliant rocket scientist, John Whiteside Parsons (above), drew attention when he attested to be the avatar of the Antichrist and made no excuses for involvement in certain occult practices: He always said a prayer to the god Pan before each test-rocket lift-off. Parson died in 1952 from an explosion while working in his home laboratory, still interested in creating a machine that would travel through time and space. There is a lunar landscape named Parsons Crater in his honor, fittingly located on the moon's dark side.

He will ascend to power on a platform of peace.
By peace, he will destroy many.
—Book of Daniel

ANTICLERICALISM

MEXICO GOES TO WAR WITH THE CHURCH

Sometimes irreligious attitudes can be as dangerous as religious ones. In the 1920s, the Mexican government tried to pry loose the Jesuits' more than three-century hold on the populace. They added new anticlerical amendments to the constitution in an attempt to end the Catholic Church's sway. The law forbade Jesuits to teach, banned most Catholic sects, prohibited churches from holding property, and even stripped clergy of protection and their ordinary civil rights—there were no trials by jury for priests or nuns. More than forty Catholic clergy were executed and countless more defrocked, manhandled, and rounded up by troops for deportation. Thousands more fled for their lives across treacherous terrain into the United States. Some priests, vowed to celibacy, were forced into impromptu marriages. Hundreds of clergy were "betrothed" to whatever women were willing—or sometimes to nuns yanked from convents—and then watched by the mob until the priest consummated the marriage with the new shotgun bride. Before the purge there were more than 4,500 priests in Mexico; by 1934 only about three hundred were given licenses to worship, and even then only inside certain churches left standing.

In the 1990s, the Mexican congress passed a "Law on Religious Associations and Public Worship," which returned some church rights, though religious organizations still could not run TV or radio stations. According to a 2000 census report, nearly 90 million Mexicans consider themselves Roman Catholic. There are now more than ten thousand Catholic churches in Mexico, with fourteen thousand nuns and priests back in operation.

THE MERRY MEXICAN MARTYR

Miguel Pro, a Jesuit priest in Mexico, suffered more than most during that time, though he was known to keep a cheerful attitude and has since been canonized. In 1926, at the height of the government's anticlerical campaign, when many clergy scrambled to flee or went into hiding, Father Cocol, as he was known, came back to his native Mexico. To bypass the ban on wearing

religious clothing in public, the priest came up with clever ways to administer sacraments. At night he would head into the slums of Mexico City disguised as a beggar to baptize and marry the faithful. He faked his way into prison dressed as a police officer to serve communion to inmates, or he'd sport a business suit to serve the remaining Catholics among the rich. His fame spread and irked then Mexican president Plutarco Calles, an outspoken atheist, who enacted laws of such strictness that even wearing a crucifix on a necklace could bring five years in prison. Eventually, Father Pro was implicated in a bombing plot against the government, with Calles ordering the Jesuit's execution by firing squad. As a last request, Father Pro only asked to kneel and pray, and then refused a blindfold. At age thirty-six in 1927, with arms outstretched, a volley of bullets nailed him to his imaginary cross, and when the soldiers realized he was still alive, one of them fired a final shot into his head at point-blank range.

APHRODITE

LUSTY GODDESS

According to Greek mythology, the sky god Uranus had his genitals extracted with a sickle by his son Cronus. These were thrown into the sea and apparently floated for an undisclosed period, until becoming the seed that gave birth to one of the most beautiful goddesses of all, Aphrodite. She emerged from sea foam with the body of a supermodel, and had a devout following through much of antiquity. She was supposedly vain, and easy to anger, though admired by all as the epitome of physical beauty. It's no wonder she was so popular, since she held dominion over love, lust, and fertility, even if she had the power to rob men of their reason through her skill at seduction and temptation. Temples to Aphrodite stretched across the civilized world; priestesses there were noted for wearing antiquities' equivalent of the latest in boudoir fashions. Men of all ages looked forward to her feast day with great excitement. To gain sexual prowess, and to pay proper respect to the deity, sex with an Aphrodite priestess was considered a prudent necessity. Donations afterward, of course, were accepted, making these temples, in some sense, spiritually sanctioned brothels.

Ishtar was the Babylonian version of Aphrodite. On Ishtar's feast day every woman in the city was required to show up at the temple and expected to have sex with strangers. A man could select the consort he wished with the ritual request: "In the name of Ishtar I choose you." It was considered a sin for a woman to refuse.

In the sixth century, Christians replaced Aphrodite's feast, by then thought to be October 1, with the feast of St. Gregory the Illuminator, which required less hands-on participation from celebrants. Aphrodite's son, Eros, or in the Roman version Cupid, was the god of erotic love and is still commemorated on Valentine's Day.

APOCALYPSE

THE END OF US ALL

Since ancient times, prophets made predictions forecasting doom and the end of the world—the apocalypse. Jesus, when quizzed on the subject by disciples, said that the stars shall fall from the heavens, though he cautioned that no one knows the hour or date, "not the angels, nor the son, but only the Father." His apostle Peter was more specific in his cataclysmic imagery, predicting the heavens will explode and that everything will be demolished in a raging fire. In the book of Revelation, we are told that a pit will open in the earth from which thick smoke will emanate and darken the sun. In addition to demonic monsters and angelic battles, millions of locusts will emerge and eat on the unfaithful, though not kill them.

Nevertheless, the apocalypse became popular from the second century—once it became clear that Jesus wasn't returning anytime soon—to scour the scriptures for clues to know exactly when it will happen.

A 2002 survey conducted by Time/CNN found that 59 percent of Americans polled believe that the events described in the book of Revelation will come to pass. Twenty-five percent thought the September 11, 2001, attack on the World Trade Center and the Pentagon was predicted in the Bible.

In 1650, the Irish archbishop James Vassler believed he had settled the problem once and for all. He spent years scouring the Bible for every reference; analogizing tangible dates, and with great fanfare announced that his final calculations showed that Adam was created on October 24, 4004 B.C., with earth's formation occurring during the preceding six days. To Vassler's reckoning, he believed mankind would be destroyed on October 24, 1996. This lack of urgency was a disappointment, and his calculations ridiculed, since fear that doomsday was right around the corner had always helped religious recruitment and encouraged more reverent prayer. Throughout history most generations actually hoped this event would take place in their lifetime. For many religions, the notion of apocalypse, although horrific, also meant that God would finally reveal himself, taking whoever was deemed "saved" into a better realm. More than 80 percent of all cults used Armageddon revelations to round up followers, a foolproof technique that always swelled ranks in the shortest time.

FOUR HORSEMEN

As described in the book of Revelation, the arrival of the Four Horsemen of the Apocalypse (right) foretells the end of the world. The first rider will be on a white horse, wearing a crown and carrying a bow, but without any arrows—signifying that the one who brings on doomsday will acquire power through nonviolent means, as in an election, or bloodless coup d'état. (*See also*: Antichrist) The second rider will be on a red horse, brandishing a battle sword, the symbol for war. The third horse charging through the gory end days will be black and its rider carrying a balancing scale, to exemplify who will die by famine, drought, and epidemics. The last, a rider on a pale-colored horse, is Death, who holds a pitchfork to round up the unsaved and herd them to hell.

APOSTATE

RELIGIOUS CRIMINALS

In December 2007, a fifty-seven-year-old Canadian man murdered his sixteen-year-old daughter for her refusal to wear a hijab. The young girl was deemed an apostate for not wearing the traditional head garment of many Muslim females; this was considered the breaking point that led to her strangulation. An apostate is one who denounces his

or her religion and often goes to lengths to bad-mouth it; in the case of the sixteen-year-old, she said the head scarf simply messed up her hair. In Syria, for example, there are an estimated two hundred to three hundred such killings each year, for actions deemed blasphemous and against religious beliefs. The Amish will shun a member and cast out

an individual in contention with their rules and consider that person as good as dead, even if he or she is an immediate family member. In the Catholic Church, an apostate will often be excommunicated and no longer allowed to partake in sacraments.

JULIAN THE APOSTATE

Julian was the last non-Christian emperor of Rome. He ruled for only a few years until he died at age thirty-one in 363 A.D. from war wounds. When he took

over he brought back the old Roman gods and celebrated their feast days, trying to keep the ruling class exclusively non-Christian. However, he didn't persecute any sect or cult and instead was tolerant of whatever and however people wanted to worship. He even attempted to rebuild the ruined Temple in Jerusalem to show the Jews he was open-minded, yet with the next wave of rulers, all Christians, he was forever dubbed Julian "the Apostate." Others, however, herald Julian as the last ancient champion of polytheism. One of the rituals Julian wanted reestablished was the custom of "reading" goat entrails prior to battle. A goat was slaughtered, then its guts put into a bowl and studied by a general to determine if the attack time should be amended. Julian misread his interpretation of goat innards and got mortally impaled by a spear while in retreat. (*See also*: Haruspicy)

APOTHEOSIS

FIRST-CLASS TICKET TO HEAVEN

Certain individuals have been so heroic, or notable, they have been transformed from mere humans into divine beings while still on earth, and in their present bodies. Many monarchs claimed to have been changed by apotheosis, which made their commands and subsequent lineage unbreakable because of their heavenly transformation. In Egypt, a pharaoh was a living god. In Rome, only dead military leaders could achieve this transformative status, while in England and throughout Europe, the term *divine rights of kings* held sway for sometime and made the king or queen's judgments unquestionable. Up until World War II, Japanese emperors were in power thanks to divine links, as it was throughout many Chinese dynasties. In Eastern religions certain children are still examined to find that special transformative mark of holiness. (*See also*: Dalai Lama)

ST. THOMAS AQUINAS

THE LEVITATING THEOLOGIAN

Thomas was born in Sicily in 1225 to parents of means, with his mother's lineage traced to Roman emperors of centuries past. This advantage allowed young Thomas to attend a monastery from kindergarten age, where he was expected to eventually become an abbot. He seemed

set for a respected and titled religious career, one a family of nobility would expect of at least one son, when Thomas suddenly dropped out of the University of Naples just months short of ordination. It seems Thomas fell in with a new order of monks, the Dominicans, referred to belittlingly as the "Black Friars," for the outfit of ankle-length, black hooded robes, and shaved heads, leaving only a ring of hair to imitate the crown of thorns. This was such a potential scandal that Thomas's family intervened. He was kidnapped by his brothers and held in his parents' castle for a year while they tried to dissuade him from associating with this upstart sect devoted to chastity and many then-unpopular deviations of Christianity.

After Thomas resisted, his two brothers had a plan, and on one summer night they hired two of the most notoriously beautiful prostitutes in all of Italy. Certain that no man could resist their gorgeousness, the brothers opened the lock of Thomas's heavy wooden door and shooed the women inside. However, the brothers soon jumped back. Legend has it that two guardian angels transformed into giant lions, which growled and roared, protecting Thomas from temptation. As for the working girls, nothing is mentioned, though a joke among clergy is that the women ran from the castle to the nearest convent and immediately signed up to become nuns.

As regards the individual nature, woman is defective and misbegotten, for the active power of the male seed tends to the production of a perfect likeness in the masculine sex; while the production of a woman comes from defect in the active power.

—Thomas Aquinas

Thomas was nicknamed "Dumb Ox" in school because of his stocky and stout frame, and for an odd habit of delaying a response to the simplest question. He was considered dimwitted by classmates for the way he stared at someone—at a considerable length—before answering. The pope eventually intervened and Thomas was allowed to join the order of his choosing, and permitted to continue studies at leading institutions in Cologne and Paris. Aquinas quickly made a reputation as an astute scholar, theologian, and articulate defender of his faith. From "Dumb Ox" to being heralded as "Doctor of the Church," Aquinas's writings helped define Roman Catholic thought. He believed the entire goal of life was to seek a beatification vision, defined as "the immediate sight of God in heaven." From the beginning, Aquinas's fellow monks reportedly heard Thomas actually conversing with God, and it became commonplace to see him levitating, sometimes only a few inches off the ground, with others claiming to find Aquinas lying on the ceiling while reading a book. (*See also*: Levitating Clergy) His death came in 1274 at age forty-nine, after a nearly two-month illness, with some believing he had been poisoned, either through his daily meals or put in the wine of his chalice when he served mass. His relics remain in a tomb at the Church of the Jacobins, in Toulouse, France, and have not been tested for traces of poison. Fifty years after his death he was canonized.

ARIUS

THE HERETIC WHO COULDN'T FIND THE BATHROOM

During the Council of Nicaea, in 325 A.D., it was decided that the Christian God consisted of three beings: God the Father, the Holy Spirit, and Jesus existing simultaneously in one as a Trinity. Arius, a priest from Alexandria, Egypt, made a big commotion about this doctrine, protesting that the idea was all wrong. He argued that God existed before Jesus and therefore Jesus was not co-equal with God the Father. At one time Arius had a considerable following, even after he was condemned for his refusal to recant his position and became the Catholic Church's first heretic, sentenced to excommunication. In addition, all known collections of his extensive philosophical and theological texts were ordered burned, and while alive, he watched the flames turn his life's work to ashes. Copies of writings found later were immediately blessed with a splash of holy water before likewise enflamed.

To further ensure his hand would produce no more of his considered blasphemy, Arius was coaxed out of exile in 336 and ordered to Constantinople under the pretense of readmittance into the church. Arius took few precautions against assassins, and when he arrived he paraded openly throughout the city, waving to the crowds with a sense of vindication that his ideas would finally be reconsidered. However, before he made it to the church, where he thought he was to be blessed by the pope, his hands clutched at both his mouth and rear end. While attempting to make a run for a bathroom, his body suddenly seized. Blood began to hemorrhage from every orifice, and eyewitnesses claimed to see his spleen and liver pass along with his bowels. This was taken as a further sign that God was displeased with his heretical notions, even though it appears that Arius was poisoned by skilled alchemists who had schemed to kill him in a spectacular way right in front of the crowds. The wall he leaned against was marked and became a tourist site for centuries to come as a reminder of the fate for those who defy belief in the Trinity.

If forgers and malefactors are put to death by the secular power, there is much more reason for excommunicating and even putting to death one convicted of heresy

—St. Thomas Aquinas

THE FIRST HEAD ROLLS

THE INVENTION OF HERETICS

Prior to the Council of Nicaea, there was much disagreement on Christian dogma. The accusation of "heretic" was only applied to pagans. Moreover, there was little the early Christians could do to punish offenders, or those who wavered in faith. The first public and nonclandestine execution of a Christian heretic occurred nearly sixty years after the council, in 385, when the newly crowned Roman emperor Magnus Maximus, a native of Spain, ordered the death of a fellow compatriot, Bishop Priscillian of Avila, for the suspicion of magic. Born to a noble class and well educated, Priscillian had developed into a stern Christian bishop. He believed one needed more than faith for salvation, and he was opposed to a number of agreements made at the Nicaean council. Priscillian believed literally in what scripture indicated, particularly that one had to give away all title and money, which he did (with the exception of his position as bishop) to be a true Christian. He also advocated nothing less than a harsh ascetic lifestyle, and openly condemned marriage; he advocated celibacy for all, allowed men and women to pray together, and treated females as equal, which was not permitted. When favoring a Pentecostal sort of worship, with long fasting recommended to achieve ecstasy, he was suspected of casting a magic spell on his congregation. When Priscillian's companions were excommunicated for following his belief, he went to petition the new emperor, thinking he had favor since both were Spanish. However, Maximus refused to entertain a petition of leniency and instead quickly condemned Priscillian, and six in his party, to death by beheading.

SHOKO ASAHARA

BLIND JAPANESE CULT LEADER

On March 20, 1995, in Tokyo, crowds fled packed subways, gagged, vomited, and collapsed on the streets after sarin, a deadly nerve gas, was unleashed during rush hour by a religious group called Aum. Twelve people died and thousands more were injured. The group's leader, Shoko Asahara, warned that the approaching millennium would bring an apocalypse and believed this act would topple the Japanese government and install him as the emperor. Born Chizuo Matsumoto to a poor family of mat makers, Asahara was virtually blind due to glaucoma from a young age and sent to school for the sight-impaired to learn acupuncture.

He studied Buddhism, Taoism, Hinduism, Yoga, and the book of Revelation, before a trip to India and the Himalayas. Upon return,

he declared he had achieved enlightenment. In 1987, he changed his name to Shoko Asahara and formed a new Buddhist sect, Aum Shinrikyo, Religion of Supreme Truth. He attracted not only a number of Japan's intelligentsia, but also more than forty thousand who came to believe his words prophetic and inspired. At first his primary message was that political change was needed before Japanese society could spiritually advance. He combined technology and ancient wisdoms to teach his adherents to control their lungs, to up to five minutes without needing oxygen, and to slow the heart rate to a nearly undetected beat. When he first started the sect he was well regarded enough to have an audience with Tibet's Dalai Lama, and even donated funds to the Tibetan cause.

However, as the 1990s progressed his visions incorporated doomsday urgency, especially after he began to dabble in LSD, among other hallucinogens. Appointed by Asahara as "Minister of Health," a former heart doctor and circulatory specialist, Ikuo Hayashi, developed the gas in the sect's secret "Light Weapons Compound." The nerve gas was used a year before the subway incident to saturate a neighborhood home to judges who were about to pass a real estate verdict not in favor of the sect. Eight died after a refrigerator truck dispersed the gas. For the Tokyo Metro operation, small pouches of sarin were dropped and punctured by sharpened umbrella tips.

Tracking the subway assailants to Asahara's complex, the police found a sarin lab and Asahara in a hidden room, sitting lotus-style in deep meditation. During his trial he never spoke and only mumbled, such that he was subjected to psychiatric tests, though they concluded he was mentally stable. In 2009, Asahara and seven others were on Japan's death row, awaiting execution by hanging. The sect has changed its name to Aleph and still has about 1,500 members.

ASH WEDNESDAY

YOU'RE NOTHING BUT DIRT

On a Wednesday, six weeks prior to Easter Sunday, Catholics, Lutherans, and Anglicans, among those in other Christian religions, have their foreheads marked with a cross of ashes. The soot comes from the leftover blessed palm fronds from the previous year's Palm Sunday. The purpose of this ritual is to remind people what was noted in the book of Genesis: "For dust thou art, and unto dust shalt thou return." In addition, it signifies that there is little time left for repentance, self-denial, and prayer before one's existence inevitably turns into an unredeemed pile of soot.

The early Christians took the forty days prior to Easter seriously (literally imitating Jesus' fasting in the desert), and dressed in sacks, covered themselves with ash smudge and adhered to strict self-denial. By the ninth century, it became customary for the entire congregation to participate, especially as the church shifted focus from the implied joy of baptism to a life that emulated the suffering of Jesus.

The Romans had a similar February celebration for the fertility god Lupercus, which turned into a citywide spring break, noted for public drunkenness and orgies. The Romans then recommended fasting for forty days afterward. Converted pagans refused to give up this holiday and Christians allowed a comparable feast, which came to be known as Mardi Gras, or "Fat Tuesday," from a French medieval custom of cutting loose before Ash Wednesday. Generally, sins of intemperance, gluttony, or fornication didn't count during this celebration. The

Romans referred to it as a *carnival*, derived from the Latin *carnis* and *vale*, meaning "farewell to the flesh"—and the last party before abstinence and repentance.

HOLY DAYS OF OBLIGATION

In 1247, Catholic decrees were passed that made attendance at church mandatory every Sunday and on thirty-six additional feast days. Failing to attend was as grievous a sin as committing murder. In 1911, Pope Pius X reduced the number to eight, removing Ash Wednesday as an official Holy Day of Obligation. However, it remains a day of fasting and abstinence and still requires all Catholics over the age of fourteen to forgo eating meat on that day, and on every Friday during Lent. Officially, Catholics are recommended not to eat meat on all Fridays throughout the year, although giving up some other item will also count.

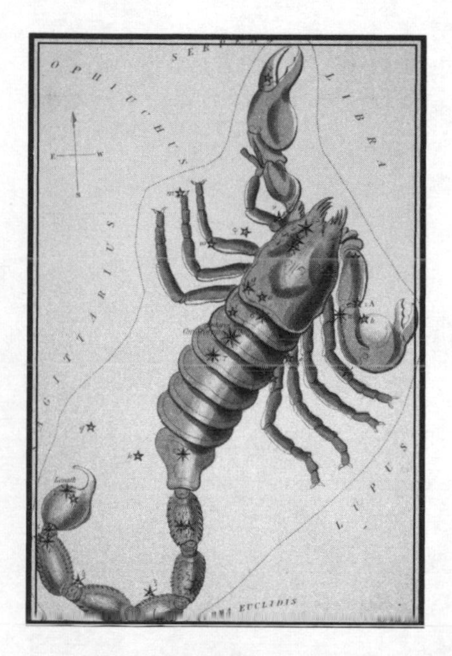

ASTROLOGY

THE FUTURE IS IN THE STARS

As early as the third century B.C., texts revealed the ancients' intense interest in stars as a means to guide human events. Celestial arrangements were cast in the sky by gods to give humankind clues and coded messages. Stars were imagined as attached to strings, dangling from an inverted dome God had placed over the earth. According to historian Philostratus, it was customary in many cultures to press one's fingers firmly to their closed eyes until they saw dots or stars. This was practiced daily as a religious reverence to heavenly bodies the way Christians would later use the sign of the cross. As sixteenth-century stargazer, mathematician, and eccentric Tycho Brahe noted about the galaxy and the symmetrical order seen in everything, "by looking up, I see downward"; stars had, and still hold, a natural wonder. Until the late 1800s, those with knowledge of stars, whether for scientific or religious interest, were considered astronomers.

Now astrologers are different from astronomers, and focus on interpreting star movements to help predict future events. Astrologers read into the hidden meanings of constellations to offer guidance on when best to partake, or refrain from pursuing various human endeavors. Horoscopes, one tool of astrology, were first employed four thousand years ago and correlate the time of year a person was born in relation to star formations that were present during one's precise moment of birth. This particular alignment of celestial objects has stamped certain

indelible personality traits into a person's character. These tendencies cannot be entirely overcome, but only guided by knowing how the stars relate to them now.

Jeanne Dixon (left), a noted astrologer, supposedly predicted the John F. Kennedy assassination years before it happened. She used charts and a crystal ball. Nancy Reagan had employed Dixon's services until the clairvoyant predicted Ronald Reagan would not become president in 1976—advice not welcomed—and she switched to astrologer Joan Quigley for consultation. Dixon lost her prophetical political pull, not so much from Mrs. Reagan's slight as from predicting President Nixon would become the greatest president of all.

THE PAGAN DAYS OF THE WEEK

Many star constellations were named after zodiac signs, and all the planets in the solar system, with the exception of earth, have the names of Roman gods. The months from January to June were also derived from Roman deities: January was named for Janus, the god of beginnings; and June for Juno, the goddess overseeing the financial health of the empire. The English names for the days of the week likewise have religious connotations, though mostly from Germanic or Norse gods. Tuesday was named for Tyr, or "Tyr's Day," with Tyr depicted as a one-handed god. Thursday was Thor's day—the god of thunder, noted for wielding a hammer and lightning bolt. The phrase TGIF, or "thank God it's Friday," is apt since it was derived for a day honoring the goddess of love, Fria, who was noted as always coming out soon after sunset.

ATLANTIS

LOST CITY UNDER THE SEA

Plato, a Greek philosopher, was the first to mention Atlantis, namely in one of his dialogues, *Timaeus*, written in 360 B.C. He described an ancient though advanced civilization that inhabited an island in the Atlantic Ocean, somewhere near the "Pillars of Hercules." The Atlanteans were ruled by descendants of the god Poseidon and had achieved astonishing technological success. According to Plato, "in a single day" in 9600 B.C. the island sunk into the sea during an earthquake, taking with it the wealth of its secret wisdom. However, it was believed a few

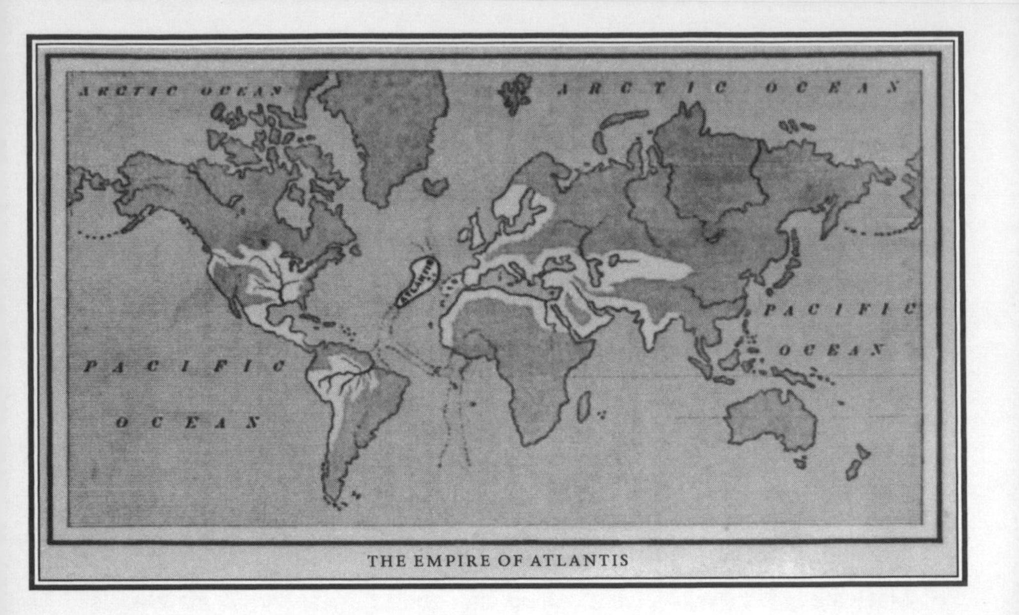

THE EMPIRE OF ATLANTIS

of its survivors managed to encrypt various messages and documents and hide them throughout the ancient world. Over time more than one hundred secret societies formed and claimed knowledge of, or direct descent from, these once-enlightened people. The forerunner of New Age movements, the Theosophical Society in the late 1800s (*See also*: Madame Blavatsky) was primarily responsible for revitalizing interest, and by adding new mythology concerning its origin and hidden wisdom. Edgar Cayce, a celebrated "past life regressor," claimed he had once been a priest in Atlantis. Cayce reported that the Atlanteans discovered electricity fifty thousand years ago and harnessed atomic energy—but saw its danger and so instead decided to power everything with a "Great Crystal."

Archeological records confirm that at the end of the Ice Age, the oceans rose three hundred feet, more or less coinciding with the time frame quoted by Plato. Some speculate that a huge landmass about half the size of Australia was located between Spain and Central America, and that its mountaintops now form the Caribbean Islands. New Age Atlantisologists assert that it was never "lost," and that it remains, due to its even more progressively advanced technology, in plain view, and that only the enlightened can see it.

BERMUDA TRIANGLE

LAND OF THE LOST

Atlantisologists have considered an area of the Atlantic Ocean known as the Bermuda Triangle—an imaginary pyramid-shaped section drawn from the longitudinal points of Miami, Bermuda, and Puerto Rico—as a possible resting place of Atlantis. More than twenty low-flying planes and one hundred ships have vanished while traversing the area. The disappearances have been attributed to the crystal power source that Atlantis used; it still beams powerful waves into the heavens, thus affecting the properties of tides and normal nautical phenomena. Today many scientists believe the unexplained disappearances are caused by methane hydrates—giant gas bubbles that arise from continental shelf shifts, which create aberrant atmospheric conditions.

SRI AUROBINDO

"SUPERMENTAL" YOGI

Born in India in 1872, Aurobindo Ghose was educated at England's Cambridge University, where his writings won literary prizes. He returned to India and before long was incarcerated for his involvement with India's independence movement. While lingering more than a year in prison he experienced a spiritual awakening, claiming visitations from recently deceased gurus, especially Swami Vivekananda, an advocate in the revival of yoga practices. After acquittal he retreated to the then-remote providence of Pondicherry and devoted himself to studying yoga. His writings, particularly his massive, 1,100-page tome *The Life Divine*, laid the foundation for much of the yoga practices known today.

Aurobindo believed humans were born completely ignorant, with no idea of their true self. The path to enlightenment lay in transcending the physical realm and reuniting with a higher, perfect consciousness. According to Aurobindo, this was essential not only for an individual's enlightenment but for nothing less than the natural evolution of mankind. In order to connect with this ever-present level of divinity during life, the mind, body, and spirit had to advance in unison through a practice he called Integral Yoga. Eventually, he foretold, a new species of humanity, a "supermental being," will emerge with higher levels of consciousness, and will be as different as contemporary man is from other animals. If societal constraints do not allow these new beings to evolve, then the hope for mankind does not appear good. Toward the end of his life, he preferred solitude, and died in 1950 at age seventy-eight. His body remained without signs of decomposition for four days, according to eyewitnesses, and was enveloped in a blue and golden hue. (*See also*: Incorruptible Saints)

B

BÁB

MARTYRED BAHA'I PROPHET

In 1844, amid the squalor of the marketplace in the dusty town of Shiraz, Iran, one twenty-four-year-old man, a seller of goat cheese, threw down his apron and declared himself to be the new messiah. He immediately changed his name from Siyyid Ali Muhammad to Báb, meaning "Gate." Nevertheless, he had an understated charisma and always appeared well kempt and extremely serious. He was noted to speak only when absolutely necessary and seemed preoccupied with his own thoughts, usually ignoring most questions. His writings were considered interesting, but his fame soared when he claimed to be the prophet that Muhammad predicted would arrive.

Within a few years, Báb attracted tens of thousands of followers who believed in his divine authority, so many that he was soon arrested by religious leaders and put on trial for apostasy. He was ordered to perform miracles to prove his claim, and was asked sixty-two questions. At first witnesses were amazed at his ability to recite original verse as answers, but then Báb got tired of the process and merely stated, "I am that person you have been awaiting for one thousand years."

What to do with Báb became a sticky issue: The clergy present wanted capital punishment, but the crown prince noted his popularity and didn't want to create a martyr or instigate a rebellion. The government tried to prove by medical examination that Báb was merely insane. He was given twenty lashes on the bottom of his feet, but still wouldn't recant his divinity. In 1850, only six years after he began preaching, he was tied with his face flat against a wall before a firing squad. After the smoke cleared Báb seemed to have vanished, and it appeared the bullets had merely frayed the ropes and allowed his escape. However, his miraculous getaway was short-lived, for he was found after a brief search to be still on the compound, and so he was restrung up. After more guns were added to the firing line, he was killed, and his body thrown in a dump to be eaten by animals.

Shortly thereafter, he was elevated to martyr status, with followers thinking he had been the reincarnated prophet Elijah, or John the Baptist. But others, more than twenty, vied to be declared the true prophet, and Báb just the "gate" that foretold of their arrival. Bahá'u'lláh, a follower of Báb, rose above the pack and became known as the founder of the new Baha'i religion. Bahá'u'lláh synthesized Báb's and a few other religious scholars' ideas, declaring that the time had come for the creation of one global society. Bahá'u'lláh spent the last twenty-four years of his life in prison, eventually dying behind bars in 1892 for his prophetic claims.

The Baha'i faith, strives, among other goals, for one world religion and one world government, one standard form of measurement, and one court of law, but with all citizens dedicated to service, which Bahá'u'lláh considers the only path to salvation for mankind and spiritual fulfillment. Membership in the Baha'i religion exceeds five million.

In the Baha'i religion there are nineteen months to the year; each consists of nineteen days, with four extra days left over that are dedicated to a period of gift giving. A follower has to attend any of the religion's star-shaped churches on the nineteenth day. In addition, a member is not allowed to join a political party, but can vote, and must spend nearly all free time at community service. There is no heaven or hell, but after death the soul takes its place in a sort of celestial game board, either moving a few spaces back or forward, with the goal of getting closer to God.

BABEL, TOWER OF

ORIGIN OF LINGUAL DISCORD

Babel was the first city rebuilt after the Great Flood destroyed most of mankind. Here builders began constructing an enormous tower, which would "reach the heavens" and perhaps even provide access to God himself. However, God was not pleased with this engineering achievement. In 1913, French archeologist Henri Genouillac claimed to find the ruins eighty miles south of Baghdad, in the original capital of Babylon, called the City of Kiss: The tower had an eight-mile-diameter base and reached two miles into the air. However, according to Genesis, when God saw this spectacle he said, "Now nothing will be restrained from them, which they have imagined to do," and He figured the best way to slow down progress was to make everyone speak a different language. Afterward, the workers at the Tower of Babel blabbered in seventy-two

different tongues. Today there are 6,912 dialects currently spoken on the planet. (*See also*: Pentecostalism)

HOW NOAH REPOPULATED THE WORLD

According to biblical scholars, the Great Flood took place in 2348 B.C. when God decided to wipe out humanity and start over again. Noah was instructed to build a huge boat and took two of every creature (a tremendous task since there are currently at least ten million different species), along with his family. Noah's family was the only survivor of the catastrophe. Therefore, in theory, everyone is a descendant from Noah and his wife—though Noah had a lot of time to repopulate the world, since he lived to the age of six hundred. In the 1980s the Turkish government opened Noah's National Park and accepted as true the claim that Noah's five-hundred-foot-long ark came to rest on "Doomsday Mountain," more than six thousand feet above sea level. More recently, archeologists said they believe they have found the original—a three-hundred-foot-long version—resting on Iran's Elburz mountain range, at thirteen thousand feet. There are more than two hundred flood myths held by various cultures, though they likely derive from an ancient attempt to explain why seashell fossils are commonly found on dry land. Geologists think the Atlantic Ocean might have breached the Strait of Gibraltar at the end of the Ice Age, around eleven thousand years ago. This possibly wiped out civilizations living near the Mediterranean Basin and deposited boats in unlikely places. Nevertheless, God vowed to never use a flood to destroy humanity again, supposedly giving us the rainbow as a sign of his promise.

ROGER BACON

PHILOSOPHER, SCIENTIST, ALCHEMIST

Bacon was born in England in 1214, and to wealth, but lost status during changing monarchs. Nevertheless, he was distinguished for a brilliant mind and was educated at Oxford. After he joined the Franciscan order and became a friar, Bacon got into trouble when he taught at various universities. He discoursed on all types of philosophies, especially Aristotle's book *The Secrets of Secrets*, which at the time was banned by papal edict. Bacon began to disagree with Catholic theologians and believed man could understand God better through science. For many years, he lived under house arrest and was prohibited from publishing his works. Although Bacon approached science through observation, physical perception, and the importance of mathematics, he also had a keen interest in magic and alchemy as a way to transcend hidden truths. The transformation of common metals into more valuable commodities was the primary objective of alchemy, and was a preoccupation of many scientists during his time. However, Bacon was more interested in concocting an "elixir of life," a drink that would help Christians act more moral and gain eternal salvation by becoming literally ageless. His theory was that if everyone had unusually long lives, there was a better chance to fulfill all the requirements for salvation.

A BRAZEN HEAD

During medieval times, commoners were suspicious of exceptionally smart people and so they thought Roger Bacon got his ideas from a talking statue. According to legend, the friar created a bronze bust that frequently came to life and could discourse intelligently on any subject—though the sculpture has never been found.

Although Bacon never admitted to owning a talking sculpture, he did claim expertise with the Philosopher's Stone, a substance thought capable of turning base metals into gold. If it weren't for Bacon's brilliance in other areas, the radical friar would have been burned at the stake for heterodoxy. Bacon's actual science experiments were considered marvelous, though frightening. Once he refracted light through a lens, producing what appeared to be rainbows, which were once thought only the handiwork of God. Bacon was more a mystic than a scientist in the end, writing once that the goal was for man to experience rapture. A year before his death in 1292, he was allowed once again to roam freely, but by then Bacon was old and frail, an obscure figure among his peers.

But there is another alchemy, operative and practical, which teaches how to discover such things as are capable of prolonging human life for much longer periods than can be accomplished by nature.

—from *Opus Tertium*, Roger Bacon

JOE BALLARD

WEATHER PROPHET

On November 13, 1833, Joe Ballard was eight years old and a slave in Salisbury, North Carolina. At three o'clock in the morning he leapt

from his bunk and ran outside to see the sky as bright as midday, with a thousand shooting stars crisscrossing the sky. People thought the world was coming to an end, and that all the celestial objects had been shaken by God's wrath—even falling from the heavens. One eyewitness, Preacher Samuel Rodgers, observed: "Some really thought that the Judgment Day was at hand, and they fell on their knees in penitence, confessing all the sins of their past lives, and calling upon God to have mercy." What Joe and the entire eastern United States actually witnessed was one of the greatest meteor showers ever to take place, later called the Great Leonid Meteor Storm of 1833. Two years later Joe was outdoors again to witness the passing of Halley's Comet.

After the Civil War, Joe Ballard remained in Salisbury and worked long hours as a blacksmith, well into his eighties. Dubbed "Our Weather Prophet" by city officials, Joe was thought to have a divine insight in regard to predicting local climate changes. Some believed his witnessing of two especially unique meteorological events gave him this special gift. It was said the Weather Prophet could foretell (with an accuracy not even modern-day weather analysts achieve with radar and computer models) when it would rain, for how long, when the first snow would fall, and when it was best to plant corn without fear of frost. He died February 25, 1917, and was buried with Masonic honors; however, Joe left little behind of his philosophy or the techniques he used to read the heavens.

MIRACULOUS WEATHER

The greatest "weather" miracle that remains baffling was Moses' parting of the Red Sea, during the Hebrews' Exodus from Egypt. The escaping Israelites

were cornered, pushed against the sea with an army of angry chariots in hot pursuit. Currently, more than 60 percent of Americans surveyed believe the ocean parted, literally, with the raised command from Moses' staff. There's still dispute as to the exact point where the Israelites crossed the sea, though it's generally supposed they hid among a marshland, or the Reed Sea, a lagoon, near the present-day city of Suez. Contemporary meteorologists have cited the possibility of volcanic eruptions that could have sent tsunami-size waves at just the right moment, while others point to unusual wind and tide factors, still common in the area. Nevertheless, "weather," as defined on many homeowners' insurance policies, can be excluded from coverage under the still-used legal phrase "Acts of God."

MODERN WEATHER DOOMSAYERS

Technology makes predicting weather easier, but forecasting remains imprecise, which leads many, despite what's said on the evening news, to pray for clear skies for a child's outdoor birthday party, or a sunny wedding day for a bride. However, contemporary meteorologists, especially at the National Hurricane Center, currently hold the attention of a far greater segment of the population than any antiquated weather clairvoyant. The words of scientists as they interpret a hurricane's path, for example, have the power to evacuate millions by their predictions. It's accepted that these specialists have the ability to decode an array of technological equipment that remains incomprehensible to the masses, thus making them powerful authorities. In New Orleans in 2008, for example, their words caused the evacuation of two million people, at a tremendous cost, because of a prediction that proved false. In 2005, a wrong prediction had 2.7 million fleeing Houston, which led to the deaths of 130 people. Among the casualties were thirty elderly who, after being packed into a church bus, died when their oxygen tanks caught fire. (*See Also:* Divining Rod)

After a prolonged drought, Georgia governor Sonny Purdue implored residents to pray for rain. In November 2007, on the steps of the state capitol, the governor had the few hundred in attendance holding hands, as he pleaded: "God we need you. We need rain." It drizzled in parts of Georgia later in the afternoon, though not enough to end the water shortage.

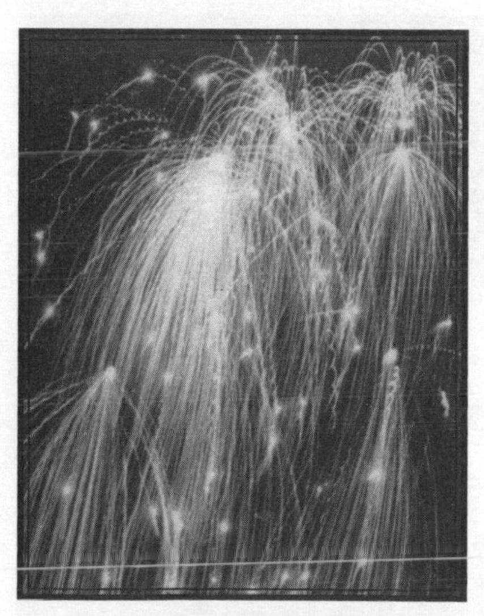

ST. BARBARA

PATRON SAINT OF FIREWORKS

A third-century Greek merchant wanted to keep his daughter protected from the outside world and so fashioned a lovely tower where she lived in nearly complete seclusion. He constructed it with two large windows, from which he hoped she could observe the world at a safe distance. However, while he was gone on business she had a third window added to her quarters, which to her secretly represented the Trinity. When the father returned and eventually discovered she had been swayed by Christian ideas, he drew his sword to lop her head where she stood. Fortunately, angels quickly opened a hole in the roof to whisk Barbara from the room. Her father then saw Barbara waving from a faraway hilltop, standing between two shepherds, and he raced to retrieve her. One herdsman tried to hide Barbara, but the other was unfaithful. He betrayed her whereabouts, though he was turned into a stone statue for doing so and his sheep were transformed into a swarm of locusts. Barbara was retrieved and tortured to recant her faith, but each morning it was found that all the previous inflicted wounds were healed. The father had enough, and cut his daughter's head off, to finally put an end to these foolish beliefs in miracles. However, retribution came quickly; as her father galloped away from the execution, he was knocked from his horse and killed by a bolt of lightning on an otherwise bright and sunny day. Since this miracle was attributed to St. Barbara, she was made the patron saint of gunpowder—and later of artillery specialists. Many pray to St. Barbara for the accuracy she evoked against her father. The saint still receives prayers at the onset of firework displays. Many pyrotechnic workers wear St. Barbara medallions to help avoid mishaps.

DEATH BY LIGHTNING STRIKE

In antiquity, those killed by lightning were considered not only unlucky but "unclean," and were frequently denied burial on consecrated church grounds. However, it's estimated that on average more than ten people die each year while attending outdoor religious events. The most bizarre tragedy occurred on March 25, 1906, to Reverend J. B. Lentz, the pastor of a Seventh-Day Adventist Church in Carson, Iowa. While a thunderstorm raged outside, he was at the pulpit giving a sermon. Standing before the entire congregation, the reverend was suddenly struck and killed by a lightning bolt. The electrical surge hit the building and somehow transferred through the wiring. The deadly charge burst from a chandelier above his head, and it was powerful enough to set his body on fire. Many fled, though a few stayed to save the church by carrying the poor, smoldering pastor outside into the rain.

BEDE, THE VENERABLE

THE MONK WHO INVENTED TIME

It was the unassuming, steady-handed English scribe Bede who popularized A.D. as an abbreviation for *Anno Domini*, or Year of Our Lord. The international calendar that is accepted today, called the Gregorian calendar, or Christian calendar, came into being in 1582, but it relied heavily on Bede's calculations centuries earlier. Bede had been concerned about getting the celebration of Easter correct, and led to his method of counting time for following lunar and seasonal changes, which he explained in his masterpiece, *The Reckoning of Time*, in 724 A.D. He decided to count modern time starting with year 1 as the day and time Jesus was born, making everything that happened before that date B.C. as a way to write about histories and distinguish eras—even if in fact it was later proved he miscalculated the birth of Jesus by four to six years. Bede used no zeros, however, meaning that according to him, no one ever lived in the year 0. Bede was an inconspicuous Benedictine monk; nearly the only thing known about him is that he was placed in a monastery at age seven. He rarely ventured outside the abbey's walls, and there were no incidents of note on his cloister records. It seems he was dutiful in attending mandatory choir practice when not writing.

RESETTING SPIRITUAL CLOCKS

Time was not standardized until the mid-1700s, such that George Washington actually had two different dates for his birthday each year, depending on the various calendars he chose to use. The Jewish calendar began counting time from what was thought as the first day of creation. That makes the year normally considered to be 2010, for example, to be instead the year 5,770 according to Jewish time. Muslims began fixing year 1, day 1 of time as the date July 16, 622 A.D., or the "Year of the Hegira." Muhammad was driven from Mecca to Medina on that day. However, whatever date it is today, it is numbered so because of the calculations of a pious monk. (*See also:* Unarius Science)

In medieval times, and in most calendar versions through the eighteenth century, a day ended with sunset. A month began with the sight of a crescent moon in the sky, which Bede considered as "the knife of time," slicing one month from the next. According to fellow monks, Bede knew he was dying and called a novice to his "cell" to give away his prized quills and inkwells. When he died at age sixty-three in 735, his tombstone was simply marked BEDE. However, the next morning VENERABLE was etched before his name, supposedly chiseled by angels during the night.

THE MYSTICAL BEDE

In addition to the calendar, Bede also wrote *The Ecclesiastical History of the English People,* which remains a well-known historical work. His insights were often profound, as when he compared the life of man to a small bird: "This sparrow flies swiftly in through one door of the hall, and out through another. While he is inside, he is safe from the winter storms; but after a few moments of comfort, he vanishes from sight into the wintry world from which he came. Even so, man appears on earth for but a little while; but of what came before this life or of what follows, we know nothing."

BEGUINES

MEDIEVAL RELIGIOUS SORORITY

Beginning in the twelfth century, all-female communes, known as the Beguines, sprung up throughout Europe. Although there were convents for Christian women, the best often required sizable entry donations, and they were dominated by males and other church authorities. The Beguines instead gathered in clusters of small cabins or huts, consisting of usually thirty to one hundred women. Unlike nuns, however, they took no formal vows, but were thought to be as pious as nuns since they avoided heterosexual sex and attempted to disassociate with men whenever possible. Although men believed the Beguines would eventually die off without them, they at first refused all alms and earned money through manual labor. By the fourteenth century, some communes were housed in large dormitory buildings as a central gathering place, in addition to outlying cottages, though still there was no hierarchy, or official doctrines. (From 1230 until 1310, a Beguine commune in Ghent, Belgium, had more than one thousand women.) Individual ownership of property was discouraged, and although they followed many Christian rituals, they often worshiped in the nude. Beguines rejected marriage in favor of polyamory, or having multiple female partners. Persecution began in Paris in 1310 when outspoken Beguine mystic Marguerite Porete was burned for heresy. After she refused to recant her views on "divine love" or cease circulation of her book *The Mirror of Simple Souls*, the church's stance toward Beguines changed. In addition, many communes, once industrious, had now begun to practice astrology and card reading, and started begging for money. Thousands were then accused of heresy and executed, although many others continued to operate in secrecy during the Inquisition. In the 1700s, the few remaining groups became Benedictine nuns and called their commune buildings convents. By the 1960s there were only twenty-five Beguines left. The last of the Beguines, one Sister Marcella of Belgium, died in 2008 at age eighty-eight in a nursing home.

WOMYN COMMUNES

In the 1970s, there were a number of attempts to form all-female communes, founded by radical lesbians. In rural locations from Florida to Oregon, groups numbering as many as one hundred attempted to establish self-sufficient, egalitarian, and matriarchal societies, all with the common principle of prohibiting men. Some folded due to financial sustainability, while others found greater conflict in deciding what feminist principles to uphold, for example whether a transsexual should be admitted. Others who rejected monogamy as "possession" still found changeable relationships among a limited number of women problematic. This eventually

caused jealousy and resentments, and one reason many of the utopian "back-to-the-land" experiments failed. WomanShare, founded in 1974 by three women in southern Oregon, wanted to make "a home and family of lesbians." It still operates on twenty-three acres, though younger lesbians usually only stay for a weekend, with many of the older and more dedicated dwindling in numbers. Another example is the all-female community called Alaphine, located in northern Alabama, which has mostly elderly lesbians as residents. One sixty-year-old, who identified herself as a "radical feminist separatist lesbian," admitted that her two daughters could not visit, since they are not lesbians. In 2009, when one member arrived carrying her newborn grandson, an e-mail alert went around—"A man was on the land!"

CONRAD BEISSEL

VEGETARIAN UTOPIA FOUNDER

When Conrad Beissel, a baker by trade, migrated from Germany to Pennsylvania in 1720, he joined whatever new religious sect flourished (*See also*: Dunkards), though he favored hermitage as the path to spirituality. By 1732 he was dissatisfied with the sects and worship practices he had tried, so he decided to start his own religious movement, establishing a spiritual commune he called the Ephrata Community, or the Camp of Solidarity, near Lancaster, Pennsylvania. He attracted followers by claiming revelations and that he had been personally instructed by God to foster a new society. In addition, Beissel had a striking posture, and when he delivered insights with his curt German-English accent, he seemed self-assured and believable. None doubted his piousness, as his reputation and character were admired for being impervious to seduction; he was seemingly uninterested in any manner of courtship. His charisma rose until people were soon willing to do whatever he said. One stern requirement he demanded of all was absolute celibacy—a necessity, Beissel believed, in order to receive God's message undiluted by lustful distractions. Subsequently, the commune assigned men and women to separate buildings. He likewise attested that vegetarianism was the best diet to cleanse the mind and body and make one better suited to hear God's instructions. Officially, his utopia became the first completely vegetarian commune in North America, even if, as some say, Beissel and his group ate lamb, once a year, on Easter. Beissel loved music, wrote hymns, and supported the commune through a printing enterprise. They grew their own food and limited mingling with the neighboring communities. He also urged complete silence—except for the sound of his eerie organ, which Beissel played when inspired. His extremely cloistered and austere lifestyle went on after his death in 1768, since Beissel had set up the finances so well that the movement continued. Nevertheless, there were no spiritual leaders of Beissel's caliber to attract new recruits, and so his utopian experiment limped along until the buildings deteriorated. Without a new generation to carry on—there was no sex allowed, even to produce children—devotees had disbanded entirely by 1814.

VEGETABLES AND RELIGION

Jainism is the only large-scale religion that forbids all followers to eat meat. Hindus and Buddhists strive to eat a vegetarian diet because they do not want to consume the violent spirit, or "negative karma," of a slaughtered animal. Christians, Jews, and Muslims limit certain types of meat or recommend abstinence on certain days. Five million people in the United States claim to have an exclusively vegetarian diet, though mostly for health and social reasons, not religion. Some sects adhere to strict dietary regimes and believe some foods are required for spiritual sustenance. Vegans, the most strident class of vegetarians, will not eat meat or dairy products, and only consume fruits or nuts that have fallen from trees. In 2002, one vegan tried to sue his employer that required workers to be vaccinated. He claimed the policy was religious discrimination since he refused to get an injection of a mumps vaccine, protesting that it contained ingredients grown in chicken embryos. The California Supreme Court heard the case and refused to accept veganism as a religion. In 2004, vegan parents Joseph and Lamoy Andressohn were charged with aggravated manslaughter when their malnourished five-month-old daughter died. The child weighed only six pounds and had been fed only wheatgrass, coconut water, and almond milk. Likewise, the court dismissed religious motivation, and after jail time, the parents received ten years probation. One vegan writer, Jeffrey M. Freedman, noted, "For me, it is a prayer, a petition asking why animals and people suffer greatly in a Universe created by a benevolent and loving God."

REFORMING CARNIVORES

St. Blaise, a fourth-century Armenian monk, is the patron saint of sore throats. He got that position because he saved a child who was choking to death on a fishbone. Before this, he had an unusual relationship with animals. In the summer of 314 he met a poor weeping widow on the road who was devastated that her pig had been snatched away by a wolf. St. Blaise found the wolf and scolded it for trying to eat one of God's friends. The wolf released the swine and in turn, St. Blaise showed the carnivore how to get as much nourishment from milk. It's unknown if the wolf remained a vegetarian. In any case, the widow was very grateful; she slaughtered the pig and offered a plate of the barbecued pork to St. Blaise.

DAVID BERG

SEX-CULT FOUNDER

David Berg was believed to be a prophet, popularly called Moses David, though he was known among his faithful as simply "Mo." Berg was raised among a devoutly evangelical family, and, following in his father's footsteps, he was trained as a minister by the Christian and Missionary Alliance. Due to differing beliefs and for issues concerning sexual misconduct, Berg left to form his own religion. In 1968, the forty-nine-year-old Berg founded the "Children of God," or "the Family of Love," and later referred to his group as simply "the Family." Berg claimed to be devoutly Christian, though he believed the essential message of the entire New Testament was boiled down to one concept—love. For Mo, this was to be displayed both figuratively and literally, as in having sex, whenever and with whomever, as long as it was heterosexual. (Except for females—they could be bisexual.) Certain hippies, then called "Jesus freaks," were instantly attracted to his ideas of a spiritual revolution against what Berg called "the System." In order to recruit converts, Mo sent the prettiest girls to hand out brochures and taught them how to practice "flirt fishing." They were instructed by the prophet to entice the potential converts with promises of sex, and even have sex if it seemed that a prospect was ripe for conversion to the sect. The more members the better, since the group was supported by tithing (that is, taking a percentage of the income from followers' paychecks). By the mid-1970s Mo had decided that his sexy recruiters made more for the church when they left the spiritual brochures behind and were able to find more "sinners" by working as outright prostitutes. He eventually became awash in legal woes concerning his ideas on rape, and he was formally accused of pedophilia in 1974.

"THE DEVIL HATES SEX—BUT GOD LOVES IT."

When questioned about his views on rape, Berg replied: "Even under the Law of Moses or even under the early days of the Patriarchs, if a guy got caught [screwing] one of the local girls out in the field, if the family didn't like it, all they could demand was reparation, pay them some money for it." He believed that no man-made laws about sex, including rape, should be punished by a court, and that God's laws allowed all of that behavior. He said a rape victim should not complain and would be given "credit" by God.

However, he escaped U.S. prosecution by moving operations to Southeast Asia, Europe, and Argentina. Berg never did time for pedophilia, though his daughters and granddaughters testified to his sexual abuse against them. River Phoenix, the actor, was a child when his parents were members of Berg's sect. He later admitted that he lost his virginity at age four while living on Berg's compound. Toward the end, Berg lived in seclusion, though he sent out nearly three thousand epistles on his beliefs, called "Mo Letters," which his supporters deemed as divine revelation, no different than the Bible. His "Law of Love" doctrine encouraged sex regardless of marital status, as a spiritual wife-swapping, of sorts, to promote unity. Berg's notion of "sacrificial sex" fostered his idea that God gave man the ability to engage in sex to combat loneliness and help those in deep need for love. In 2008 the group was known as the Family International (TFI), though cleaned up and disallowing sex with children, and is currently headed by Berg's second wife, Karen Zerby, known as Mama Maria. Members then and now were excommunicated if they did not adhere to secrecy concerning sexual conduct; by some accounts there are still nearly eight thousand followers. Berg often resorted to apocalyptic prophesy, without much accuracy, as in his claim that the Second Coming of Christ would occur in 1993. In his later years he was a fugitive and on Interpol's worldwide search list, though he supposedly died in 1994, at the age of seventy-five. (*See also:* John Humphrey Noyes)

THE BIBLE

THE GOOD BOOK

Bible comes from the Greek word for "book," any book. *Biblia* was the Greek term for the bark of a linden tree, then used as paper. It didn't come to refer exclusively to the combination of the Old and New Testament until the fifth century. The Catholic Biblical Federation conducted a survey and found that 75 percent of Americans read the Bible at least once each year. Gallup polls revealed that more than a third of all Americans believe the Bible to be the literal "Word of God," though even among that group 56 percent said it is not easy to understand. No doubt, since it underwent numerous translations, with Jewish scholars attesting to archeological data that places the Torah, the most revered scriptures of Judaism, at more than three thousand years old. But oral Hebrew tradition believes it was created way before that by God, at least two thousand years prior to creation, and served as a blueprint and an agreement, or "covenant" between God and man.

The origins of Bible stories have been examined thoroughly through the ages. Many scholars find similarities in the Bible to writings taken from older cultures and religions. For example, Genesis, the first book of the Bible, has been compared to the Epic of Gilgamesh, written by a Mesopotamian in the seventh century B.C. and now considered the oldest known example of literary fiction. The story includes a man and woman in a garden, expelled from paradise, and even has a story about a flood that washed away much of known civilization due to man's evil. The Old Testament was written in Hebrew and Aramaic between the 3rd and 1st centuries B.C., and was translated into Greek. The most complete surviving text was written in Greek, called the Septuagint, meaning that seventy translators worked on it, and dates to 300 B.C. Religious authorities in varying periods, both Hebrew and Christian, kept or rejected some chapters, books, or parts for various reasons, until the testaments contained sixty-six books, thirty-nine in the Old and twenty-seven in the New Testament. There are at least thirty-six known authors. The first complete version was translated into English in 1380 by John Wycliffe, and became the first book ever printed on a press in the 1450s by Johannes Gutenberg. The King James Bible became the "Authorized Version," deemed so by "forty-seven learned men" in 1611. Many words from older sources were changed or mistranslated. One of the key words changed was *slave*, found often in early texts and used to describe a disciple or follower of Jesus: This was altered across the board and became *servant.* The entire book was revised again in 1870 when a committee of eighty American and English scholars from various denominations decided on the "true" translation of many archaic words. One 1631 edition was circulated widely until an important typo was discovered: That printing, eventually dubbed the "Wicked Bible," listed one of the Ten Commandments as "Thou *shalt* commit adultery." (*See also:* Gideon Bible)

RECYCLED SCRIPTURES

In 2007, Americans spent nearly half a billion dollars buying more than 25 million bibles, even though estimates cite nearly 600 million copies are still in circulation. People have always been reluctant to discard an old Bible. Placing them in trash cans seems irreverent to many, or bad luck to others. A torn or faded American flag has an outlined procedure for disposal, and is ultimately supposed to be burned. However, for many, igniting a Bible seems too dramatic, though numerous religious leaders suggest keeping, giving away, or burying it. Catholic priests have a certain prayer to bless a Bible before putting it into its grave.

BITTER HERBS

A TASTE OF THE BAD OLD DAYS

Many religions have developed lists of certain foods, some merely symbolic and others recommended to help promote health. These foods have thus become part of the religions' canons and traditions. (*See also*: Pharmacologism) Rastafarians, for example, cannot eat fish that measure less than twelve inches. Mormons are discouraged from eating chocolate, and from consuming various prescription drugs. Catholics at one time avoided meat every Friday, but now do so only on certain feast days or during Lent. Orthodox Jews must refrain from a number of foods, including pork and clams.

Eating bitter herbs was an instruction given to Hebrews to commemorate Passover and the Exodus, when Israelites escaped Egypt and slavery. Bitter herbs are to be reminders of that harsh bondage, and originally might have been such plants as hawkweed, sow thistle, and wild lettuce, which grew in the Sinai Peninsula during the time of Moses. Sometimes dandelions and chicory served as bitter herbs and were to be eaten with unleavened bread, or flour made without yeast. Today a side of kosher horseradish will satisfy the requirement of bitter herbs.

BODY OF CHRIST

When Jesus broke bread at the Last Supper and gave it to his disciples, he began a tradition that is still practiced today, with more than 634,000,000 pounds of Eucharistic wafers consumed each year worldwide. (A wafer weighs less than two grams.) A Eucharistic wafer can only be made from wheat and water. In 1237, a bishop stationed in the boondocks of Greenland implored the pope to allow another grain as an ingredient, since no wheat was available. He was told to forgo serving communion until the proper recipe was followed. The missionaries in North America in the 15th and 16th centuries wanted to use corn, a native staple, to make their communion hosts. They were not permitted, either, and so every traveling missionary always carried a handful of wheat seeds in his satchel. During the bubonic plague, some exceptions were made because rumor had it that wheat was the source of the disease (actually, it was infected rat urine and fleas that had saturated stored food supplies), and priests didn't want to get sick handling it. Some then used wafers made from thin slices of potatoes, which is believed to be the forerunner of the potato chip. Today, 1 in 133 Americans suffer from celiac disease and are allergic to wheat, making communion truly "bitter" to them. The faithful who suffer from this predicament have petitioned the Vatican to offer a wafer substitute, and now must put in a special request for a holy but gluten-free host.

BLACK MAGIC

THE DARK ARTS

Hexes and curses intended to kill or harm are categorized in occult writings as dark or black magic spells. The intention is to summon unseen forces to maliciously alter the destiny of another. The spells are guarded, and in the wrong hands are considered more dangerous than a loaded gun. For the inexperienced, the spells often backfire and adhere to the principles of "the Threefold Law." (*See also*: Wicca) According to this canon, the fee the dark powers charge is seven times the evil done on others, and is often extracted from the curse-caster and their lineage. Many hexes target a person's money, but usually it's health; a curse can be as specific as bringing an affliction to a precise body part. For example, if tourists in Hungary and other parts of Europe refuse to get their palms read or fortunes told, a gypsy will quickly curse the penny-pinching traveler to impotence.

May you wander over the face of the earth forever, never sleep twice in the same bed, never drink water twice from the same well and never cross the same river twice in a year.
Original Gypsy Curse

In voodoo, black magic spells involve making a figurine or doll of the person intended to be cursed. To activate these effigies, however, one must have the victim's hair or drop of blood. Pins are then stuck into the doll; these can affect any part of the cursed person's actual anatomy—anything from an ingrown toenail to heart problems. In Australia, if an aborigine points a kangaroo bone at you, it's usually in order to cast a curse. In Russia, Rasputin was held in high esteem by the ruling czars for his ability to see the future and cast curses to alter it. But on his deathbed, he was more than a bit upset when members of the ruling family turned against him: Rasputin was shot, thrown through a hole in a frozen lake, and reportedly castrated. His last hex was against the entire ruling Romanov lineage, and within a year they had all been murdered. (*See also*: Rasputin's Lucky Charm)

In 2004, Masoud Haroub Saidi, a member of the Tanzanian parliament, got tired of the slow-motion pace of politics and the corruption he considered rampant. He got worldwide attention when he threatened to use a curse to make political rivals "drop dead like locusts." He went on record: "When we [Muslims] want to stop things from happening we use Halbadiri—an Islamic death curse."

BLACK MASS

SATAN'S SACRAMENT

The black mass was originally intended as a protest against the Catholic Church's insistence that only certain men could perform a mass. There was rampant abuse among clergy, with the less than pious often charging a fee to say mass, hear confession, or bless fields and such. It originally had nothing to do with Satanism or the worship of the devil, but as a remonstration that only priests could perform the transformation of bread into the body of Christ. In medieval times, defrocked priests supplied hosts to be desecrated at these ritual ceremonies. The Lord's Prayer was recited backward, and sacrifices both animal and human often supplied blood to be mixed with wine. In the seventeenth century, the black mass included placing a naked woman on the altar, with

a chalice balanced between her breasts. Sometimes the host was colored black, and in some cases the priest wore a long hooded red robe and carried a dagger. (*See also*: Church of Satan)

THE MASS OF ST.-SÉCAIRE

In France from medieval times to the mid-1800s, certain priests were paid to secretly perform a version of a black mass known as the Mass of St.-Sécaire, meant to inflict pain and death on an enemy. For the ceremony to work, the liturgy of a regular mass had to be recited backward, beginning at exactly 11 P.M. and ending at the stroke of midnight. The priest employed a prostitute as altar server and used a black, triangle-shaped host for the Eucharist. Instead of wine, water taken from a well—preferably one where an unbaptized baby had been disposed—was used in the chalice. Upon leaving, the priest had to make the sign of the cross using his left foot into dirt, thus cursing the targeted victim to suffer incurable consumption. If Christians were to witness the mass, they were warned that it would cause immediate blindness. If a priest was caught serving this mass it was considered such an unholy crime that only the pope could pardon it. Nevertheless, many took the chance, since the going rate paid to priests was considerable, as much as ten thousand dollars in today's currency. It was a ritual that supposedly seldom failed to conjure up dark forces. The Vatican denies that such a liturgy or St.-Sécaire himself ever existed.

A detailed picture of black mass rituals was revealed in court documents during the trial of the French aristocrat Gilles de Rais, who had once fought alongside Joan of Arc. He kidnapped and killed hundreds of children for the black masses conducted in the basement of his castle. He was hanged in 1440. In the 1600s a French priest practiced black masses that also included orgies: The Catholic abbot Étienne Guibourg was encouraged by Catherine Monvoisin (called "La Voisin" and known for her expertise in poison potions) to perform a mass. The ceremony included kinky sex to help speed the death of King Louis XIV's mistress. In 1680, La Voisin was burned at the stake; the abbot, at age seventy, was sent to prison, where he died three years later. (*See also*: Devils of Loudon)

For centuries, to prevent the stealing of hosts for use at black masses, the church required the bread to be placed directly into a person's open mouth. Now priests and church deacons are allowed to put the communion wafer into a recipient's hand, though many are trained to watch carefully to make sure he or she swallows it and does not carry it off for other purposes.

MADAME BLAVATSKY

RUSSIAN PSYCHIC BELOVED BY THE NAZIS

Elena Petrovna Gan, better known as Madame Blavatsky, co-founded the Theosophical Society in New York City in 1875. This was the first "New Age" type of organization in the United States, with a focus on the discovery of universal mysteries through psychic powers. After visiting Tibet, Egypt, and numerous exotic locales, the Ukrainian Blavatsky arrived in America with an unusual grasp of many world religions that at the time were little understood by the general public. However, the organization soon decreed it was formed to create a "Universal Brotherhood of Humanity." The group fostered examination of Buddhism, Zoroastrianism, and scriptures, though they believed these and many religions had problems and conflicts that could only be fully understood by tapping into "the psychic and spiritual powers latent in man." Madame Blavatsky could levitate, send messages telepathically, see and hear spirits, and transport her consciousness from her body to other locations, and was even reported to materialize things, or make things appear from nothing. She was certainly controversial but she had a loyal following. Soon after the group was formed, she devoted her time to writing, in an attempt to explain in depth her ideas, which influenced many thinkers. Her goal was to encourage open-mindedness.

Although she advocated peace and unity for all humanity, one notion she held concerning man's origins later became an important propaganda tool of Nazi Germany. Blavatsky believed humans evolved from seven distinct "root races." She claimed that Aryan people were the most advanced, the descendants of a superior race that once inhabited the lost city of Atlantis. The Theosophical So-

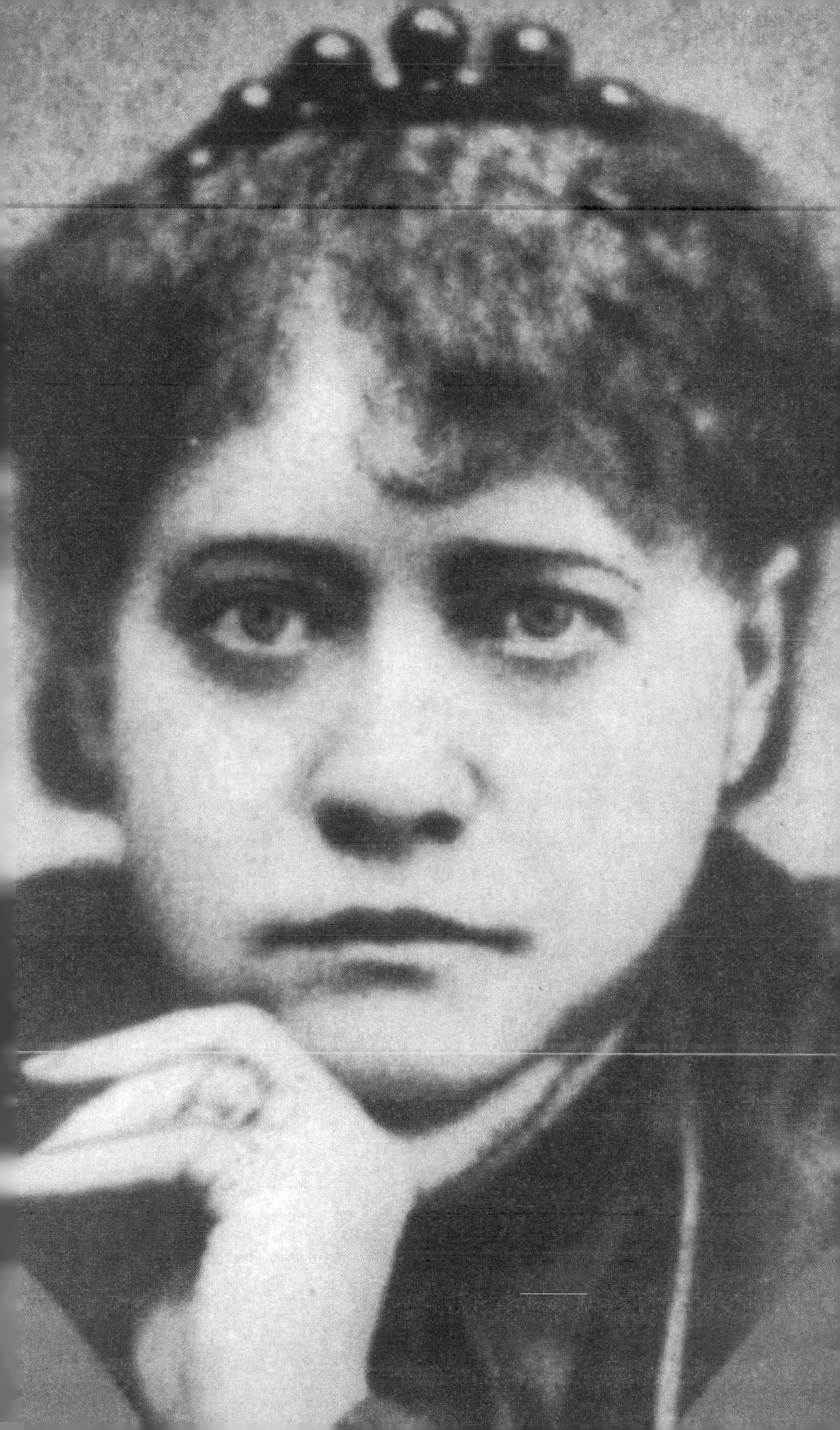

ciety logo consisted of iconic symbols, including a Star of David and a swastika, which was originally a Hindu figure signifying good luck.

She wrote books still consulted today, especially *The Secret Doctrine*, which came to her through clairvoyance and automatic writing. Eyewitnesses who watched her compose merely marveled at her pen flying across the page. Blavatsky produced more than five thousand book pages in her fifteen-year literary career. She was a stocky woman who wore bizarre outfits, usually sacklike dresses, and often led a big black dog around on a gold chain. She wore a small animal head around her neck that served as a pouch for tobacco to use for her hand-rolled cigarettes. She ate whatever she wanted, sometimes little, other times bingeing, and was known to have a red-hot temper. In addition, some claimed she liked to occasionally smoke hashish as a way to "multiply life a thousand fold." For herself, she advocated celibacy to keep her psychic powers pure, and attested proudly on her deathbed that she had indeed remained a lifelong virgin. She left the United States in the late 1870s when newspapers began to seriously assault her character. She moved to India, then England, where she died of a urinary tract infection in 1891 at age fifty-nine. Her last words: "Keep the link unbroken! Do not let my last incarnation be a failure."

Man is also triune: he has his objective, physical body, his vitalizing astral body (or soul), the real man; and these two are brooded over and illuminated by the third—the sovereign, the immortal spirit. When the real man succeeds in merging himself with the latter, he becomes an immortal entity.

—Madame Blavatsky

BLOODY MARY

ENGLAND'S MURDEROUS CATHOLIC QUEEN

At first the pride and joy of her father King Henry VIII, Mary lost status quickly and was deemed illegitimate after the king divorced her mother, Catherine of Aragon. Although outwardly conceding to Henry's demand that all subjects become Protestant, Mary was raised Catholic and remained staunchly so, especially when it gave her allies in Rome reason to help her regain her royal rights. When Henry died, his son Edward VI took charge, but he died shortly after. Mary then stepped in to claim the crown. By then the Church of England had been entrenched as the official religion for nearly twenty years; however, Mary attempted to reestablish Catholicism as it was before. While at first popular among the country's closeted Catholics, she fought off numerous Protestant plots to kill her, until she struck back with a vengeance. It's recorded that she burned nearly three hundred Protestants at the stake, earning the moniker "Bloody Mary." She reigned for nearly five years as monarch before she died of ovarian tumors at age forty-two in 1533. The morning of her death she had a Catholic mass said in her chambers, received last rites, and agreed to turn the throne over to her sister Elizabeth. Mary requested that Elizabeth keep England Catholic, but it was a stipulation that Elizabeth abandoned immediately.

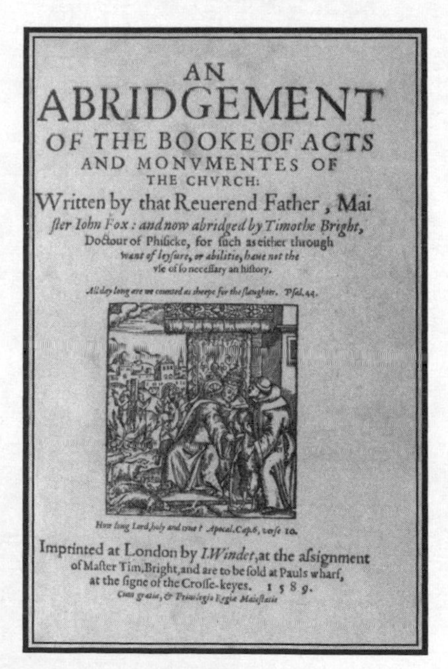
AN
ABRIDGEMENT
OF THE BOOKE OF ACTS
AND MONVMENTES OF
THE CHVRCH:
Written by that Reuerend Father, Mai
ster Iohn Fox: and now abridged by Timothe Bright,
Doctour of Phisicke, for such as either through
want of leysure, or abilitie, haue not the
vse of so necessary an history.

All day long are we counted as sheepe for the slaughter. Psal. 44.

How long Lord, holy and true? Apocal. Cap. 6, verse 10.

Imprinted at London by *I. Windet*, at the assignment
of Master Tim. Bright, and are to be sold at Pauls wharf,
at the signe of the Crosse-keyes. 1589.
Cum gratia, & Priuilegio Regiæ Maiestatis

FOXE'S *BOOK OF MARTYRS*

John Foxe chronicled Queen Mary's brutality in his famous *Book of Martyrs* while Mary was on the throne. Foxe had risen in prestige under Protestant rule and was a deacon when Mary came to power. He managed to escape his abode just as the Bloody Queen's soldiers were en route on horseback to arrest him. He wrote his

ARYE the QUEENE.

first list of English Protestant martyrs while he was abroad in 1554. When he returned to England after Elizabeth I, or "Good Queen Bess" (*See also:* Martyr Factory) restored Protestantism, his books became standard reading, though Foxe earned no royalties since copyright laws didn't exist. Nevertheless, he became a martyr of a different passion, dying in 1587 at his writing desk, at age seventy.

BOOK OF THE DEAD

THE MUMMIES' GUIDE TO THE AFTERLIFE

Egyptians listed certain hymns, mantras, and answers to trick questions that would be needed in the afterlife. Known by Egyptians as *The Book of Going Forth by Day*, these collected pieces were usually placed in the casket or tomb as a general reference guide to the Underworld. In 1842, when some of the original text was first published by a German Egyptologist, it came to be called the Book of the Dead. It wasn't an actual book in Egyptian times, but rather was written on papyrus scrolls and had no strict versions, unlike the manner in which Bible scriptures were translated. These coffin texts were laboriously handcrafted, sometimes specifically designed for the deceased. A personalized copy cost a fortune, though there were bargain copies, in which the deceased's name was filled into a blank. The pages had elaborately painted images, with many pop-up messages that gave the dead clues on which incantation to use to overcome obstacles that various gods had devised. For example, some versions suggested using a ladder, while others might recommend invoking a flying spell, very similar to video games that rely on the player to decipher what strategy to employ to get to the next level. (*See also*: Mummy)

DEVOURER OF THE DEAD

To the Egyptians, the matter of the afterlife was an ever-present and fearful concern. The prospect of getting to an ideal and heavenly version of Egypt was remote. Demons, traps, and at least forty-two different judging deities made such chances extremely difficult, and thus many spent a lifetime learning the incantations in order to crack the code. If you failed these tests when dead, you would be found by the Devourer of the Dead, a creature with a sharp-toothed crocodile head, the powerful front legs and clawed paws of a lion, and the stout hindquarters of a hippo. When you were ultimately judged, your heart was placed on a balancing scale and had to be so pure that if it weighed more than a feather the Devourer immediately snatched you away. Once you were savagely torn apart, it would then swallow your soul, which would reside forever in the foul intestines of this monster.

ISAAC BULLARD

WILDERNESS PROPHET, LOVER OF BEARSKIN CLOTHING

In 1817, the residents of the small town of Woodstock, Vermont, were roused by strange chants emanating from the surrounding woods. Isaac Bullard, a self-proclaimed prophet, had come to town. The Canadian-born Bullard, once a mild-mannered farmer, was on a mission after receiving a direct command from God. Bullard believed the advancements of modern society were sure to lead to humanity's damnation. Bullard claimed he was called by God while he had been seriously ill, though ill of what, he refused to elaborate. To cure himself he shunned all man-made items, and after a forty-day fast, he had a vision. He began to preach that God had given him a command to interpret the Bible in a new way and had sent him forth on a mission to start a model community in the wilderness. He cast off traditional clothes, gave away all possessions, and dressed in bearskins. He also grew an imposing pointed, flaming-red beard that complemented his piercing eyes and booming baritone voice.

A reporter from the Pittsburgh Gazette *was dispatched to Woodstock to get a glimpse of this new prophet and described Bullard and his followers in the November 4, 1817, edition: "So incessant were their professed addresses to and communications with invisible beings, with whom they pretended at times to hold converse . . . their fame went abroad."*

Bullard mostly scared the devil out of people rather than enlisting followers. Still, a few locals from each town joined his sect and called Bullard the new Prophet Elijah, until his number of followers reached more than two hundred. Bullard cared little for titles, though he was the de facto leader, and he was a stern disciplinarian. He promised to lead his disciples to a Promised Land and set up a commune. By example of their righteous living, they alone would be spared on Judgment Day. Bullard was certain his ideas would start a global sensation and be imitated by all. However, by any measure Bullard's path to God was rigid, for he demanded a life of severe self-denial. His followers had to wear coarse, skin-irritating bearskins, as well as endure long fasts, broken periodically by simple meals of gruel. He even thought his flock might be corrupted by eating this plain fare with utensils, and so he forbade knifes, forks, and spoons, and made them suck up the mush with a straw while standing. In his reading of the Bible, Bullard ascertained that bathing and cutting hair was wrong. So he encouraged his congregation instead to roll in the dirt as an act of repentance. Naturally, the sight of wild-haired, mud-caked men, women, and children dressed in bearskins heading toward a town drew crowds.

What really got the media's attention during Bullard's journey, and in the various sightings of his group, was their strange chanting. Bullard would spontaneously stop and face his traveling congregation to pose his favorite rhetorical question: "My God, what wouldst thou have me do?" Without fail, his group began to sway and scream at the top of their lungs: "Mummyjum, mummyjum, mummyjum." Seemingly as if spirit-possessed, they repeated this phrase until they were stricken with exhaustion and fell to the ground. What the phrase actually meant no one knew, and when Bullard was pressed to explain, he spread his arms over his followers and indicated that their lifestyle and beliefs were the answer, or "the mummyjum." Bullard led them on a long journey to the Arkansas Territory, outfitted with nothing more than their bearskin clothes, a few oxen, and a resolution that the power of "mummyjum" would make them prosper. They never asked for handouts, though were routinely beaten, horsewhipped, and run out of one town after another. However, they persisted, convinced that their example of a religious commune would one day spread across the planet.

Little is known of what actually became of Bullard and his followers; they were last seen in 1824, approximately seven years after Bullard began preaching. Legend had it that Bullard walked across the Mississippi River and ascended into heaven. Others think he was scalped by Indians. In 1824, a traveler came across the last of the surviving Mummyjummers, found desperate and seemingly deranged, living in three primitive huts along the banks of the Mississippi. Reportedly, "mummyjum" was the only word they uttered, no matter what the query. The prophet Bullard left no writings to further explain his visions. The incantation he invented remains, though it has yet to be deciphered.

a

COUNT CAGLIOSTRO

ALCHEMIST, TELEPORTER

Count Cagliostro remains one of the most enigmatic and mysterious figures in occult history. Most scholars think he was merely a street urchin, petty criminal, and expert forger named Giuseppe Balsamo, born in 1743 in the slums of Palermo, Sicily. However, admirers of Cagliostro attest he was not this notorious con artist, but instead an enlightened time-traveler. Although mostly lauded, the count was eventually alleged to be a charlatan by enemies, who feared that he held the fascination and the ear of too many of Europe's leading monarchs. According to Cagliostro's own testimony, when questioned by the court of Louis XVI about the heist of a valuable necklace, the count said he was of noble birth, though orphaned at a young age. During his youth, he traveled from Mecca to Cairo, then remained through his teen years on the island of Malta. There he was inducted into the Sovereign Military Order of the Knights of Malta, where he learned the secrets of Kabbalah, alchemy, and magic. He often worked as a physician throughout Europe, and was credited with miraculously healing hundreds of peasants. His reputation was such that crowds of the infirm, carried on stretchers or wobbling on crutches, lined the roads wherever he went. He said he learned these curative powers from the discovery of ancient secrets, though he never named the exact source.

Cagliostro traveled with a beautiful young wife—said to be fourteen years old when he married her, though he claimed she was actually fifty and that he had merely transformed her into her more youthful form. His repute became legendary and he inspired Mozart's *The Magic Flute*, Goethe's *Faust*, and possibly Alexandre Dumas's *The Count of Monte Cristo*. In addition, Benjamin Franklin was told that Cagliostro was the best physician in France, during the American diplomat's stay in Paris in 1776. It was thought that Cagliostro had the skill to bring back the dead, and once, according to Tolstoy, the count resurrected a woman by making the portrait of the deceased climb out of the frame. The count told of how he had been born many times before and had walked on the shores of Galilee with Jesus. He also had a magic cloak that allegedly transported him and his wife at will, and they were noted to arrive at record time in England, Germany, Russia, and Italy. He also opened a number of Egyptian Rite of Freemasonry lodges throughout Europe and repopularized the ancient symbol of the Ouroboros, an icon that depicts a snake biting its own tail.

In the end, Cagliostro appears to have been arrested in Rome in 1789 by the Inquisition. He was primarily accused of advocating Freemasonry, which was banned by papal canon. After confessing under torture to be whoever they wished him to be—namely Giuseppe Balsamo—the count was sentenced to death, as the crime of heresy demanded. Some suggest that by Cagliostro's skill at hypnotism, the pope suddenly commuted the sentence to life and placed him in a prison for the well-to-do. However, after a failed escape attempt, the count was locked in the darkest dungeons of Italy at the Fortress of San Leo. It was reported he died there at the age of fifty-two in

1795, though those who knew him and witnessed the count's alchemy doubted this was possible. Napoleon sent an investigative team, members of which had known Cagliostro in person, and when they found the corpse presented—and even though they were in disagreement if the body was indeed the count's—they were instructed to certify the death. Spiritualists who have tried to reach Cagliostro through séances have been unsuccessful. There have been no sightings of the count since the 1950s.

OUROBOROS

Plato had described a self-eating circular creature as the first living thing made by the Creator. He admired its efficient design, in that it needed nothing other than itself to survive. After devouring itself, it would be reborn, continuing the cycle for eternity. It came to symbolize immortality and was particularly used by alchemists to evoke any number of magic spells; it was also considered a purifying glyph to contemplate. Although similar creatures were depicted in varying mythologies, represented as either a dragon or a snake, it was found in early Egyptian art and was a favorite of Cleopatra. Since the concept is spread among so many different cultures—including the Aztecs, Native American mythology, Hinduism, and Norse symbolism—psychologist Carl Jung believed it rose from the universal consciousness of humanity. The Ouroboros is considered the oldest known mystical symbol preserved to this day.

CAIN

THE FIRST MURDERER

Cain was a farmer and the first son of Adam and Eve. He also goes on record as committing the first homicide, the premeditated murder of his brother, Abel, then employed as a shepherd. Cain, in some accounts, was at least five years older, with Abel possibly a young

Doré

teen while Cain seemed to be in his early twenties. No one knows how long Cain felt animosity toward his sibling. It was perhaps from Abel's birth, since Cain was no longer the main focus of the very first parents' undivided attention. Being the first farmer had its problems, too, with no manuals on how to plant or raise crops properly, while Abel got to roam the hills, merely following his sheep. Nevertheless, the brothers were instructed to present the fruit of their labors to God. All versions of this story say God was pleased with Abel's gift, probably a nice fat lamb, while Cain's bushel of puny vegetables was less welcomed. Cain's reaction was a "fallen countenance"—understandably so, though God pulled Cain aside and instructed him to do better and cheer up. Cain instead remained angry and became jealous, hatching a plan to take out his frustration on Abel. The same devil that had tempted Eve was still afoot, and while Cain hoed and hoed—a labor he apparently considered a waste of time—the devil encouraged Cain to go through with his fantasy of killing Abel. Shortly after, Cain invited his brother to take a walk together out into a field, and Abel, as young brothers are still inclined to do, was eager to spend some quality time with his older sibling. Cain then beat his brother to death with a club. Cain's punishment was banishment from the land adjacent to the Garden of Eden; he was no longer allowed to partake in agriculture and was cursed to wander the earth. When Cain heard God's sentence, he thought it was extremely severe, and he argued that he had only killed one person. He also worried that others might murder him when they heard what he had done. God then gave Cain a mark to protect him, for reasons still unknown.

THE MARK OF CAIN

Centuries later, those trying to add a religious justification for the African slave trade thought the mark of Cain had to do with skin color. This concept was deeply entrenched; many Protestant churches banned the ordination of blacks until the 1960s—citing the mark-of-Cain theory. In one Hebrew text, "The Zohar," the mark of Cain was a letter, similar to a *Y*, without the upper right-hand line. This was branded onto Cain's forehead, but others think it might appear as a birthmark, or a freckle under the armpits, or near the groin. (*See also:* Six-Six-Six)

JOHN CALVIN
SPIRITUAL DISCIPLINARIAN

John Calvin was a French Catholic and a lawyer in his mid-twenties when, around the year 1533, he suddenly had what he claimed was "a sudden conversion." He abandoned Catholicism and sided enthusiastically with the Protestant Reformation. A prolific writer, an eloquent speaker, and no doubt intelligent, he became more convinced that nearly all of his ideas on theology, or any subject for that matter, were always and entirely right. He reduced the seven Catholic sacraments down to two, leaving only baptism and the Eucharist offering. He discarded the typical Catholic mass, closed monasteries, denied papal authority, and deemed that the clergy were not above normal men, and that all were equal (though not women). He believed that knowing Jesus and absolute reliance on the literal interpretation of the Bible were all one needed for a moral life.

Calvin also was resolute on the notion that before creation, God had predetermined everything, deciding beforehand how the minutest of events were to occur in the future. Human souls were either "elected" from that time for salvation or, regardless of how exemplary a life a person might have led, were indelibly marked, such that at Judgment Day they would be automatically damned. If one followed his way, then by default, in Calvin's reasoning, those few were obviously pre-chosen to be among the elect. He stressed that only the "fruit of your labors" might indicate a person's eternal fate. Continuing this thought, he observed that society had run afoul because it was not governed by politicians who adhered to these principles of faith. In 1536 he was given an opportunity to create an experimental government in Geneva, ruled by his ideas. He quickly formed a democratic theocracy. With adherents to his brand of religion in seats of power, he had new laws passed. Among them, offenses previously considered as sins were instead made into civil crimes, punishable by law. For example, if a person worked or participated in pleasurable activities on a Sunday, he or she was physically punished. No citizen was to dress extravagantly, or dance. In addition, taverns could only serve liquor if it included Bible readings, and no customer could be served at a restaurant unless first heard to say grace. Blasphemers—they had their tongues pierced. He went further. For instance, when he was insulted once, he had the man dragged through the streets by chains, begging for God's forgiveness. Another "reformer" who came to his town preaching the denial of the Trinity, which to Calvin was heresy, was burned "slowly." Calvin allowed no images of saints or angels; that was idolatry. He also believed in citizens' rights to choose their way of life free of monarchy—they'd pick his theocracy, preferably—and these were among the notions that motivated nearly every person on the *Mayflower* to flee persecution in Europe. Many of them became the leaders of early American colonies. Persecution, in Calvinist logic, further proved they were among the chosen. By the time of the American Revolution, in 1776, more than two-thirds of the three million European immigrants to America had been trained in Calvinist doctrines. "Don't Tread on Me," the Revolutionary War slogan, was a central part of Calvin's way of thought. Prestigious historian Leopold von Ranke noted that "John Calvin was the virtual founder of America," since many biblical interpretations believed by Calvin were incorporated into the laws of the U.S. government.

Humans are unable to comprehend why God performs any particular action, but whatever good or evil people may practise, their efforts result in the execution of his judgments.

—John Calvin

When Calvin died in 1564, at the age of fifty-four, from unknown causes, though "in his sleep," he was fairly destitute and had requested no ceremony or monument. His tombstone, located in the Cimetière des Rois, Geneva, Switzerland, simply bears the initials J.C. After his passing, adherents to his doctrines were called Calvinists, while many other reformer sects, such as Presbyterians and French Huguenots, adopted many aspects of Calvinism. Today any church with "Reformed" in it is likely Calvinist, making for more than ten million believers worldwide who adhere to John Calvin's theology, considering themselves among the "elect."

"LET YOUR WOMAN BE SILENT IN CHURCHES."

After St. Paul wrote these words in a letter to the Corinthians, secular punishments were devised to enforce this literal Bible interpretation. In Calvin's time, one torture device, known as the scold's bridle, proved how truly unequal women were considered to be. This iron birdcage-like apparatus was placed over a woman's head if a husband considered his wife sinful. *Scold* was a word for "nuisance," and a woman could be subjected to wearing the humiliating head cage for the sins of nagging her husband, talking back, gossiping, or using a curse word. Some had a spike designed to pierce the tongue to prevent speaking, as well as a loop to be used to lead the sinful wife about by a chain, tied to a post in town, or to a hook in the house. Some models also featured a bell to alert a napping husband of his wife's whereabouts, possibly so he could prevent himself from getting strangled unexpectedly.

CANONIZATION
HOW SAINTS ARE MADE

Canonization—the process of elevating a person to saint status—usually commemorates individuals of extreme holiness. A saint, or canonized person, resides in heaven and is capable of intervening there on our behalf. A saint is also allowed to be the object of public worship. Miracles and martyrdom were the usual requisite items for a saintly résumé, with the most outlandish miracles always bumping candidates higher on the clerical waiting lists. It wasn't until 1962 that the Catholic Church allowed the names of non-miracle workers to be deemed saints; these included people of good character, and potential role models.

Pope John Paul II broke records as the most prolific saint maker in history, with more than 280 beatified from 1978 until 2000 during his reign.

In the first thousand years of Christianity, saints were championed at the local level, where the holy person lived, or visited, providing many potential eyewitnesses who attested to miraculous happenings or were able to point to the exact place where martyrdom had occurred. From the twelfth century on, the Catholic canonization process usually ended after a lengthy investigation, with only a papal approval or veto to make the verdict final. In those days a splendid saint (*See also*: Levitating Clergy), such as the original flying nun, might be bypassed if towns or regions already had too many saints, or, in the case of the airborne nun, when there was already a patron saint of aviation on the books. Others were sainted for entirely political reasons: French king Celestine V was canonized in 1305 when the reigning monarch, Philip IV, put a strong arm on fellow Frenchman Pope Clement V. Joan of Arc was pulled from the bottom of the list in 1920, almost six hundred years after she was burned at the stake, when the Vatican wanted to reestablish relations with France, shortly after World War I.

After 1900, the process leading to canonization had to begin with a local bishop who first carefully inspected artifacts belonging to a prospective saint, such as a robe, sandal, or bone shard. The bishop or his assignees visited the house, site of miracle, or place of death and tried to intuitively sense something saintly about it. Once this happened, the Vatican had various committees, especially the "Congregation for the Causes of Saints," scrutinize things further before the candidate was offered to the pope for consideration. There are more than ten thousand saints listed in the ancient book *The Catalog of Saints*. Some have not been Catholics. Martin Luther King, Jr., civil rights champion, though not sainted yet, is one expected to eventually be included at least in *The Catalog of Martyrs*, a list of notable persons of faith.

CANONIZATION REVOKED

In 1802, the bones and vial of blood belonging to a thirteen-year-old girl were found in a Roman era catacomb. From letters inscribed on tiles used to name the tomb's occupant, it was thought to be the relics of one St. Philomena. Miracles were soon attributed to her and people began to report visitations from the young saint, said to be a fourth-century Greek princess who died as a martyr to protect her virginity. She was canonized in 1837, even though nothing was known of her life. It was later discovered that the tiles identifying the tombs in the section of the catacomb where her remains were found had been rearranged, and that the vial of blood was actually perfume. It was doubtful that a person named Philomena existed and she was removed from the church calendar of liturgical veneration in 1961. However, in 2005 the priest in charge of the Italian church where St. Philomena's bones were stored employed scientists to investigate the integrity of the saint's relics. At the "Conference of the New Philomenian Studies," it was announced the bones belonged to a girl who had indeed died violently, possibly martyred around 200 A.D. The conference urged continued veneration to St. Philomena.

CATHERINE OF SIENA

THE SAINT WHO MARRIED JESUS

Catherine was born the twenty-fourth and final child of an Italian dye worker and daughter of a local poet in 1347. It's no wonder that when her turn came for raising a family she refused, claiming she had already been betrothed to Jesus by a "mystical marriage." She explained that instead of wedding rings, she and Jesus had exchanged hearts. As insurance to avoid a secular marriage, she made herself unattractive. She cut to the scalp patches of her hair and scalded herself with boiling water to cause disfigurement. Catherine left home to work among the poor and preached that "love of God" was the only requirement for salvation. After attracting a huge following, Catherine was at first investigated for heresy, though eventually she won favor with the pope and monarchs. In addition to attesting to her marriage with Jesus, she fasted for long periods and survived by eating only the Eucharistic wafer. She could levitate and was impervious to fire. In 1375, after receiving her breakfast communion at a mass in Pisa, she suddenly had all the stigmata marks of Jesus' crucifixion. Witnesses saw laserlike bolts come from a statue of Christ on the cross and cause wounds to appear on her hands and feet, and one to appear near her heart. Afterward, she redoubled her

persistence in self-flagellation and extreme fasting, until she died of a stroke while in Rome in 1380 at age thirty-three. Catherine, noted for humility, asked before her death that the stigmatic wounds be made invisible; the prayer was apparently granted by God, since no markings were found on her remains.

News of her death spread quickly, and citizens from her hometown of Siena were dispatched to collect some relic of Catherine. Without permission from church authorities, they nevertheless managed to get access to her corpse, and finding it too risky to carry her whole, they decided to take her head and one thumb, which they tried to smuggle out of Rome in a sack. When stopped by guards, the bag miraculously only contained a bushel of rose petals, though the contents turned back into Catherine's head after passing that unexpected inspection. Catherine's head and her thumb remain on display at the Church of San Dominico in Siena, Italy. She has a foot, without stigmata in it, for the devout to contemplate in Venice, with the remaining body lodged at the Minerva Church in Rome.

More than three hundred people have been recognized by the Vatican as having received stigmata on their palms, feet, and side, even if it was later discovered that Jesus most likely was nailed through the wrists, and not his hands. Ninety percent of the stigmatics have been women, even if some have made the marks "invisible," as Catherine did. The first record of stigmata belongs to St. Francis of Assisi, who died of stigmatic complications, including continuous bleeding, at age forty-four in 1226. Eight additional saints have had stigmatism, and more than one thousand noncanonized cases are still under review. This condition is apparently on the rise, with more than five hundred ordinary people affected in the twentieth century alone.

THE LATEST NEWS IN STIGMATAS

In 2008, the stigmatic body of Padre Pio, canonized a saint in 2002, was exhumed and put on display in Rome. He was born to peasant farmers in southern Italy in 1887, but even by the age of five he was declaring he would become a priest. His mother frequently stopped him from inflicting physical penances upon himself. His stigmata first manifested in 1910, appearing first in the palm

of his hand. He described the mark as "the size of a penny." He prayed for the visible markings to disappear, which they did until 1918, when they bled again. They stayed visible until his death in 1968. He noted that the heart wound was worst of all and bled "especially from Thursday evening until Saturday." He was also reported to levitate while naked.

St. Rose of Lima, the first canonized saint of the Americas, didn't wait for supernatural stigmata. Instead she wore a metal-spiked crown of thorns pressed into her forehead, just above her eyes; she decorated it with a rose garland. She had avoided marriage and was allowed to become a nun, as Catherine of Siena had done, though St. Rose disfigured her face permanently with lye. Her goal and nightly prayer before she lay on a bed of broken glass: "Lord, increase my sufferings." She died from the complications of self-mortification at age thirty-one in 1617.

EDGAR CAYCE

THE "SLEEPING PROPHET"

In 1883, when Edgar Cayce was an eight-year-old Kentucky farm boy, a mysterious white light appeared before him. He remained undisturbed when an angel showed itself and offered to grant Cayce one wish. The young boy merely asked to be given the power to help people. The request was granted, however strange, for it seemed Cayce could only communicate with the angel realm while he slept. He discovered this anomaly of clairvoyance soon after the initial heavenly visit. When unable to memorize a small set of words for a quiz, he eventually fell asleep and used the book as his pillow. After a five-minute snooze, young Edgar not only knew the list of lesson words but could also recite and spell every single word in the entire textbook.

Dubbed the "Sleeping Prophet" by the press, Cayce amazed the most discerning doubters, who witnessed his self-induced sleep and hypnotic trances. Many claimed that during his sleeping sessions Cayce spoke fluently in numerous foreign languages and used vocabulary beyond his grasp. From 1901 to his death forty years later, Cayce gave more than fourteen thousand psychic readings. He easily and accurately predicted the sex of thousands of unborn children in an age before sonograms. He also offered unorthodox cures for diseases ranging from epilepsy and cancer to the common cold. He even foretold the 1929 stock market crash, World War II, and presidential assassinations.

Cayce was dedicated to studying the properties of the strange powers he was given. He traveled to the astral plane and the spirit realm, writing forty-nine thousand pages of treatises that primarily combined Christian ideas and Hindu philosophies. A proponent of reincarnation, he viewed God as a great "Creative Energy" and believed Jesus was reincarnated nearly thirty times before taking the form of the Christ. He also concluded that all living souls coexisted from the moment of creation but were splintered when born into the realm of physical matter. Cayce was never interested in forming a religion or sect. Instead he lived a simple life: gardening, writing, always dismissing the idea that he was a prophet. He did, however, believe his clairvoyant gift was teachable and available to everyone.

Cayce's subconscious is in direct communication with all other subconscious minds, and is capable of interpreting through his objective mind and imparting impressions received to other objective minds, gathering in this way all knowledge possessed by endless millions of other subconscious minds.

—Dr. Wesley H. Ketchum

Often complaining of headaches throughout his life, Cayce died of a stroke or cerebral hemorrhage in 1945, at age sixty-seven.

THE LEFT PAW

Cayce predicted that Christ's Second Coming would occur in 1998, and that Antarctica and the North Pole ice cap would somehow change places in 2000. He thought that a "library of wisdom," supposedly hidden by survivors of Atlantis, would one day be discovered under the left paw of the Sphinx at Giza;

he promised this information would surpass all current knowledge. Cayce claimed to know this because he was once a priest during Atlantis's heyday, having been known during that reincarnation as "Ra-Ta."

CHURCH OF THE UNIVERSE

WORSHIPING HEMP AND NUDITY

Walter A. Tucker formed a religion in 1969 that advocated nudism and the worship of the cannabis plant: the Assembly of the Church of the Universe. On a 360-acre wilderness tract in Ontario, Canada, he established a back-to-nature religious commune that centered on growing marijuana, smoking it, eating it, and using hemp products. While forming his church, Tucker incorporated theological aspects from a wide range of religions and concluded that "God is God," meaning you can worship or believe in whatever you wish. Old Testament texts, however, were quoted to base his belief that cannabis was the Tree of Life. Said Tucker, "[marijuana] is the most sacred plant in the world; it's the most user friendly plant you could ever smoke." All members of the church were required to use cannabis at least once a day, and to wear hats made from hemp when in public. On nudism, he reasoned: "You are willing to stand before God exactly as he created you, without shame." The original Canadian church commune, Clearwater Abbey, was raided by police and closed in 1982, as were other locations. As recently as 2008, Tucker was convicted of drug trafficking, after selling three dime bags of the "sacred herb" to an undercover agent who had posed as a hemp devotee. At age seventy-five, Brother Wally, as the bushy-bearded, hemp-hatted Tucker was known, was sentenced to one year in jail and had his house and church building confiscated. There are approximately thirty thousand members of the Church of the Universe.

CIRCUIT RIDER

GOD'S TRAVELING SALESMEN

In the 1800s, a traveling preacher, or "circuit rider," made the rounds to the faithful on horseback, with saddlebags laden with Bibles. It was principally a concept developed by the Methodist Church to bring its message to the masses. These riders, assigned as much for their horsemanship and survival abilities as for scholarship in scriptures, jumped off their horses to preach anywhere. A log cabin, a farm field, or a wilderness town's muddy street corner was as good as any place to give their interpretation of God's word. Even though Methodism was considered a religion of the then-hated British motherland, these equestrian preachers' unpretentious approach helped the Methodist Church to become one of the largest Protestant denominations in the United States. In 1784, the church had forty-three sanctioned circuit riders, but by 1850 the number had swelled, with more than four thousand riding preachers on the books.

The founder of the American Methodist movement, Francis Asbury, logged more than 270,000 miles on horseback and reportedly gave more than sixteen thousand sermons. He died in 1816 at the age of seventy.

CIRCUMCISION

SENSITIVE RELIGIOUS CUSTOM

On the eighth day after birth, all Jewish males have the penis foreskin removed and folded back to produce a bulge and give the organ a small "head." Abraham commanded it be done among the male descendants of his people; it was likely a custom he learned from Egyptian culture, where it was practiced among royalty. For Egyptians it stemmed from worship of various snakes, noted for shedding their skin and symbolizing rebirth. It was and still is practiced by Muslims and many Christian religions, though worldwide more than 70 percent of the male population is not circumcised. In Britain and most other European countries, as well as New Zealand, Japan, and most of South America, it is not performed at birth. During the 1880s, it became popular in the United States, where it was heralded as pro-hygienic and as a way to stem sex drive in boys as they got older—primarily believed to discourage masturbation, considered a sin.

Female circumcision is traced back to African religions and often includes removing the clitoris and portions of the labia. Since antiquity, it was done to females usually after marriage as a way to curtail sex drive and keep them monogamous. Highly controversial, it remains a religious initiation into womanhood throughout many Muslim countries and among African religions. In 2009, more than ninety million girls had the procedure done, usually at around the age of ten. In Sierra Leone, for example, more than six thousand girls from the ages of two to eighteen endure the procedure every day.

The Prophet Muhammad, when meeting a female circumciser, offered advice:

"Do not cut too severely as that is better for a woman and more desirable for a husband."

COURSE OF MIRACLES

CHANNELING JESUS

In 1965, Helen Schucman, a Jewish research psychologist, began to hear a voice in her head and copied what she heard in shorthand. This was not a muse of the kind to which many attribute their writing inspiration, but rather a "distinct voice" that identified itself to Helen as Jesus. For eleven years she "scribed" a vast amount of material, which her colleagues, fellow psychologists William Thetford and Kenneth Wapnick, edited and transformed into a spiritual study program called A Course in Miracles. In summary, Jesus told Helen that everything is an illusion, including suffering, while the path to enlightenment is through love and forgiveness. Schucman did not seek to gain undue notoriety from her channeling experience, though her partners did, and subsequently had more than 1.5 million people take the course. Helen died in 1981 at age seventy-one of cancer. The priest who gave her eulogy said, "This woman who had written so eloquently that suffering really did not exist spent the last two years of her life in the blackest psychotic depression I have ever witnessed."

COVENANT, ARK OF THE

ELUSIVE LUGGAGE OF THE TEN COMMANDMENTS

In 1440 B.C., three months after leading the Exodus from Egypt, Moses not only had the harsh wilderness and desert to contend with, but also a lot of complainers. At least they had plenty of food and water

in Egypt. Moses then went alone to the summit of Mount Sinai, where God told him the Ten Commandments, or rules that, if obeyed, would save his people. Moses claimed that God wrote them on stone tablets. Geologically, Mount Sinai is made of granite and volcanic stones, so it's assumed the first commandments were inscribed on the more easily etchable volcanic slabs. Nevertheless, after returning to the Israelites' camp with such a windfall, Moses saw his people worshiping a golden calf, and in utter desperation he smashed the tablets.

Moses climbed the mountain for a second time, and apparently received a replacement set of inscribed stones, once again written by God, which were then placed into a special container. The ark, or box, was similar to an Egyptian design Moses had seen while in the pharaoh's court, an item of furniture often used as a storage trunk. His was made of wood, covered with gold, and had four rings so it could be carried by poles. It was probably about four feet long, two feet wide, and two and a half feet high. For nine hundred years the Israelites kept close tabs on the ark that contained the Ten Commandments, a miracle walking staff, and a sample jar of bread or manna that was sent from the heavens to feed the Jews during Exodus. The ark had powerful properties, capable of everything from parting rivers to killing scorpions, and if aimed at enemies it acted like a divine cannon, blowing armies to smithereens. According to the book of Joshua, the walls of Jericho crumbled after Jewish priests paraded the ark around the perimeter of the fortified city of Canaan for six days straight. Afterward, when they sounded ram horn-trumpets, the mortar loosened and the stone walls tumbled, thereby adding a sonic-wave-like arsenal to the ark's might. When the Philistines captured it for seven months and installed it in front of their idol Dagon, the statue kept falling. While in enemy hands, the ark caused the plague death of fifty thousand Philistine citizens, making it a potent biological weapon as well. When King David had it returned with great fanfare and a parade, one of his servants tried to steady the ark from falling off an ox-drawn cart, but the servant was killed instantly for merely touching it.

The ark was last seen intact during Solomon's reign, though according to the book of Kings the magic staff and jar of bread were missing. When Jerusalem was destroyed by Nebuchadnezzar and the Babylonians in 586 B.C., the ark was forever lost. By deciphering the more than forty-five references to the ark in the Bible, biblical scholars surmised it went with the son that Solomon had with the queen of Sheba and was hid in Ethiopia. New researchers think it might be in Zimbabwe, among the Lemba people, who interestingly called their high priests "cohens." (DNA tests made on Jews with the last name Cohen are claimed to match this African tribe.) Another theory proposes that Solomon had a tunnel system under the temple and that the ark was removed and hidden in a cave someplace in Jerusalem. During the Middle Ages, rumors abounded that the ark was retrieved by the Knights Templars and brought back to London, then buried in an undisclosed field near Warwickshire, England. However, it's most reasonable to assume it was looted by Babylonians, the gold scraped off and the wooden storage trunk discarded. Too heavy to carry, and containing words Babylonians couldn't read, the original stone tablets with the commandments were mostly likely left behind among the rubble.

WAS MOSES TRIPPING?

Mount Sinai (also known as Mount Herob) was noted for its hardy acacia trees, a type of gum tree that contains powerful psychedelic toxins in its bark. Moses not only had a conversation with a burning bush on Mount Sinai, but he received two sets of tablets while among the acacia trees, where God's rules were literally "carved in stone." Professor Benny Shanon of the Hebrew University of Jerusalem stated, "As far as Moses on Mount Sinai is concerned, it was . . . and this is very probable, an event that joined Moses and the people of Israel under the effect of narcotics."

CREATIONISM

RELIGION VS. SCIENCE

In 1859, Charles Darwin published *On the Origin of Species*, and among other ideas in that revolutionary book, he suggested that all carbon-based life forms evolved slowly over time through the process of natural selection. Further additions to his theory purported that humans had ancestors that were possibly once tree-dwelling apes. Many religions have creation stories that feature one god, or a cast of them, who created man and the world just as it is. However, for those denominations that base nearly all precepts on strict interpretation of biblical scriptures, the idea of evolution cannot be found in any text. According to Genesis, God created the first human, Adam, from dust, cast in the image and likeness of himself. For those placing all faith in the literal meaning of the Good Book, there is no such thing as human evolution.

From the infamous "Monkey Trial" in the 1920s, when John Scopes, a high school science teacher in Tennessee, was convicted of violating new laws that prevented evolution theories to be taught in schools, through 2005, when stickers on science textbooks were added that warned "Evolution is a theory, not a fact," the debate on which notions of creation should be taught in public schools has continued. In a survey of 480,000 scientists in the United States, only seven hundred believed in creationism, defined as the denial of evolution theories and the literal belief of creation as presented in the book of Genesis. However, according to a Gallup poll, 45 percent of the general population believes "God created human beings in their present form at one time within the last 10,000 years," while 37 percent think man could have evolved from other life forms but were "god guided." Only 12 percent chose the survey answer "Human beings have developed over millions of years from less advanced forms of life, but God had no part in this process." According to a 2007 study, the United States ranks seventeenth in the world in science education. (*See also*: Big Bang and Akasha)

ALIEN SEEDS

One idea that discredits neither evolutionists nor creationists is the alien visitation proposal, or the "Ancient Astronauts Theory." This hypothesizes that intelligent alien life forms left offspring here on earth and taught man the rudiments of civilization. From then on, remembrances of these ancestors were worshiped as gods, and humankind has been long awaiting their return. The

extinction of humanlike species such as Neanderthal man or *Homo erectus* was not a weeding out by natural selection but rather simply failed genetic engineering experiments of these early space travelers. When these "star gods" were able to seed the planet with creatures similar to themselves, they left and hoped for the best. Others (*See also:* Raëlism) think these superior extraterrestrials return periodically to guide us. Although it stretches credulity, many creation myths of countless religions could plug in this "Astronaut Theory" to fill in the illogical gaps. For evolutionists, certain unexplained deficiencies such as the "missing link" between man and ape are believed to be solved according to this premise.

CRISTO REDENTOR

GIANT JESUS OF BRAZIL

The 120-foot-high statue perched on a mountain in Rio de Janeiro, known in Brazil as *O Cristo Redentor*, was the vision of a visiting missionary, Father Pedro Boss, in 1859. He woke one moonless night in a cold sweat and went outside to clear his head. He looked up at the dark mass looming above, an odd hunchbacked-shaped outcropping on Corcovado Mountain called Pináculo da Tentação, or the Peak of Temptation, and prayed for help. In the next instant he witnessed an electrically charged crackle in the sky that he first thought to be light-

ning. The bolt of energy suddenly formed into a neon-blazing outline of a giant crucifix.

A humble man who never expected visions from God, Father Boss kept this divine hallucination to himself. Nevertheless, he went about crafting the proposal to erect a giant cross—obtaining the advice of masons and architects as well as church superiors—before sending a request for financial backing to Isabel, princess imperial of Brazil. She said she had to "sleep on it" but after two years, the queen rejected the notion, denying money to construct it. In the 1920s Father Boss's idea was put forth again, and this time successfully funded by donations from Brazil's Catholics. After a contest, an Art Deco–style Jesus was chosen and constructed piece by piece on the rugged site, taking five years to complete.

Jesus' head weighs thirty tons, and each of his outstretched hands weighs eight tons.

Two engineers were in charge, Catholic Heitor da Silva and Jewish Heitor Levy. Levy reportedly converted to Catholicism during the process, and he put the names of his family members in a batch of concrete where the heart would be. In 2007 the statue received 100 million Internet votes to become one of the "New Seven Wonders of the World."

ALEISTER CROWLEY

"THE WICKEDEST MAN IN THE WORLD"

The man born to English aristocracy in 1875 as Edward Alexander Crowley changed his first name to Aleister to disassociate himself from his strict Puritan upbringing. By his early twenties, he took on another name when he joined an occult group, the Hermetic Order of the Golden Dawn, to "Perdurabo," a name meaning "I Will Endure." After rising quickly in the ranks of the Golden Dawn mystical cult, fractures developed and his use of black magic and other negative curses summoned up to vanquish enemies caused his own banishment two years later. He traveled widely and became interested in using incantations to contact angels, particularly his guardian angel, which he called "Aiwass." When

he conjured this spirit he became entranced and through a three-day, sweat-dripping psychic frenzy, he dictated *Liber AL vel Legis*, or the *Book of Law*, in which Crowley attacked organized religion in favor of doing what your higher self believes is right. He elaborated on the process and offered a set of incantations that he often performed in a theatrically macabre, though darkly reverend manner. In 1907, Crowley started his own magic cult, Argenteum Astrum, or the Order of the Silver Star, and later the Ordo Templi Orientis, and believed himself to be the reincarnation of occultist Eliphas Levi, who died the year Crowley was born. He also accessed knowledge of his other past lives, including one as the eighth-century Count Cagliostro, Pope Alexander VI, and John Dee.

A poet and prolific writer, Crowley sought adventure and pleasure and seemed determined to extract every experience possible in his allotted time in human form, regardless of the consequences. He remained persistently dedicated to meditation and magic. Sex was also a big part of his rituals, as were blood sacrifices; during one episode he baptized a toad that he had transformed into Jesus of Nazareth and then crucified it, surely solidifying his reputation as a Satanist. In turn, Crowley was vilified in the press as a "beast" and the "wickedest man in the world." However, he believed one had to battle the devil, the force that blocked the search for God and the "Secret Chiefs," or bearers of ancient wisdoms. He thought the only way to subdue the devil was by following an elaborate set of rituals and incantations.

In the end, Crowley went through two inherited fortunes and became addicted to heroin and cocaine, remaining a displaced wanderer, quickly wearing out his welcome wherever he went. Today he's considered the father of modern occult, synthesizing various texts and credited with defining and differentiating between stage magic and what he called paranormal "magik," defined as "the science and art of causing change to occur in conformity with the will." *K*—or the eleventh letter of the alphabet—held significance and from ancient times had been associated with the demonic forces that need to be vanquished before one's Guardian Angel can be summoned. (*See also*: New Aeon, Church of the)

Crowley died in 1947, at age seventy-two, at a London boardinghouse. It was believed his last words were "I am perplexed," though others say it was "Sometimes I hate myself." Nevertheless, the doctor who had refused to renew his morphine prescription died within twenty-four hours of Crowley's passing.

THE BOOK OF LAW

I am the Snake that giveth Knowledge & Delight and bright glory, and stir the hearts of men with drunkenness. To worship me take wine and strange drugs whereof I will tell my prophet, & be drunk thereof! They shall not harm ye at all. It is a lie, this folly against self. The exposure of innocence is a lie. Be strong, o man! lust, enjoy all things of sense and rapture: fear not that any God shall deny thee for this.

—Aleister Crowley

THE CRUSADES

HOLY WARS THAT KILLED MILLIONS

In the seventh century, Muslim armies spread across the Middle East converting great numbers to their faith by conquest. They captured Palestine and then took control of Jerusalem. In 1009, when an imam destroyed Christian relics, most notably the Church of the Holy Sepulchre, a fervor spread among the common people of Christian Europe. Papal blessings were already given to soldiers in Spain, long at war against invading Muslims. Starting in 1099 there were eight church-sanctioned crusades organized with the intent of recapturing Jerusalem and halting Islam expansion, which by then had taken control of two-thirds of the Christian world. The Christian poor saw it as a sacred religious war, while knights and nobles viewed it as an opportunity to attain war booty.

The religious zeal was not exclusive to adults. In 1212 A.D., a French peasant boy, twelve-year-old Stephen of Cloyes, claimed that Jesus appeared to him in a vision. He began preaching that Jesus had told him to form an army of children and march with their innocence to recapture the Holy Sepulchre. Thirty thousand children gathered from all over France and paraded with candles and banners, while singing religious songs toward Marseilles, where they were offered ships to take them

to the Holy Land. Once at sea, they were instead sold as slaves at various Muslim markets along the way. Another fifty thousand children marched from Germany, though most perished while crossing the Alps, with the last, numbering around three thousand, also boarding ships only to be betrayed and sold into slavery. Pope Innocent III used their deaths as a recruitment tactic, saying, "These children put us to shame; while we sleep they rush to recover the Holy Land." By the end of the last crusade, in 1272, nine million lay dead, victims of barbarous carnage inflicted in the name of God.

CRYSTAL BALLS

THE ANCIENT ART OF CRYSTALOMANCY

A traditional crystal ball is a transparent quartz sphere that fits in the palm of the hand and weighs approximately five pounds. The first ones

CRYSTAL SEER
Sees Your Life from the Cradle
to the Grave

appeared around 700 A.D. and were used by fortune-tellers, or diviners, who employed a technique called scrying. Before crystal balls, scrying was practiced by gazing into a bowl of water, or by interpreting shapes formed by smoke from a fire, and was thought to give clues to the future. Those who could decipher otherworldly signs through scrying were usually treated well, put on the payroll of monarchs or persons of wealth and power.

Smaller, polished, and transparent stones, now called crystals, go back even further in history and were used by the Druids for the same purposes. The crystal stones the Druids used, a natural gem made of beryllium, gave off an energy—similar claims are made for crystals today—that allowed the wizard to fall into a trance and gape, mesmerized, into the gems held up to sunlight or flame. The stones foretold battles and guided nearly every aspect of their decision making. Some current crystal specialists state that a ball itself has no powers, but through an intense and concentrated look into it the globe can send the user into an altered state, the orb turning into a sort of projection screen. Here the skilled diviner can discern stanches of images, and glimpses at events, which have come from an astral plane or an alternate dimension. This is possible to the diviner because in the astral level time exists differently—the past, present, and future occur simultaneously.

The British Museum now has the crystal ball that is thought to be the famous "shew-stone" used by Queen Elizabeth's diviner, John Dee. Dr. Dee was a prodigy, teaching mathematics at universities by his early twenties, and was an expert in astronomy, navigation, and cartography. He also believed in magic and eventually abandoned all study to spend the last years of his life attempting to communicate with angels, primarily through crystal-gazing. To Dee, both math and magic were equally legitimate paths to reveal the underlying secrets of the universe. With the help of a known seer, Edward Kelly, he learned to scry, and translated a lost angelic language, called "Enochian." Dee wrote a dictionary of this language and other books, which he claimed were inspired from communication with spiritual entities.

SEEING HIS OWN FUTURE

When Dr. Dee met Edward Kelly, Kelly had already had his ears cut off, a typical punishment at the time for forgery and counterfeiting, which he had been convicted of. Kelly was a persuasive character nevertheless, and he gazed into the ball for nearly ten years under Dee's payroll while Dee transcribed messages Kelly divined from the "skew-stone." Kelly's association with Dee paid off well for a while, especially after the pair moved operations from England to Bohemia. There Kelly was given land and heralded by royalty as the "Marshal of Bohemia." Seeing his future success elsewhere, Kelly tired of Dee's spiritual quest and focused again on alchemy. To break off the relation with Dee, Kelly said the angels wanted each of them to swap wives. Convinced the angels were real, Dee obliged, though afterward Dr. Dee eventually had second thoughts and they parted ways. Kelly's promises of making gold brought affluence, but the Bohemian royalty got impatient for the product and demanded Kelly make it happen, and quick. To motivate him, they locked Kelly in a tower and only released him when he promised to make a pile of gold promptly. When this didn't happen he was imprisoned again, and he tried to escape by tying bedsheets together to climb out of a very high window. Edward Kelly subsequently fell and died of injuries at age forty-two, in 1597.

CULTS, COPYCAT VARIETY

IMITATING EVIL

In the 1980s, the Ripper Crew thought of themselves as a satanic cult that would eventually attract millions and become more famous than the Manson family had. (*See also*: Charles Manson) However, the founders proved less than charismatic, ultimately enlisting only four members. In a plan to gain notoriety, the group drove around the streets of Chicago in a van and snatched up and killed eighteen women. Their methods were extremely gruesome: While the women were still alive, the members cut off and ate the women's breasts while reading passages from the Bible, until finally murdering the victims. Once captured, the Satan worshipers caved, and were quick to testify against the others to save themselves—at that point no longer wishing to copy the Manson family. Two of the four cult members were sentenced to death, and the other pair received 120 years to life.

Another copycat cult originated at a fifteen-acre farm in rural Ohio in the late 1980s; Jeffrey Lundgren started his own version of the Reorganized Church of Jesus Christ of the Latter-Day Saints, hoping to be considered a prophet of equal status to Joseph Smith, Jr., founder of the Mormons. After convincing six families to join his movement, he changed his name to "Jeb7." He exercised total control: beating children that misbehaved, training his men for war, and having bizarre sex with everybody. He particularly liked to have naked women cover his body with excrement as he gave long speeches about the nearing doomsday. One day he decided that God was still angry with his sect, because of "man's sins," and came up with a sacrifice he believed mimicked biblical offerings. He ordered five members of one family to dig a pit in a barn. When it was deep enough, Jeb7 had them read from the Bible and pray before he shot them. When the end of the world didn't come as planned, Lundgren and his remaining group went into hiding, but they were arrested a few years later, in 1990. Jeb7 was sentenced to death and was executed in 2006, while his followers each received life in prison.

In 1998, four teenagers and two young adults called themselves the "Kentucky Occult Teen Killers." Out joyriding, they encountered a married couple and their two children, ages six and two years old. The Satan cult marched the family into the woods and with children on the parents' laps, shot them all many times. Though shot in the eye, the two-year-old boy survived, the only one to do so. He is currently living with relatives in Sweden. The cult ringleader, then nineteen-year-old Natasha Cornett, claimed to be the daughter of Satan. She had a hundred knife slashes on her arms when arrested and claimed Satan spoke to her only after self-inflicted cuts. They testified that they got their idea of a cross-country murder spree after watching the movie *Natural Born Killers*. The cult members are serving life terms.

Inspired by newspaper accounts of the Kentucky Cult, three bored Tennessee teenagers, dubbed "the West Memphis Cult," killed three children in a botched satanic ritual. They stripped, beat, and mutilated their victims in the woods, and all cult members are now serving life terms.

CULT OF PERSONALITY

CELEBRITY WORSHIP

In 1956, Nikita Khrushchev was the first to use the phrase "cult of personality." He was indicating his disapproval of former Soviet dictator Joseph Stalin's massive and successful media campaign to elevate himself to the status of an infallible god. Hitler, Mao, Saddam Hussein, and North Korea's Kim Jong-Il were other notable dictators who made their omniscient presence a part of daily life through the proliferation of their image. Accompanied by messages and slogans of their goodness, wisdom, and power, portraits of these watchful leaders were everywhere—and required by law to be in every home. They were impossible to escape. Vast segments of their countries' populations began to venerate them as if they were gods. In free societies the closest equivalent might be celebrities. Adoration of entertainment personalities or sports stars is so widespread that today many popular celebrities' images permeate society's consciousness far greater than any religious icon could ever have done. The number-one item on most search engines relates to a celebrity. For example, Google compiled an annual list of the most popular globally viewed images. It counted an "aggregation of billions of search queries people typed into the Google search box," and surmised that in 2008, Heath Ledger and the Jonas Brothers were in the top ten; in 2007 images of Britney Spears, Paris Hilton, and Anna Nicole Smith were viewed religiously many millions of times.

CHURCH UNIVERSAL AND TRIUMPHANT

TANK-OWNING CULT

In 1958, Mark L. Prophet, a follower of theosophy and former member of the Mighty I AM movement, claimed that he was visited by an entity from another dimension. He was singled out by the spirit of a deceased holy man, El Morya, to be its "secretary," or revealer of a new message. Soon after this revelation, Prophet opened a spiritual center, the Summit Lighthouse in Washington, D.C., where he declared himself the divinely selected leader of all the varied and splintered I AM groups. (*See also*: I AM) More than ten years later, Prophet still only had a mere hundred followers, and the movement seemed about to collapse when he did not ascend "into oneness with God," as he had promised, but died suddenly of a stroke in 1973, at age fifty-four. His wife, Elizabeth Prophet, then changed the name of the group to Church Universal and Triumphant (C.U.T.) and went on an aggressive campaign to solicit followers, claiming communication with Mark, who in death was renamed "Lanello." Elizabeth was instructed to find a secret manual called *Operation Christ Command*, written by Mark before his death. It became the group's survival guide for the predicted and imminent apocalypse.

THE MYSTERY OF MASTER MORYA

Madame Blavatsky, founder of the Theosophical Society, claimed to have made contact with El Morya, whom she placed among one of the "seven rays of light," alongside Jesus, St. Germain, and Buddha. El Morya supposedly came to London in 1851, when he was 125 years old, and once saved Blavatsky as she contemplated leaping off a bridge. She referred to Morya as her Master, and she his slave. Although no archeological records prove El Morya existed, she had his portrait committed to memory and described El Morya as a "handsome Oriental prince." He also was seen at the turn of the twentieth century in Paris and America, though eluded cameras and paparazzi. No wonder, since Blavatsky had noted "he traveled in great secrecy, accompanied by Oriental attendants, and met but a few people." He was a figure who has appeared to many spiritualists since, and is even thought to have been the emperor of Atlantis in 222,000 B.C.

By the late 1980s there were more than ten thousand dedicated C.U.T. supporters, even if Elizabeth, then known as "the Mother of the Universe," boasted to have more than 150,000 followers. She made national news when she proclaimed that a Russian missile attack would take place in March 1990. By then she had a forty-thousand-acre ranch in Montana, designated as the church's inner retreat. She constructed an elaborate system of tunnels and bomb shelters, and charged twelve thousand dollars to anyone who wished to be saved from the impending doom by taking refuge in her underground sanctum. Some rooms reportedly were large enough to hold seven hundred people. Based on Elizabeth's communication with Lanello, C.U.T. devotees stockpiled weapons and even purchased armored tanks in preparation for Armageddon. The shelters, however, had no toilets and excrement was removed by buckets. Children were strapped to bunks deep inside while heavily armed men patrolled the grounds.

When the prediction failed to materialize, some former members sued the church for "slave labor," and its religious tax-exempt status was revoked, when the government cited a clause that churches were not allowed to own tanks. Nevertheless, once it gave up its heavy artillery, tax-exempt status was returned and the church survived, even after Elizabeth stepped down in 1999 due to Alzheimer's disease. Some claim C.U.T. remains flush, primarily from inheritance bequeathed to the church by devoted followers, and operates thirty international centers that offer guidance as prescribed by the Ascended Masters. Although members must now pledge "to shun the black arts including the practice of Satanism, witchcraft and black magic," and "avoid deliberate association with discarnate spirits through such activities as psychic channeling, automatic writing, and spiritualism," the C.U.T. faithful believe God is in "elemental spirits" and is often manifested as fairies and elves. Elizabeth Prophet died in October 2009, after suffering the complications of Alzheimer's, at age seventy.

D

SANTO DAIME

PSYCHEDELIC AMAZONIAN RELIGION

The son of Brazilian ex-slaves and devotedly Catholic, Irineu Serra was in his early twenties when he left home to work as a rubber-tree tapper deep in the Peruvian rain forest. At six feet, five inches in height, Irineu was a muscular black man with a gentle demeanor who made friends easily. In time, he was invited to attend an ayahuasca ceremony conducted by a shaman of Inca descent. Ayahuasca, a powerful hallucinogenic beverage brewed from Amazonian vines and plants, had for millennia been considered a sacred concoction by the indigenous people. Used for prophecy, healing, and cleansing of the spirit, the tea called "rope of the soul," or "vine of the dead" allowed one to visit the astral plane and garner insights unattainable from ordinary sensory perception. The psychoactive properties of the drink caused an outer-body-type-of-experience that included shape-shifting visions and teleportation, and for those unprepared to imbibe, horrific demonic nightmares that lasted for days. For the enlightened, the brew was a portal to hidden knowledge about one's soul and yielded answers to universal mysteries.

Irineu took to the drink immediately and on the third "journey" had a conversation with a woman who appeared in the light of the full moon. He was sitting on a hammock when the moon soared free from the sky and descended toward him until it was only an arm's length away. A woman appeared from the brilliance and identified herself as Clara, or Queen of the Forest. Clara became his personal guide and instructed Master Irineu to write hymns and doctrines that would be the foundation of a new religion, Santo Daime. It took twenty years from the first visions, countless more hallucinations, and dedicated study of the special plants, including harvesting and preparation techniques, before he had his first followers. He believed life's goal was to find the "superior I inside" through the use of ayahuasca. His doctrines blended the worship of Jesus, saints, and spiritual entities from indigenous and African cultures, advocating harmony with nature and fellow man as well as incorporating notions similar to Hindu karma, including reincarnation.

Catholic authorities in Brazil prosecuted sects devoted to obtaining spirituality through the use of ayahuasca, but eventually Irineu, through his formidable presence and healing powers—even among politicians—gained permission from the government to continue. When he died, "doing his passage," in Santo Daime idiom, in 1971, at age eighty-two, his more than five thousand followers splintered.

ALTERNATIVE TOURISM

The Amazon region still caters to a booming business in what's called ayahuasca tourism for those coming from all over the world to seek transcendental experiences. Advocates say the ritual and controlled intake of ayahuasca is much better than the compulsive and indiscriminate use of various street drugs and alcohol. In 2009, the most widespread of the Amazonian-linked hallucinogenic religions was União do Vegetal (UDV), founded by another Brazilian rubber tapper, Mestre Gabriel, who coincidentally also died in 1971. UDV claims more than ten thousand members and has more than 130 church-centers in the United States alone. In 1999, when the U.S. Drug Enforcement Administration raided a UDV compound and found thirty gallons of the special tea, considered by authorities as an outlawed Schedule I drug, the group brought its case to the U.S. Supreme Court. In 2006, the high court allowed the use of this psychoactive brew and wrote in a unanimous decision that banning it would "substantially burden a sincere exercise of religion."

KING DAVID

PHILANDERING HOLY RULER

The Israelites squared off on the battlefield facing the more formidable Philistines. David was just a boy, though sent by his mother to bring his brothers a bag lunch while they awaited orders to attack. A proposition was made by the Philistines to let victory or defeat—winner takes all, losers become slaves—be decided by the outcome of single combat between each force's best man. The Philistines had Goliath, a virtual giant and an undefeated brute, while the Israelites had none comparable, nor were any clamoring to volunteer to meet that challenge. When David walked out alone, he was met by hoots and howls of laughter from the Philistine camp. However, David was an expert (or extremely lucky) with the slingshot, and staying out of Goliath's reach, he nailed the giant with a rock to the forehead. Once the giant was down and gagging, David raced in and severed Goliath's head with the fallen giant's own sword. He held it up for the Philistines to see, and they retreated in terror.

Afterward, David became a general and eventually gained more popularity than King Saul, who attempted to kill the boy warrior. When Saul died, and after Saul's heirs were murdered, David, by about the age of thirty, was made king of Judea. He continued racking up victories and conquered Jerusalem, making that his home base and the place where the original Ten Commandments were housed. (*See also*: Covenant, Ark of the) He ruled Judea and Israel for forty years, and is credited with writing psalms, though David was also noted for using his power to sleep with any woman he wished. The most detailed biblical account involves Bathsheba, whom he eventually married and who bore his heir, King Solomon. One morning, while David was up on the palace roof terrace, he spied a woman taking a bath and told his servants to bring her to him. When David was informed that she was married to one of David's top soldiers, he ignored it and took her to his bed. He eventually had Bathsheba's husband murdered. In total, David had at least seven wives, which in scriptures are mentioned by name, but it was alluded that David slept with hundreds more before dying in 967 B.C. at age sixty-nine. (*See also*: Absalom)

BIBLICAL HAREMS

Even though adultery was a crime punishable by death, monogamy was not a requirement in biblical times. One ancient orator, Demosthenes, summed up the status quo this way: "We have mistresses for our delight, concubines for the day to day needs of our body, and wives to breed legitimate children and serve as faithful housekeepers." David's son King Solomon broke the record with seven hundred wives and three hundred concubines. His son Rehoboam took eighteen wives and had sixty concubines. Abraham had three, and Jacob had two at the same time. In total, more than forty biblical patriarchs had multiple-occurring matrimonial arrangements.

DÍA DE LOS MUERTOS (DAY OF THE DEAD)

MOCKING DEATH IN MEXICO

Spanish missionaries found a way to incorporate an Aztec festival for the goddess Mictecacihuatl, queen of the Underworld, known as "the Lady of the Dead," by encouraging converts to celebrate the ancient festival on All Soul's Day. For Christians, that is the day when souls in purgatory are remembered, though in Mexico the festivities have retained a homegrown twist and interpretation. Ceremonies honoring the dead among indigenous people of Mexico and South America trace back more than 2,500 years. For Aztecs and other Mesoamerican cultures, reality was considered a dream; not until death was one fully awake. Considerable time was spent conversing with dead ancestors, displaying their skulls, making their favorite food, and setting a place for them long after they had died. The original Aztec monthlong death celebration was never completely eradicated by missionaries, though it was condensed to a two-day affair that is still practiced. The ceremony retains the lighthearted and near-mocking attitude to the final exit, once horrifyingly sacrilegious to the early missionaries. Small altars are made and adorned with foods that the dead liked; skulls, wooden skeleton masks, and dances also mark the feast. Pillows are placed randomly throughout towns so that spirits might rest. Many families still spend the night in the graveyard and picnic at a deceased loved one's burial site, often bringing special candies and a bottle of tequila, which make it easier to stay the night, telling stories about the deceased. Mexican writer and Nobel laureate Octavio Paz remarked: "Rather than fearing death, a Mexican chases after it, mocks it, courts it, embraces it, sleeps with it; it is his favorite plaything and most lasting love."

During a celebration in 1487 at the Aztec Tenochtitlan Pyramid in Mexico City, as many as eighty thousand prisoners were offered as human sacrifices to appease Queen Mictecacihuatl, a record-breaking number for Día de los Muertos. Further down the Yucatán Peninsula, Mayans thought naturally occurring sinkholes were secret passages to the Underworld and the water god, Chaac. One excavation found more than forty sacrificially killed bodies stacked head to toe, with most of the victims being less than twenty years old.

DEISM

BELIEF IN THE DIVINE MACHINE

To the deist there is a supreme higher power, but it has no role in everyday affairs. Through the use of reason and studying nature's laws, God is revealed, and intelligent design can be observed to function like the intricate gears of an old fashioned windup pocket watch. Deists believe organized religion, by relying on sacred texts, ritual, and superstition, causes more societal harm than good. The founding fathers of the United States, such as Thomas Jefferson, George Washington, Benja-

min Franklin, and Thomas Paine, were well versed in deist writings, particularly in philosophers Descartes' and Locke's view that through the learning of science God would be revealed. The U.S. constitutional concept of separating church and state was a deist notion. More than half the signers of the Constitution—although officially members of various church congregations—were more aligned with deist thought than Christian dogma.

These laws may have originally been decreed by God, but it appears that he has since left the universe to evolve according to them and does not now intervene in it.
—Stephen Hawking, from *A Brief History of Time*

OFFICIAL RELIGIONS OF U.S. PRESIDENTS

Thomas Jefferson was the only president who claimed openly to be a deist, noting, "Say nothing of my religion. It is known to my god and myself alone." Abraham Lincoln and Andrew Johnson cited no church affiliation. Episcopalians have furnished the most U.S. presidents: George Washington, James Madison, James Monroe, William Henry Harrison, John Tyler, Zachary Taylor, Franklin Pierce, Chester Arthur, Franklin Roosevelt, Gerald Ford, and George H. W. Bush. The Unitarian Church counted John Adams, John Quincy Adams, Millard Fillmore, and William Taft as members. Baptist presidents included Warren Harding, Harry Truman, Jimmy Carter, and Bill Clinton. Martin Van Buren, Theodore Roosevelt, and Calvin Coolidge were Calvinists. James Garfield, Lyndon Johnson, and Ronald Reagan were members of Disciples of Christ Church. Methodists who became chief executives included James Polk, Ulysses Grant, William McKinley, and George W. Bush. The Presbyterian presidents were Andrew Jackson, James Polk, James Buchanan, Rutherford Hayes, Grover Cleveland, Benjamin Harrison, Woodrow Wilson, and Dwight Eisenhower—though Eisenhower was raised as a Jehovah's Witness; he was baptized to his new faith while in the White House. There were two Quaker presidents, Herbert Hoover and Richard Nixon. John F. Kennedy was the only Catholic, and there has yet to be a Jew, Buddhist, Muslim, or Hindu in the White House's top seat. President Barack Obama's father was a Muslim, but he has said his presidency will be "guided by [his] Christian faith."

DEVILS OF LOUDON

MASS DEMONIC POSSESSION OF FRENCH NUNS

Toward the end of his life, twentieth-century writer Aldous Huxley became fascinated by the exploration of alternate levels of human consciousness. He became particularly obsessed with the story of sexual hysteria and mass possession at a French convent in the 1600s, and published a nonfiction account of these happenings, *The Devils of Loudon*, in 1952.

The mayhem was said to be the result of a very handsome and well-connected parish priest named Urbain Grandier, who was already suspected of fathering a few illegitimate children in the small town of Loudon, France. He was also in hot water for writing a paper against celibacy. Mother Superior Jeanne des Anges of a local convent became obsessed with the handsome Grandier. She began to claim that Grandier appeared as an angel during the night, and tempted her to have sex with the apparition. Before long other nuns in the convent began to have the same dreams, and soon moans and screams of nuns experiencing sexual orgasms filled the convent's halls. Exorcisms on the nuns were performed before a crowd of more than seven thousand who witnessed the religious women speaking in tongues and gyrating sexually.

Grandier was arrested and brutally tortured. Convicted of heresy and accused of casting demonic spells, he was burned at the stake in

1634. Before he died, he cursed his accusers—and nearly all the priests and exorcists involved with his downfall died shortly afterward, many suffering from delirium. The nun, Jeanne des Anges, left Loudon and went on a tour of Europe, where she was welcomed by aristocrats and church dignitaries. Afterward, a local hired actresses to dress like nuns and set up a small stand near the convent. For a penny, these "nuns" lifted their garments and called sexual obscenities to the crowd.

THE PRICE OF CELIBACY

According to an American Psychological Association study, more than 50 percent of Catholic priests, after vowing celibacy, have at some point broken it. Since 1965, two hundred thousand nuns and priests have left their orders primarily because they were unable to continue being celibate. Researcher A. W. Richard Sipe, a former priest, found that 23 percent of Catholic clergy are homosexual. Other estimates count as many as half of this group as sexually active. At seminary training, students are told "not to look into a woman's eyes," with prayer and cold showers suggested as the solution to quell temptation. There is no mention about altar boys. In 2006 alone, the Catholic Church paid $615 million stemming from incidents involving clergy found guilty of or suspected of child sex abuse.

DIONYSOS

DRUNKEN DEITY

Many religions thank certain gods for gifting humanity with pleasing intoxicants. Evidence that rituals were performed for these deities dates to when man first learned to ferment grapes and mead. More minute attention to cultivation of these intoxicating plants are considered one of the primary reasons mankind went from nomadic life to agricultural, ultimately allowing for the development of civilization; staying in one place to wait for the wine to ferment was a powerful catalyst to stay put. After time, religions sprung up believing the altered state acquired by drinking to excess was divinely sent, and that it was the possession of their minds and bodies by an outside spirit that caused them to feel different. Some of the first religious rituals centered on these gods, beseeching them to help grow the spe-

cial plants and protect those crops most of all. Even when the more enlightened Greek society flourished it included a god in charge of drink, and invented a wine-inspired mythology about Dionysos, the god of wine.

According to mythology, Dionysos was discarded from his mother's womb, and then Zeus stuck unborn Dionysos into his thigh for the remainder of the gestation. This odd birth gave him handsomeness that was as beautiful as that of a woman, as well as a history of acting unpredictably and odd. Subsequently a flourishing cult formed to celebrate the inspiring madness Dionysos dispensed through wine. It was his spirit that allowed for flashes of uninhibited genius that one got while getting drunk. For the most serious followers, worship of the god was shown not merely by getting high, but also by continuing to drink until passing out, which they believed transformed them during the comatose-like state, even rendering them holy. Dionysos also gave followers a long-lost formula to guarantee no hangovers, which records indicate included poison ivy as an ingredient. Many of the ceremonies to this god were performed in the mountains so that the effect of the high altitude increased inebriation. Dionysians had their special drinking mugs, everything from hollowed-out bullhorns, giant wooden goblets, and chalices to long rods and walking staffs filled with wine, corked with a pinecone.

In Roman times, Dionysos, called Bacchus, had similar cults, equally rowdy and dedicated to drinking to access. The followers, or Bacchanalibus, believed that insights and appreciation of nature's true beauty were possible only when intoxication removed the illusory preoccupation and distractions of daily life. The Roman followers of Bacchus described this state as one of "enthusiasm," or *enthusiasmus*, which meant being in a spiritual realm of ecstasy, and under the possession of the god. In the second century B.C. Rome passed laws forbidding the sect, believing many plots were hatched under Bacchus's influence. Spartacus, the leader of the rebellion of Roman slaves in 70 A.D., was a dedicated worshiper of Bacchus and lover of wine.

FATHER DIVINE

SELF-PROCLAIMED MESSIAH

Born George Baker in the 1870s, "Father Divine" took to preaching from a young age, having learned the trade by traveling the American South with a fiery speaker known as Father Jehovah. After his mentor died, and narrowly escaping a number of lynch mobs, Father Divine formed his own religion when he moved to a huge house in Sayville, Long Island. He called his group "the Peace Movement," and before long his residence was a center for blacks arriving by busloads to hear his sermons and get a meal or help finding a job. After some time, his unusual benevolence led his followers to believe Father Divine was actually God himself, which the good preacher at first neither denied or encouraged.

When the same racism that plagued his efforts in the South attacked his congregation in Sayville, leading to his arrest for "disturbing the peace," he bravely stood his ground. In 1931, instead of paying the fine for the summons issued, Father Divine went to trial. According to court records, the judge disdained Divine and his "gullible" disciples, quickly sentencing Divine to one year in jail. Miraculously, four days later the judge died suddenly, to which Divine responded: "I hated to do it." With even more public attention, Father Divine's sentence was appealed, and he was released from lockup four weeks later. Once out, he moved his headquarters to Harlem, and satellite branches of his Peace Movement soon spread from coast to coast. By then Divine had proclaimed he was in fact the Messiah promised in the Old Testament. Hymns and prayers were composed to worship him. He also provided food, shelter, and jobs, though he demanded followers show him love, in addition to donating their salaries and all earthly possessions.

By 1942, he moved to a mansion in Philadelphia, had a stable of luxury cars and private planes, and was attended to like a king. When his wife, Sister Penny, died, he married his white Canadian secretary, announcing to his followers that she was the incarnation of his first wife. He crowned her Sweet Angel, or Mother Divine. Supposedly, the marriage was never consummated. When he died in 1965, Divine's holdings were worth, by conservative estimates, more than $4 million. Under Mother

Divine's leadership, his religious enterprises continued. It was known that Mother Divine continued to set a place for Father Divine at dinner and always referred to him in the present tense. The movement had dissolved by the mid-1990s.

DIVINE HAIR

RELIGIOUS HAIR FASHIONS THROUGH HISTORY

Jesus is always portrayed with long, shoulder-length hair, parted in the middle, and sporting a nicely trimmed beard, but by the fourth century, looking like Jesus was frowned upon by clerical orders. It was recommended to cut your hair at least once a year, preferably on Easter. In 456 A.D., St. Patrick wanted anyone with long hair to be expelled from the church, and soon after a church deacon was permitted to forcibly subject an unruly-haired parishioner to a shearing. Beards, mustaches, or goatees were definitely out. In 572 A.D., the religious had to have their ears exposed, and again in 721, the church in Rome passed rules to excommunicate all uncombed, lock-loving persons among the congregation. The monks of the Middle Ages thought shaving the head, once a common practice applied to only slaves, was a good symbol for ascetic lifestyles, though some left a ring of hair to represent the crown of thorns.

For religious women, a veil, covering the hair and pulled down to the middle of the brow, and wrapping around to conceal the neck—just as the Virgin Mary had been depicted—was standard for nearly two thousand years, and up until recently females were required to wear at least some kind of hat in many Christian churches. Mary's fashion was merely following Jewish custom, that only adulterous women left their hair exposed in public, and if a man saw his wife on the street without a head covering, it was immediate grounds for divorce—and in some incidents, a stoning.

The Torah outlawed men from trimming the side locks of a beard and made specific rules on shaving, a practice still followed by Orthodox Jews. It was originally conceived as a way to distinguish Hebrews from pagans. (*See also*: Samson) During the Protestant Reformation, beards once again became a statement and a mark of conviction. Anabaptists (whose hairstyles can still be observed among the Amish) fashioned beards without mustaches; they served as a virtual in-your-face wedding ring: once married, a man could no longer shave, except the mustache, which had been favored by military and soldiers, and in opposition to Anabaptists' pacifist views. One of the original religious hairstyles, favored among many biblical prophets and known today as dreadlocks, came back in style after singer Bob Marley, a Rastafarian, came on the scene in the mid-1970s. Rastafarians interpreted the same Old Testament text as the Jews but believed it meant that no hair should be cut; instead, the hair should be allowed to grow as it naturally wishes. Rastas believe Jesus was black and wore dreadlocks himself. (*See also*: Rastafarians)

DIVINING ROD

"WATER WITCHING"

A Y-shaped sapling, usually from a hazel or willow tree, in the hands of a dowser, will bend and point to the ground when there is either water or oil hidden below. The liveliness of the bending and pointing branch enables the dowser to tell of the exact depth and the volume of the unseen subterranean supplies. Skeptics say this may have to do with heat, magnetic fields, or minerals emanating from water or oil sources, while believers say the stick resides in the realm of the divine. It's still a method used around the world to find potable water. As recently as 1999, a seventy-two-year-old Kansas man with a reputation for "water-witching" used his abilities with a divining rod to locate unmarked graves in a local cemetery. By the time his rod went limp he had found 272 graves that needed tombstones. City officials believed in his skills so much, they didn't bother to dig, and simply placed the markers where he said to.

DOBLE-CRUZ, ORDER OF THE

PIRATE RELIGION

A militia-style cult of Spanish seafarers operated in the Caribbean during the 1500s and became the most successful, if not the most brutal and cunning, of the early buccaneers. Known as the Order of the Doble-Cruz (Double-Cross), they were responsible for keeping all non-Spanish nations away from the Americas for nearly a century. A tattoo, consisting of two crucifixes, or two crosses, encased by an image of a sunburst was inked into the palm of their hands. They proclaimed themselves guardians of many Christian relics, including the Holy Grail, provided to them by the Spanish monarchy to bring luck and protection. The actual cup—Le Graal—was supposedly discovered by French Crusaders. It was not golden and bejeweled, but rather an ordinary pottery vessel that was eventually turned over to Rome in exchange for favorable edicts. It stayed there until the first Spanish pope, Alexander VI, took control. Since he had achieved his power through the influence of Spain's Queen Isabella, the queen instructed him to rifle the vaults and bring her the best religious icon, especially the cup used at the Last Supper. She then supposedly gave it to Christopher Columbus, who then secretly passed it to other explorers, then eventually into the guardianship of privateers, sworn for life into the Order of the Doble-Cruz. With possession of the holy cup as proof, these seamen believed they were mandated by God to monopolize the Caribbean, and cited as a motto, "God is Spanish."

By the time the swashbuckling and legendary pirate era began, the Order of the Doble-Cruz seemed to dissolve; mention of its existence faded soon after Pedro Menéndez de Avilés, an admiral of the Spanish Main, died in 1574. Menéndez established the first Spanish colony in what would become the United States, at St. Augustine, Florida, in 1565. He was said to allow the first Catholic mass to be held on the American mainland, using the pottery cup as a chalice. However, almost as soon as the mass was over he raced to the site just south of St. Augustine where three hundred French sailors had been shipwrecked, and he killed them all—as he said, not because they were Frenchmen, but because they were heretics and Protestants. The Holy Grail was thought lost when St. Augustine was ransacked and burned by English privateer Sir Francis Drake in 1586. It's assumed that when it came time to count and divide the treasure, the cup was considered worthless and tossed overboard. Grail hunters believe that members of the Doble-Cruz retrieved it and later buried it in a secret spot.

ST. DOMINIC

POPULARIZER OF ROSARY BEADS, FOUNDER OF THE DOMINICAN ORDER

In 1170 A.D., when Dominic's mother was pregnant, she had a recurring nightmare that instead of giving birth to a baby, a dog holding a flaming torch in its teeth would leap from her womb. It's assumed, regardless of the actual labor pain, that she was relieved when the future patron saint of the Dominican Republic and recipient of the first rosary was born, and not a crazed animal. Dominic was from Spanish wealth and sent to universities to study art and theology, but just as he was about to graduate a famine struck, killing thousands. At age twenty-one, Dominic gave away all his money to feed the poor, along with his prize books, which would have made him a scholar, with a lucrative career. Instead he devoted himself to religion. After spending time among traditional clergy, he rejected their comfortable lifestyle. He chose to wear burlap sacks and bound his waist and legs with iron chains that he never removed, not even for sleep. His zeal for penance was equal to his fervor for preaching and fighting heresy. He eventually founded a monastic order, the Dominicans, noted for "mortification of the flesh," or self-inflicted penance. Many from his order later served as interrogators and administrators of pain during the Spanish Inquisition.

According to Dominic, the rosary, a string of prayer beads, was personally given to him in 1214 by the Virgin Mary. Dominic had the vision after a long fast and session of self-flagellation while visiting the very first Dominican convent. The apparition of Mary handed him the rosary, which held fifty-nine various-sized beads. He was instructed that a Lord's Prayer was to be said at each large bead, while each of the ten smaller beads was for the repeated recitation of her prayer, the Ave Maria, or Hail Mary. Most biblical scholars, however, believe that it's likely the rosary evolved naturally: There is evidence that an earlier version of the rosary had 150 beads, one for each of the Bible's psalms. Nevertheless, Dominic stuck to his story that Mary herself gave him the rosary, for she had promised to make his order flourish if he promoted it.

Regardless of his fame and apparent saintliness, Dominic throughout his many travels never slept in a bed. Always seeking out the most discomfort he could find, he also took his shoes off before entering a town, preferring to walk barefoot on jagged stones. In addition to self-punishment, he demanded that his followers mimic his example of abject poverty. When dying of exhaustion in 1221, he crawled down from his sickbed to the dirt floor, where he expired at age fifty-one.

DREAMTIME

ABORIGINAL AUSTRALIANS' CREATION STORY

Aborigines, the original settlers of Australia, got there by walking across a land bridge that connected the continent to Asia forty thousand years ago. The world's first religion that seemed nearly universal among early man, including aborigines, was animism, the belief that spirits were in everything they encountered in the world. All things, humans included, were part of a dream. Native Australians believed the planet was fashioned during a period called "Dreamtime," when creator beings conceived of this reality during an hallucination. The spirits created in this fantasy then inhabited the earth and eventually took the shape of plants, animals, and landscapes, and accordingly, remain so in this state to this day. Dream serpents, for example, left their mark on the earth, which explains why most rivers snake their way through terrain, rarely making a straight line. A gully was considered a footprint of an ancient dream being.

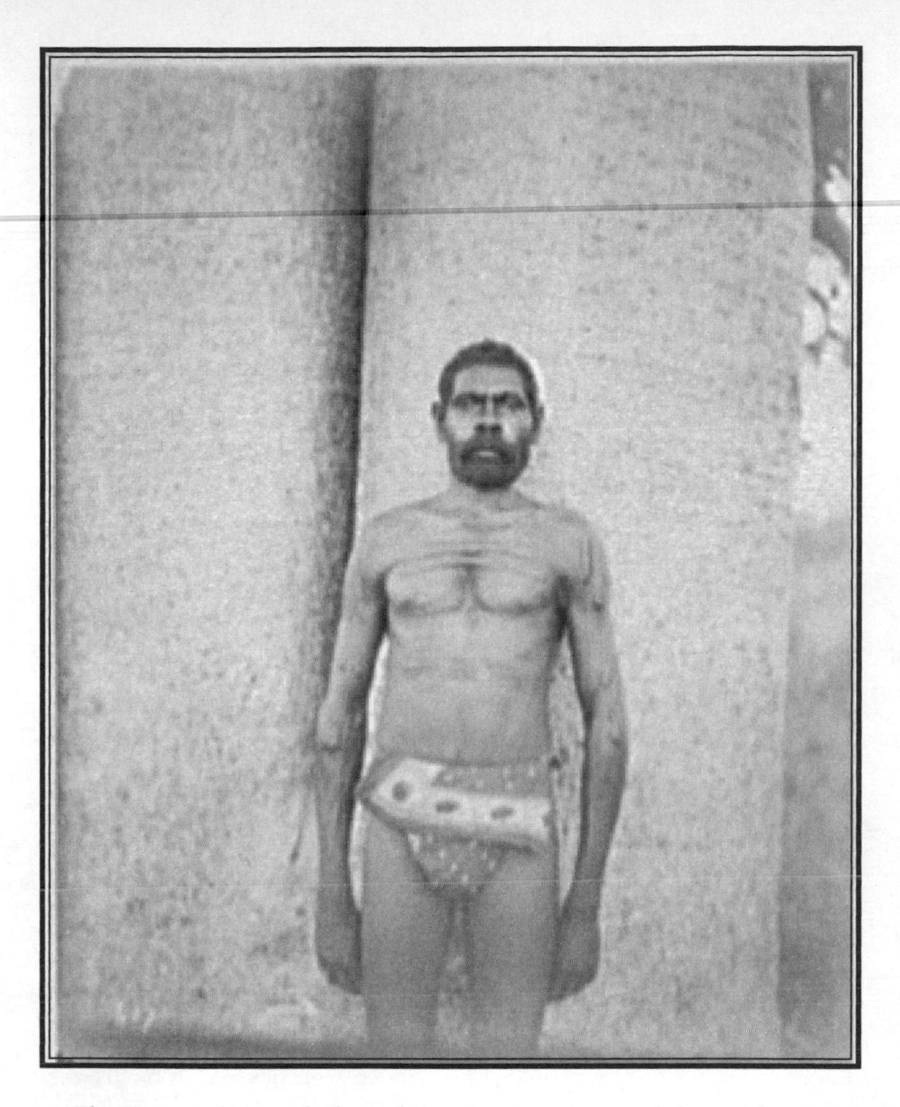

Aborigine religions believed in reincarnation, and that each person had two souls in a quarrel, similar to the id and ego theories developed by later psychologists. At death, after the corpse was placed on a raised platform and reduced to bones, the skeletal remains were painted red with ochre and carried by relatives for a year or more until a dream told where to place them. Unlike other reincarnation beliefs, the aborigines were not trying to get to a higher level, since all things were the same. To come back as a sweet potato, or the native yam, was as good as coming back as a king, since even vegetables had dreams.

DROGO

THE SAINT WHO WAS IN TWO PLACES AT ONCE

When Drogo learned that his mother had died while giving birth to him, he was racked with guilt. From his teenage years, he decided to subject himself to a life of penance for the harm he had done. In the twelfth century, Drogo abandoned his status in the French noble class and became a penniless pilgrim, walking back and forth to Rome nearly a dozen times. On one tour he came down with a flesh-eating virus that left him horribly disfigured. Although he was unquestionably pious, his repulsive looks scared children and townspeople. For the good of the community, Drogo allowed himself to be locked inside a small storage-room-sized cell, attached to a church. For more than forty years, he remained in the dark space with only a small sliding aperture through which he received his daily Eucharist, one barley roll, and a cup of water.

However, while confirmed to be in his vault, he was also seen miles away, working in a field. Other times he was spotted tending sheep, and frequently noted to be in other cities, observed drinking a cup of coffee. His miraculous ability to bilocate—defined as literally being in two places at once—confirmed his holiness, though he was kept in isolation until he died in 1185, at age eighty.

Recent quantum physics has shown that wave particles can exist in two different places simultaneously, though historical reports of the entire body achieving this phenomenon, other than St. Drogo, is rare. Unlike an out-of-body experience, bilocation is a strange physical anomaly that allows each copy of the bilocated self to fully interact with its separate environment. Another incident of miraculous bilocation occurred in 1770 when St. Liguori was witnessed both serving mass and visiting a sick pope, four days by carriage ride from where he was. This event had many eyewitnesses and was extensively recorded. Maria de Agreda was another noted bilocationist. She was most assuredly living in a nunnery in Spain during the 1600s when she duplicated herself more than five hundred times across the ocean in North America. When missionaries first reached the remotest locales within Mexico they were stunned that the natives knew so much about Christianity, and were told of Maria's many visits.

St. Drogo is officially the patron saint of shepherds, and of people with bodily disfigurement. He's also the patron saint of coffeehouse owners.

DRUIDS

PAGAN ORIGIN OF CHRISTIAN HOLIDAYS

Druids were Celtic priests, the leaders of the most prevalent religion practiced through the British Isles and northern parts of Europe for more than five centuries, until they were forced into secrecy in earnest around 1000 A.D. Celts believed in many gods and considered spirits to be in plants, certain enchanted places, and in animals. Physical reality was viewed as existing in a middle world, which was frequently invaded by fairies, gnomes, goblins, and other beings thought to exist in the lower and higher realms. Human sacrifice was used to appease gods, and many went willingly to their death, believing it brought honor.

Christian missionaries met a difficult time converting Celts until they incorporated many pagan beliefs and practices into various Christian holidays. The Celtic Yule Festival (Feailley Geu) was a twelve-day affair that was Christianized into the "twelve days of Christmas," and Yule Tidings, centered on the birth of Jesus instead of human sacrifice. In some Celtic ceremonies, a human was offered on the ninth day of rituals and feasts. (*See also:* Ritual Killings) The use of mistletoe during Christmas season likewise originated in Celtic practices; Celts considered it a sacred plant and used it to appease the fairies and gods of both fertility and immortality.

Christians said the mistletoe had at one time grown to tree size, but since it was used to make the cross, God turned it into a parasitic vine, though it still bears red poisonous berries in winter, signifying the blood of Jesus.

Halloween remains the least modified of Druidic practices; the Celts called it the "Summer's End" feast, or Samhain. This date marked the end of the calendar year and was their version of New Year's Eve. Druid feasts commemorated harvest, though a plate and a mug was set for the recently dead. It was considered a turning point in their seasons where the departed could come back, migrating easily from the upper and lower realms, and if the bad ones were not intercepted they would cause problems to livestock and bring disease. The Celtic fire festival called "All Hallowtide" scared the bad dead spirits away and purified their villages: The donning of Halloween masks was meant to frighten these spirits back to their graves. The missionaries thought to make this important Celtic holiday into "All Hallows Day," when Christians remembered all the souls stuck in purgatory. (*See also*: Día de los Muertos [Day of the Dead])

LEPRECHAUNS

LITTLE GREEN IRISHMEN

According to the Celts, leprechauns, a race of small mystical humanoids, belonging to the "fae" or fairy class of beings, once inhabited Ireland. Originally, leprechauns were mischievous and cunning shoemakers, though as the human

population encroached, they found secret spots to hide treasure and, of course, pots of gold. Many searched to find leprechauns' hidden cash, though leprechauns threw off seekers by spreading a rumor that their loot was hidden at the end of a rainbow, a place impossible to find. Until recent times, reports of leprechaun sightings were frequent; if a person came across one, these male fairies had the power to freeze them in their tracks until the fairies vanished, back into hiding. Leprechaun contemporaries included goblins, physically more grotesque, and considered dangerous pranksters who usually only worked at night. Gnomes were even smaller than leprechauns. They were often depicted as industrious and reportedly seen under new-sprouted mushrooms, where they harvested spores to make "magical" cupcakes and other baked goods. Other prehuman beings included ogres or giants, though these were mostly exterminated due to their size. Trolls were more resilient and had the ability to appear in human form, with the exception of their tails, which often gave away their identity.

DUNKARDS

FULL-SUBMERSION BAPTISTS

Alexander Mack, a German miller, or wheat grinder, established this Protestant offshoot in 1708. Originally called the Church of the Brethren, Mack's movement soon spread westward and earned the nickname "Dunkards" from the practice of dipping a convert fully underwater three times. Adherents believed exclusively in the New Testament and were encouraged to avoid carnivals, swearing, filing lawsuits, or attending secular colleges. Baptism was usually reserved until an age when prospects could make a choice whether to adhere to church principles.

Water as the symbol for rebirth and purification had been popular since ancient times. The Christian practice really took off after Jesus was baptized by John the Baptist in the Jordan River. Jesus' apostle Peter was said to have baptized a record three thousand people during

a massive ceremony in Jerusalem ten days after Jesus ascended into heaven, or approximately fifty days after he was crucified. (*See also*: Pentecostalism) It wasn't a onetime deal, and a person could be baptized or cleansed as many times as needed. Infant baptism didn't begin until around the second century. Although many Christian groups accept or disregard other sacraments, baptism remains an essential key to salvation for many. It is estimated that another person is baptized every sixty seconds.

CHURCHES THAT BAPTIZE

Full submersion is practiced by Baptists, Church of Latter-Day Saints, Churches of Christ, Jehovah's Witnesses, the East Orthodox Catholic Church, Seventh-Day Adventists, Pentecostal denominations, and Revivalist churches. Roman Catholics, Anglican denominations, and Lutheran churches use sprinkling with water to baptize.

E

EBONITES

JEWS FOR JESUS

Even if nearly all of the earliest Christians were in fact "Jews for Jesus," there was a second-century ascetic Jewish sect that believed Jesus was the messiah but didn't go so far as to accept the Virgin Mary theory, or that Jesus and the Holy Spirit were part of a Trinity. They strictly practiced most customs outlined in the Torah, in addition to encompassing some of the new tenets of Christianity. The Ebonites had gospels written in Aramaic and thought James, the alleged brother of Jesus, was the most reliable source, and came to primarily hail Jesus as a prophet. By the seventh century, and with the rise of Islam, the Ebonites in Syria and Egypt, at least, became Muslims. The contemporary Jews for Y'shua (Jesus) movement was begun in 1973 by Moishe Rosen and encouraged celebration of both Hanukkah and Christmas. The more than seventy-five thousand current members are also enlisted to frequently proselytize to mostly Jewish people, urging them to get baptized.

ECKANKAR

GOD'S CO-WORKER

Paul Twitchell had been practicing yoga for eight years when one day, in the early 1960s, while in deep meditation, he was disturbed by a loud clap. To his surprise, the unexpected commotion happened to be caused by an alien from Venus. Formerly on staff at the Church of Scientology, Twitchell followed the guidance of the extraterrestrials and in 1965 formed a new religion called Eckankar, meaning "co-worker with God." Twitchell claimed he was selected as the 971st ECK Master, though he was not given a complete list of the previous line of divine guides. Members are required to log dreams, use audio pulse and light strobe devices (for the "sacred current of sound"), practice for fifteen minutes a day at any of the 131 spiritual exercises devised by Twitchell, and chant "Hyoo," a word for God, many times during the day. Mandatory reading remains study of Twitchell's *The Key to Secret Worlds*, as well as other texts and tapes. Reincarnation until perfection is the afterlife belief of an Eckankarist, though the main focus of life is to contact your own spirit guide. These divine aliens will help a person during out-of-body experiences, time travel, and visits to the astral realm, all designed to break through twelve separate levels of consciousness, until enlightened. Today there are about fifty thousand practicing ECKs.

According to the Gospels, Jesus said, "Come and follow me." But few knew He was saying that He wanted them to go with Him into the worlds beyond.

—Paul Twitchell

ECSTASY

IN SEVENTH HEAVEN

In many religions, achieving an altered state of consciousness, or "receiving the Light," is referred to as "ecstasy." To qualify, the experience must involve a religious subject, as opposed to demon possession, ordinary psychotic madness, or hypnotic trance. In the complete state of ecstasy, the ecstatic can no longer react to his or her physical surroundings, and though awake, they stare wide-eyed, open-mouthed, or appear as if about to convulse. They will not answer, nor react to physical handling of their body. Instead they are focusing on images that appear absolutely real to their minds, even though no others can see them. The end result of religious ecstasy is physical euphoria and a belief that a God-related incident has been experienced. How to achieve this has been the quest of many religions, with techniques used to trigger an ecstasy including flagellation, fasting, meditation, sex, dancing, and psychotropic herbs or drugs, to name a few. The Hebrew mystics, Kabbalists, and Sufi Muslims believed ecstasy offered a taste of what it would feel like to be eternally in Seventh Heaven, the place they envisioned as the highest of heavenly concentric spheres, the realm of God and angels.

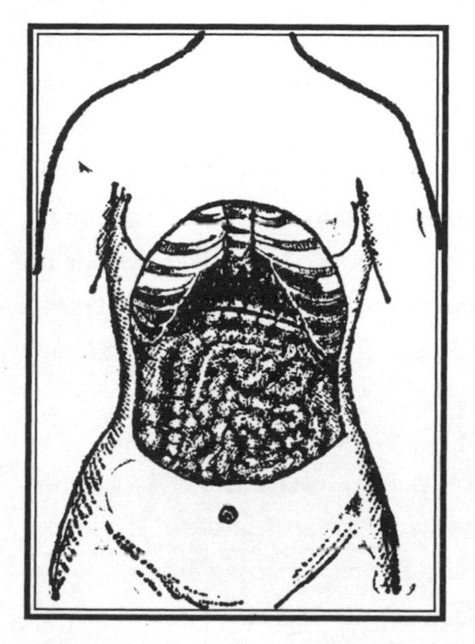

HOLDING IT IN

In Victorian times, bodily functions were generally considered uncivilized, especially among women, such that normal urination was kept to an extreme minimum. One entrepreneur marketed a St. Marina Corset, which was claimed to make for less frequent trips to the bathroom. After a day of withholding urine discharge, one would experience a near state of "religious ecstasy with relief." St. Marina was a second-century martyr, depicted wearing a belt covering her kidneys, as well as wielding a hammer and a cross. She reportedly had the power to cure urinary diseases and is currently the patron saint of nephrology. St. Marina, after refusing marriage, had her body scraped by rusted iron shards, and the wounds rubbed with lime and salt. In addition, her kidneys were singed with hot pokers and she was submerged in a cauldron of molten lead. Still, St. Marina didn't die; finally she was beheaded. The executioner who killed St. Marina was so remorseful for his act that he followed her to heaven by immediately decapitating himself.

In the United States, "ecstasy" was cited as a cause of mortality on nearly twenty thousand death certificates dating from the 1790s through the 1880s. Although most often attributed to religious fanaticism, it came to describe a person who had lost the ability to reason and had become noted for violent outbursts. Any form of mental illness where the patient was acting so wild and agitated that they required being chained to a tree or a pillar—often calling out to God or the devil until dead—was said to be a result of ecstasy.

GOD MACHINE

Technology promises to manufacture the experience of ecstasy via electrodes and by applying breakthroughs in neurosciences. It's possible to pinpoint the parts of the brain responsible for anatomically producing a religious vision. Neuroscientist Michael Presinger has developed a helmet-like machine, fitted with wires that flood the brain with pulsating electromagnetic currents. Presinger experimented with duplicating brain patterns experienced during epileptic seizures and frequencies experienced by subjects during euphoric episodes in order to jump-start mystical ecstasy. Others believe "neuro-theology" will be a new religion, though relying on science to trigger a transcendental moment, even if conveniently by appointment. Presinger's machine has usually only worked for people willing and already seeking an otherworldly experience. Geneticists differ and instead believe there might be a "God gene" that makes it easier for some to gain enlightenment. Genetic researchers at the National Cancer Institute found a gene sequence called "the spiritual allele" to be more prevalent in intensively spiritual-minded people. It's supposed that one day an unlimited number of genetically engineered individuals could be preconditioned to mature into full-time mystics.

ECTOPLASM

GHOST SCIENCE

Parapsychologists believe the dead move about as shapeless nonmatter, as energy fields called ectoplasm. These souls can momentarily materialize as a gas, liquid, or solid. The proximity of ectoplasm can be detected by a chill, and whiff of a distinctive odor, similar to semen, or as a gust of extremely foul body sweat. These clues from the other side are among the signs the "Ghost Busters," or ectoplasmic specialists employ when hired to identify and rid premises of malevolent ectoplasm. Some believe that when a violent incident occurs, such as a murder, wicked ectoplasm can regenerate and continue to kill at the exact location where the original death happened. Good ectoplasm merely observes, though it still causes chills and a sudden sniff of a bad smell.

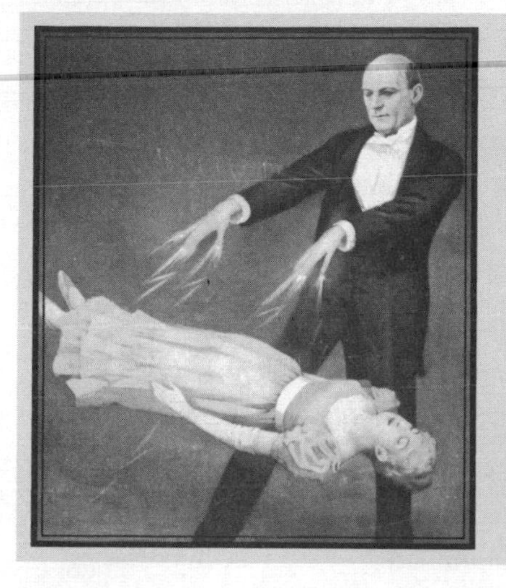

ECTONIC FORCES

In spiritualism, the power to physically move objects without touch is called ectonic force. It can be seen—and has been claimed to appear as small, static-like lightning bolts emanating from the tips of a wizard's or magician's fingers. The means to harness this force can be achieved through the intensity of the mind, or from the aid of an unseen entity.

According to a recent Harris poll, 75 percent of Americans believe in ghosts.

MARY BAKER EDDY

FOUNDER OF "CHRISTIAN SCIENCE"

Mary Baker was a sickly child known for episodic hysteria and convulsions, raised a Puritan, and encouraged to spend most free time dedicated to Bible reading and prayer. There were no early indications she would later perform miracles or become a founder of a religion. She had gone through two marriages by her mid-thirties, with her first husband dying of yellow fever only six months after the wedding, leaving her pregnant and penniless. The second husband, a dentist, she divorced for suspected philandering. Eddy became preoccupied with not being sick and feeling lethargic all the time. She sought out the famed doctor Phineas Quimby, an advocate of alternative healing. Quimby believed that the mind brought on most illness, and that if channeled properly it could also restore one to health. Mary married again, though when the third husband, Asa G. Eddy, soon died of a heart attack, she dedicated the next ten years to writing a book based on Quimby's ideas, *Science and Health with Key to the Scriptures.* In 1866, Mary was inspired to start a religion, the First Church of Christ, Scientist, based on her book. Years later, she recalled the precise moment she got the notion, and said it came to her after she took a bad spill on an icy sidewalk. She remembered how she cured herself to a state of superior health, nearly instantaneously, through the power of her mind and intense prayer. She relied on reading the Bible to activate the process but also used hypnosis and clairvoyance to interpret the scriptures, which she said had been corrupted: She claimed the new version was in her book and "uncontaminated by human hypotheses." Mary believed disease was imaginary, "a mortal fear, a mistaken belief," and she offered her controversial advice that "no true Christian Science member should ever go to a doctor, hospital, or take any kind of medicine, for to do so is to deny 'Divine Science.'" Her philosophy was not merely a mind-cure technique but offered as a goal the attainment of salvation. This was to be achieved through an intuitively grasped relationship with the Supreme Being, since her main underlying notion was that all physical reality, including suffering and illness, was an illusion.

Mary took hands-on control of building a headquarters in Boston, dubbed the Mother Church, and had zero tolerance for disloyalty, real or imagined. A disagreement, contradiction, or a mere interruption while she spoke was met with ferocity. As tough as a battleship well into her seventies and eighties, Eddy initially focused recruitment from the upper classes, and by the time she started a newspaper in 1908, at age eighty-six, the *Christian Science Monitor*, membership rose dramatically, eventually numbering more than three hundred thousand. Today, healing through prayer alone has met with numerous lawsuits, and Christian Science parents who let their children die from curable diseases have been sued successfully, such that membership in 2008 was estimated at less than twenty thousand in the United States.

CHRISTIAN SCIENCE "DOCTORS"

To become a healer in the Christian Science religion, one must take a two-week course before being allowed to offer services to "sick" patients. On completion of the training, one is likewise permitted to add the official initials C.S. after one's name. The duties of a C.S. "practitioner," as they are called, include assigning readings from the Bible and *Science and Health*. They offer no other remedies to the patient. To become a full-time practitioner, one must have documents from three individuals, other than family, that certify to one's healing powers. (They can be handwritten and signed without a notary's seal.) The skilled practitioners know that certain diseases need a different spiritual prescription, and that they might have success with a varying arrangement of readings and prayers. The drawback to this job is that prayer treatments cannot be done over the phone, or by e-mail or webcam. The patient must be attended to in person. To get a reputable C.S. practitioner to heal an ill loved one could be costly, with air travel and time; the IRS allows fees paid to practitioners to qualify as a certifiable medical tax deduction.

EDEN

ORIGINAL PRIME REAL ESTATE

Paradise derives from an Arabic word meaning "pleasure land." Eden was the first paradise, envisioned as a manicured park with green grassy fields, and mowed like the classiest golf course. A portion was set aside for a self-sustaining organic orchard and vegetable garden. Clues on where Eden was located come from scriptures, which indicate it was near the Hiddekel River, known now as the Tigris, and bordered by the Euphrates River, a tract now encompassing portions of Iraq and Iran. Early Christian Church scholars thought it was hidden in some exotic locale near Mongolia, or in India, close to the Ganges River. The arid landscape that now makes up biblical Eden was not always that way, for it was once similar to a tropical rain forest. Archeological evidence indicates it once flourished

with abundant game animals and fertile vegetation. Biblical climatologists now think the original Garden of Eden is underwater, perhaps below the present Persian Gulf. With changing climates, early man—hunters and gatherers primarily—flocked to this area. However, there was already a rudimentary agricultural society in place, populated by people called Ubaidians. The Sumerians came after them, though they used many of the names of rivers, and such that the Ubaid culture had established. The Sumerians are credited with one of the oldest known cities, Eridu, dated at about 5000 B.C. They also had the first written language and mention "Edin," for a "fertile plain." The first prime real estate eventually became less so due to overpopulation and climate change.

FIRST EVICTION NOTICE

Before eating the forbidden fruit, Adam had walked with God, and seemed to be the best of friends, but afterward the Lord pulled divine rank: "Cursed is the ground for your sake; in toil you shall eat of it all the days of your life . . . in the sweat of your face you shall eat bread."

EIGHT-FOLD PATH

BUDDHA'S LAWS

Siddhārtha Gautama (loosely translated as "Goal Achieved by Big Cow"), also known as Buddha, or "Enlightened One," founded Buddhism, one of the world's most long-lasting religions, in the fifth century B.C. A Hindu by birth from northern India, he abandoned luxury and princely titles at age twenty-nine, when during one venture outside his secluded palace he saw a crippled old man, a decaying corpse, and a monk. He abandoned his wife and children to become an ascetic, fixated on finding a way to end human suffering. While sitting under a fig tree he came to understand the scheme of the universe and decided to teach others how to achieve "awakening." His ideas urged "Four Noble Truths," stating that suffering is the primary nature of the universe, and that all suffering is caused by desire. He preached that the "Middle Way," or avoiding extremes, was the most prudent course to attain enlightenment. In order to eliminate physical and mental cravings, Siddhārtha said it was best to follow a step-by-step guide, known as the "Eight-Fold Path."

EIGHT RIGHTS FIX THE WRONGS

THE PATH BUDDHA OFFERED

1. *Right Knowledge*
2. *Right Aspiration*
3. *Right Speech*
4. *Right Behavior*
5. *Right Livelihood*
6. *Right Effort*
7. *Right Mindfulness*
8. *Right Concentration*

Here Buddha established a list on how best to think and act. He urged people to perceive differently, think of themselves less, and have good actions and noble thoughts. Desires, to the Buddha, were considered as sorts of "sins," or what led to sin, but they could be modified or eradicated through meditation. In theory, Buddhism has no theology—worship of deities—and believes salvation does not come from any god, but solely from within. One Buddhist cannot save another Buddhist, or help one to achieve Nirvana. This end phase is nothing like a Christian heaven, but rather a state of nonexistence, free of suffering and desire, considered a reward gained by only the wisest. Likewise, once attaining Nirvana, or the "annihilation of self" and a return to the "Great Void," the human soul does not need to reincarnate again. Siddhārtha died at eighty after drinking a cup of mushroom tea and dining on a pork dinner called *Sukaramaddava*, in 483 B.C. His body was cremated, and the ashes placed in his hometown in Nepal and at seven other locations, marked with a mound and topped with an umbrella. A later king dug up the ashes and divided them among eighty-four thousand other locations, supposedly scattered near the site of other Buddhist temples.

Buddhist temples are usually built in a pagoda design that displays images of Siddhārtha, with rituals performed, meditation practiced, and mantras and prayers said. If not to achieve enlightenment, many still regularly attend to bring good fortune. The monks overseeing Buddhist temples cannot own anything except a begging bowl, and

get their meals and necessities through donations. The first Buddha begged, and so it's considered good karma to give to panhandlers for either religious or secular reasons. A Buddhist monk or nun can marry and have children, as long as monastic duties come first. Buddhism does not actively seek recruits and prefers gaining converts by attraction rather than promotion. There are approximately 350 million Buddhists worldwide.

REINCARNATION AND BEER

Buddhist monks in Thailand began collecting beer bottles in 1984. Eventually amassing more than one million bottles, and to prove that recycling is a form of reincarnation, the monks fashioned a twenty-building complex and a temple made entirely of mortar, brown Chang beer, green Heineken, and Thai Red Bull bottles. Although no alcoholic beverages are served on the grounds, visitors can even use toilets made from old "Heinies."

WHY IS BUDDHA FAT?

The original Buddha was an ascetic, fasting often, and was depicted in early art as extremely emaciated, eyes closed and in prayer. However, most know Buddha as a happy, big-bellied guru sitting in a lotus position. Origins of the fat Buddha are clouded in folklore, and most claim the images are of others who achieved enlightenment. One story maintains that a jolly monk named Angida collected venomous snakes endangering a village, removed their fangs, and while releasing them, laughed heartily. Others say there was one monk born so handsome that even the angels tried to flirt with him, such that he transformed himself into an obese body in an attempt to limit suitors. Although there is no Buddhist text concerning how Buddha should be displayed, most consider the big-Buddha image as representative of abundance and comfort. Rubbing the belly of a Buddha statue supposedly brings prosperity and good luck.

A BURNING DESIRE

Through the 2,500 years of Buddhist history, a persistent streak of monks have done more than mediate, being prone to self-mutilation. Though generally frowned upon, it is accepted as one's own path toward enlightenment. That was the conclusion drawn from the report of a Thai Buddhist monk in 2007 who unexpectedly experienced an erection during meditation. He castrated himself on the

spot, explaining later that he did so to remove that "most distracting of desires." A famous incident occurred in 1963 when Buddhist monk Thich Quang Duc sat in a lotus position in the middle of an intersection in Saigon, South Vietnam. He did it to protest the ruling regime's treatment and persecution of monks. To make his point fully known, the monk then poured a can of gas over his head and set himself on fire. The crowds watched and the cameras clicked as the monk, never moving from his meditative posture, was consumed by flames and turned to ash. A number of monks followed and likewise immolated, until five months from when Thich Quang Duc first started the trend, the then-oppressive regime was wiped out by an unexpected army coup, considered by many a divine intervention.

EIGHTH SPHERE

MAXIMUM-SECURITY PRISON FOR EVIL SOULS

In the unseen world, there are many levels where angels and divine beings reside. In the occult, rising from theosophical doctrines, the Eighth Sphere is considered the place where the evilest souls—those not even deserving the vilest reincarnation—are sent for atomic pulveriza- tion and ultimate end to a soul's existence. Dante, the fifteenth-century Catholic writer, envisioned the justice dished out to those kind of souls more sinisterly, and conceived an ironic, separate suffering to match

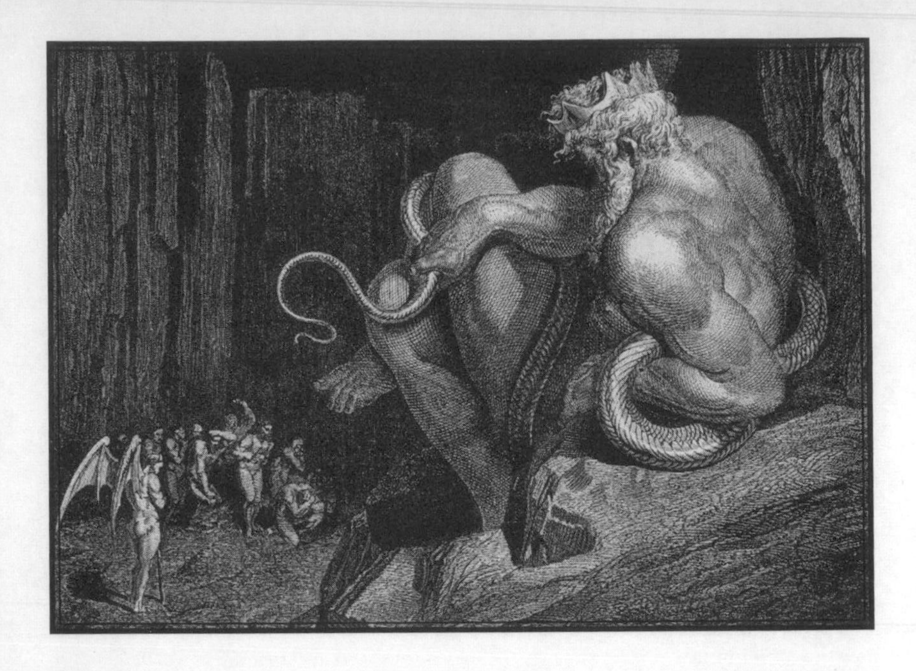

the sin, or crime. For example, obsessed lovers are locked together in an eternal embrace, and are among the lustful souls blown here and there by an endless whirlwind. Murderers tread forever in a river of blood. For Dante, his last of nine circles of hell, or "the Pit," is the fate for betrayers, or those who stole from the helpless, pillaged trust funds and such, or generally had an icy heart while alive. The ninth circle is a frozen sea where the perpetually shivering and condemned are constantly tortured by a three-headed Lucifer. So excruciating was the pain in Dante's hell that the occult Eighth Sphere seemed an act of mercy.

ELAN VITAL

MEET GOD AT THE ASTRODOME

An Indian family claimed they had groomed their son to be a guru, especially after the boy, Prem Rawat, began preaching at the age of three. In 1971, the thirteen-year-old guru, then called Maharaji Ji, or M-Ji for short, arrived in Los Angeles to start a religion dubbed the Divine Light Mission, though at first he attracted few followers. However, by 1973 twenty thousand had heard rumors of the wonder boy and filed into the Houston Astrodome to worship his enlightened presence, seated as he was high on a throne above the crowds. There he formally announced he was the "savior" of humanity. Most of his ideas were Hindu in theology, though he offered a shorter list of M-Ji's five must-do activities to achieve divinity: no procrastination, or don't put off till tomorrow; meditate on the holy name M-Ji gave you; never disbelieve; never avoid attending one of his appearances; and always trust in God. In short, he was seen favoring luxury cars and a lush lifestyle, and was considered a dubious incarnate, even more so after he married an airline stewardess and claimed she was a goddess. However, even after Maharaji Ji's own mother slammed him as a "drunken, carousing, meat-eater" and tried to install her second son as leader, M-Ji persisted, and renamed his group Elan Vital, or "life force." There are approximately 250,000 devotees of Elan Vital, following the "Six Keys of Knowledge." The first five are offered via CD and tapes that take about ten hours to view. The sixth key, however, is revealed only by a "special session" with the guru.

ELIJAH MUHAMMAD

CONTROVERSIAL NATION OF ISLAM LEADER

In the 1930s, a mysterious man, known simply as Farrad, began walking the Detroit slums, claiming he had come from the ancient city of Mecca with a message. He quickly attracted a following and gained attention when urging all "so called Negroes" of America to realize they were actually descendants from the "Original Black Man," and members of a superior race to whites, or "blue-eyed devils." He also denounced Christianity as a tool to keep blacks enslaved and said their real religion was Islam. Four years later, Farrad suddenly disappeared, but before leaving he named a close follower, Elijah Poole, as the leader of his movement, dubbing him Elijah Muhammad and putting him in control of the Nation of Islam's (NOI) organizational headquarters in Chicago. Poole stated that he believed Farrad was Allah reincarnated.

WHERE RETIRED PROPHETS GO

Wali Farrad Muhammad's true identity remains in question, even if FBI records indicate he was Wallace Dodd Ford and had served three years in California's San Quentin prison for drug dealing before arriving in Detroit. Although NOI considered this part of a smear campaign, he was of mixed race, though it was uncertain if he was half African, Polynesian, or Indian, and it remains unknown exactly where he was born. After a murder, with a suspect citing a Farrad pamphlet, *Secret Rituals of the Lost-Found Nation of Islam,* as motivation—"The unbeliever must be stabbed through the heart"—Farrad was told to leave town. His sudden disappearance in 1934 didn't make sense to many; NOI members claimed he was killed by white policemen, while others say he was assassinated by mosque members vying for power. The official church stance on the founding prophet is that he went back to Mecca to prepare for ascension, while U.S. government reports indicate he may have lived quietly in New Zealand until dying in the 1960s, at which point the FBI file on Ford was closed.

Elijah's reach for more converts was aggressive and successful, primarily through publications passed out on street corners and the message of empowerment he offered to marginalized blacks. Included in his mythology of black supremacy was a story of how in ancient times a mad scientist named Yakub genetically created the inferior white race, which the original black men had kept imprisoned in caves. Somehow these "white monsters" eventually escaped, rose to power, and sought to enslave Africans wherever found. He further believed Christianity should be rejected and all followers must change their surnames to X, symbolizing their unknowable African family lineage. Like so many fanatics, he also preached a coming apocalypse and predicted North America would be engulfed in a great fire, at which point blacks would once again regain full control. He dictated strong codes, even demanding avoidance with all whites for fear of contamination. NOI members were to adhere to Islam's practices of forgoing pork products, liquor, and tobacco, though he likewise banned corn bread, since Elijah considered it a food from the slave era. He set up schools, which instilled pride by teaching black history and other skills, including how to be hardworking and punctual. He also recommended dressing in a dignified style, including a bow tie in particular. Although he adopted many texts from the Qur'an, the Bible was still the most quoted book. In addition, he strengthened Farrad's sub-sect, called the "Fruit of Islam," which functioned as a paramilitary organization, teaching youth martial arts and other warfare tactics in preparation for the impending Armageddon.

In the 1960s Elijah's charismatic follower Malcolm X, after a pilgrimage to Mecca, disagreed with the Nation of Islam's direction under his former mentor. In 1965 members of Elijah Muhammad's sect allegedly murdered Malcolm X. Elijah died in 1975, with the organization run by Elijah's son for a number of years, though the Nation of Islam is now led by the no less controversial Louis Farrakhan. Membership numbers for the NOI range from twenty thousand to sixty thousand.

MALCOLM X

MURDERED BY THE NOI?

There is nothing in our book, the Koran, that teaches us to suffer peacefully. Our religion teaches us to be intelligent. Be peaceful, be courteous, obey the law, respect everyone; but if someone puts his hand on you, send him to the cemetery. That's a good religion.

—Malcolm X

Malcolm X, born Malcolm Little in 1925, was about to give a lecture at the Audubon Ballroom in Harlem on March 1, 1965. With his wife and four children in the audience, three men with NOI affiliation seated within eighteen feet of the podium stood up and fired handguns and a sawed-off shotgun, killing Malcolm X. Three assailants were arrested and convicted of murdering the thirty-nine-year-old black leader, though it was never learned for certain who ordered his assassination. After breaking from the Nation of Islam and forming the Muslim Mosque, Inc., he was commanded by the hierarchy of the Nation of Islam to remain silent and stop preaching. Thereafter, Malcolm was under constant threat, saying only two days before his death: "It is a time for martyrs now, and if I am to be one, it will be for the cause of brotherhood. That's the only thing that can save this country."

ENNEAGRAM

ANCIENT PERSONALITY TEST

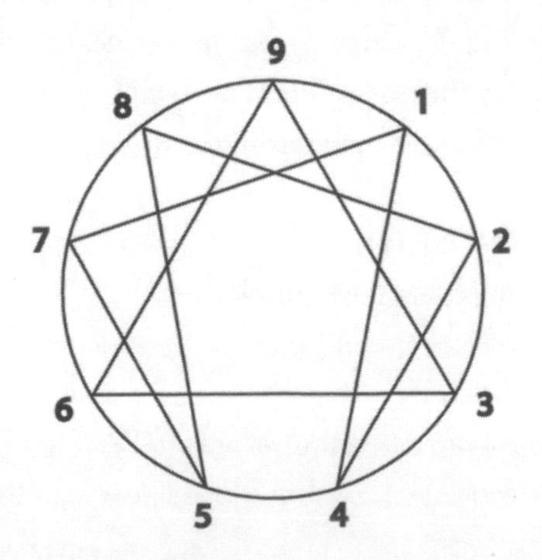

In astrology and occult practices, an enneagram is a drawing divided into nine equal points made by placing three triangle-segments inside a circle. Numbers are assigned to each point and correspond to personality types and star constellations. A reader can look at any number and sense which one means most to them, and then consult a chart that tells of the predominant personality characteristic that number represents. Proponents claim it is used to discover weaknesses and strengths, and employ a system of mathematical mysticism to help find guidance. The first ones were used by mystics in Kabbalah during the second century, and earlier versions were devised by Pythagoras. The most popular personality enneagrams used today were developed in the 1970s by a Bolivian seer, Oscar Ichazo.

EPICURUS

GREEK PHILOSOPHER OF THE GOOD LIFE

One of the most prolific writers of ancient times, Epicurus wrote hundreds of books, though all but a few fragments are lost forever. But for nearly four hundred years his ideas on how to find happiness and enlightenment were imitated, with many communes and centers espousing his thoughts flourishing through the civilized world.

Epicurus sought to free the mind of religious superstition, and he was one of the first to believe the physical world consisted of microscopic pieces of matter, which we now call atoms. Natural occurrences such as earthquakes or meteors were not the gods' doings, but merely the normal movement of atoms. He was accused of being an atheist, but he did not deny the possibility that gods exist. He explained they had achieved a state of tranquility and thus had no concern with man's mundane issues. In order to achieve this godlike state, Epicurus offered a formula on how to live, and placed great value on listening closely to what one's senses experience as a way to gain knowledge. Many think Epicurus simply advocated feeling good, with drink, food, and indulgence as the primary purpose of life, but he actually fostered moderation. If drinking caused a disturbance to one's tranquility, then it was no good; the same was true for eating to excess, or eating bad foods. He did think that everything in life should be done so to ensure pleasure, and in the end he saw a more ascetic lifestyle as the best means to attain this goal. He warned against seeking office, of the pain of ambition, and of fears that arose from contact with too many people—even marriage was a situation Epicurus urged one to avoid. Epicurus elaborated on the meaning of pleasure and often used it as a near synonym for "ataraxia," or tranquility: "Ataraxia is freedom from pain in the body and disturbances of the mind." In his later years he set up a school in a large house on the outskirts of Athens that became known as "the Garden," and he was criticized for allowing women and slaves to join his entourage. At age seventy-one, in 270 B.C., Epicurus died from urinary calculus, commonly called kidney stones. He didn't believe in the afterlife, but rather that his soul would join the universe of atoms. His name and legacy remains, such that a person called an Epicurean today is considered a consummate connoisseur, favoring only a gourmet selection of the best foods and wines. The phrase many attribute to Epicurus, "Eat, drink, and be merry, for tomorrow we die," is actually drawn from biblical text, namely Ecclesiastes and the book of Isaiah.

NO WHINERS ALLOWED

Zeno of Citium was a contemporary of Epicurus and founded a school of thought that considered all this worry about joy, pleasure, pain, grief, or any emotion, in fact, as foolishness. To Zeno, the gods determined everything anyway, so it was best to accept whatever you were dished out—stoically. The word *stoic* has come to mean a dignified acceptance. However, earlier in his life, Zeno raged against his own circumstances and raised a clenched fist at the gods. A seafaring cargo merchant, he had tumbled onto the shores near Athens, having lost his goods, and nearly his life, in a shipwreck. The frugality he came to recommend was thrust upon him as he arrived in the cultured city without status or means. But Zeno eventually stood out among the hordes of homeless and penniless. He was notably unaffected by driving rains, for example, or freezing-cold winds, or burning-hot temperatures. He began to lecture during storms, to a trapped audience of sorts, as people gathered under covered walkways near the marketplace, called *stoas*—and hence his followers came to be called *stoics*. He eventually was able to start communes and schools where he decried the idiocy of temples, laws, marriage, and the use of money. Similar to Epicurus, Zeno accepted men of all classes, as well as women and slaves, and had everyone wear the same simple undistinguished garment as he himself wore. Although Zeno said the road to personal illumination was achieved by living in accord with the laws of nature, he was accused of being lax not only on issues of governance, but on any number of taboos as well—including incest among his commune members and even cannibalism. He ultimately influenced others by believing that time, space, and reality in general were an illusion, and that humanity was no more than an insignificant play toy moved about at the whim of gods. Thus to secure freedom from this cosmic sitcom, of sorts, one must refuse to react and remain poker-faced, regardless of the destiny the gods have dealt.

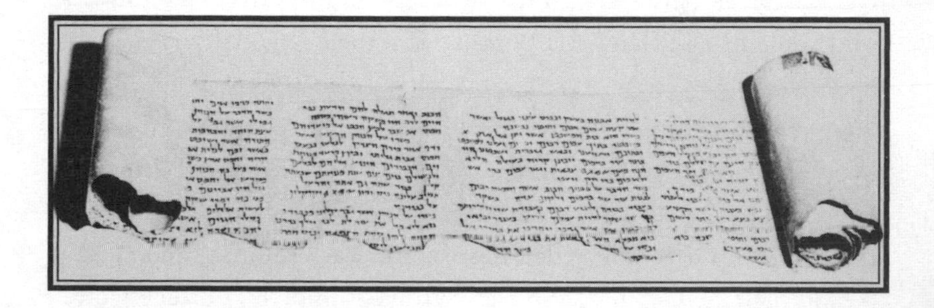

THE ESSENES

HERMITS WITH TIME CAPSULES

In 1947, a shepherd boy threw a rock into a small opening in the side of a cliff, in a mountainous area on the West Bank of the Gaza Strip, to chase out one of his lost goats. When he heard what sounded like breaking glass, he crawled inside to find a smashed ancient clay vessel. By chance, he discovered rolls of parchment wrapped in linen sealed in a stockpile of old vases. He had actually stumbled upon a series of caverns where ancient, holy text—more than 850 documents—were eventually unearthed and ultimately became known as the Dead Sea Scrolls. Some believed the scrolls were hidden there on purpose, on the site of a once-flourishing pottery factory, nearly two thousand years ago. It's believed that a pious group of seers known as the Essenes had made a time capsule and filled the clay pots with their most sacred writings. Further excavation found a writing studio with inkwells and desks nearby, where it was assumed a monastery for the Essene scribes once stood.

The Essenes were a well-known sect in ancient times and were regarded as Judaic hermits and mystics. The cult, however, believed they were guardians of all hallowed texts, and heir to ancient wisdoms, which included the ability to communicate with angels. They followed a solar calendar, as opposed to the Jewish lunar calendar; believed in reincarnation; admitted men as well as women; and customarily washed the feet of each other before eating or praying. Some believe Jesus had spent time with an Essene group, called "the School of Prophets," and that he found lodging during his preaching among Essene disciples. In Egypt, Essenes were referred to as healers, and in Greece as holy doctors. Some of the Dead Sea Scrolls were their original writings, but most were copies from other sources they deemed holy. It appears the group spread throughout antiquity, at its height claiming more than four thousand adherents. They were all but wiped out and disbanded around 70 A.D., when they were suspected of aiding a rebellion against Rome. It seems they knew their destruction was imminent and decided to hide their texts—the Dead Sea Scrolls—amid the pottery factory they correctly assumed the Romans would not destroy.

INDIANA JONES AND THE SCROLLS

Attempts at reviving Essene communes throughout the centuries never fully blossomed, as they were suppressed by mainstream Christian dogma. However, after the discovery of the scrolls a number of groups interpreted Essene writings in various ways. The International Biogenic Society was co-founded by Hungarian scholar Edmund Szekely after he translated portions of the ancient documents and centered his religion on the concept that the Essenes and Jesus were vegetarians. American explorer Gene Savoy, often thought of as the swashbuckling model for the Indiana Jones movie franchise, claimed that after studying Essene texts he had deciphered the mysteries of antediluvian wisdoms. He founded the Church of the Second Advent in 1971, based partly on Essene writings, and was reported to have discovered the secrets of immortality. However, he died at the age of eighty, in 2007.

EST

HUMAN-POTENTIAL MOVEMENT FOUNDED BY A USED-CAR SALESMAN

Werner Erhard (née Paul Rosenberg) had an awakening while driving over the Golden Gate Bridge in San Francisco in 1971. Although previously employed as a used-car salesman, he had studied Scientology, Zen Buddhism, and many of the sects that had sprung up in the 1960s, though wasn't expecting the epiphany he received. While crossing over the bridge he achieved total "enlightenment." He claimed as much, and said, "in that one moment I knew nothing and came to know everything." Reality was illusory, he concluded, and he devised a system to help make the mind perceive differently and achieve enlightenment. Thirty-two curious individuals attended the first lecture he held. By 1979, there were 160,000 enrolled in his EST, or Erhard Seminars Training. In 1985, he changed the name EST to the Forum, and by 1997 had annual sales reaching $50 million, with more than seven hundred thousand having taken his training since its inception. Higher human consciousness is achieved primarily through an expensive two-week, sixty-hour course. For example, a seminar might consist of three hundred people gathered inside a hotel ballroom. They agree to follow all EST rules, which include no talking, leaving for restrooms, chewing gum, etc., and know they might be singled out for ridicule and humiliation. Erhard believed his system was designed to get to "absolute truth," noting, "Truth believed is a lie. If you go around telling the truth, you are lying." Many began to follow the principles religiously to achieve what Erhard called "getting it"—an aphorism for salvation—at which point one will become capable, through a higher level of consciousness, of fully understanding the meaning of many things, including his philosophy.

During the 1980s Erhard was an extremely popular guru, with some followers even believing him the reincarnated messiah. When asked if he was the savior, he said, "No, I am who sent him." After being accused of sexual abuse by his daughter—which he refuted by calling the incident a "nurturing experience"—and serious trouble with IRS investigations and tax liens, the former used-car salesman fled the United States. Now called the Landmark Forum, the organization focuses on corporate training seminars.

EUTHANASIA, CHURCH OF

"SAVE THE PLANET—KILL YOURSELF"

"Thou shalt not procreate" is the main tenet of the Church of Euthanasia, founded online in 1992 by Reverend Chris Korda. Although it has no brick-and-mortar facilities, the organization is recognized by the U.S. government as a tax-exempt religion. The son of famed editor and author Michael Korda, Chris grew up to become a transgendered woman and a staunch vegan. Nevertheless, she had a vision and conversed with an alien, identified as "the Being," who warned of an eco-catastrophe about to cause the contemporary equivalent of biblical Armageddon. "On a dime," she noted, it charged her to immediately start the church and preach against worldwide overpopulation. Employing black comedy and satirical wit, the church theology is summed up in its popular bumper sticker: "Save the Planet—Kill Yourself." Reverend Korda delivers mostly "E-Sermons," in which she explains the four pillars of the church's dogma: abortion, cannibalism (or might as well be, if one is a meat eater), sodomy (or any non-procreating sex), and euthanasia. There are two saints, namely St. Jack Kevorkian, aka "the suicide doctor," and St. Margaret Sanger, a pioneer of birth control. One can believe in any deity to be a member, though one must be a vegetarian and not produce children.

EVANGELICAL CHURCHES

"BORN AGAIN" CHRISTIANS

While membership in most churches has declined in the last decade, Evangelical Christian churches have grown, and even more so during economic downturns. Evangelicals for the most part believe the New Testament as the literal word of God—to the letter. In addition, one must become "born again" through the Holy Spirit and Jesus. Evangelicals are also required to "spread the good news" by preaching and stead-

fastly seeking new candidates for conversion. The belief in *crucicentrism* is central to all Evangelical churches. This means that Jesus' death on the cross and resurrection is the central point of an individual's salvation and reconciliation with God. From the 1700s onward, Evangelical movements have usually targeted poorer and uneducated assemblies with great results and have continued that tradition with the invention of radio and TV. (*See also*: Televangelists)

The National Association of Evangelicals claims more than 66 million members, or 25 percent of the U.S. voting-age population, though with numerous degrees of difference in belief and worship practices.

WHO STARTED WWJD?

In 1896, a book about how one should imitate the life of Jesus, *In His Steps,* by Charles Sheldon, had a subtitle "What Would Jesus Do?" Sheldon was a Protestant minister aligned with the Social Gospel movement, which urged working with the poor. He posed the question in his sermons and articles, urging everyone to measure moral and ethical questions after first contemplating his hypothetical benchmark—What would Jesus do? Sheldon, however, failed to get adequate copyright protection for his writings, and no less than forty publishers distributed various editions of his book without obligation to pay him royalties. It sold more than eight million copies and was translated into two dozen languages. When Sheldon noted the windfall, he apparently asked the question he famously posed, and ultimately did nothing to reap financial rewards.

It was the same fate for the originator of the WWJD bracelet, Janie Tinklenberg, a religious instructor. In 1989 she was encouraging teens to read Sheldon's book, and knowing their love of wristbands, ordered two hundred cloth bracelets with the acronym *WWJD*. By the mid-1990s it was a billion-dollar industry, but Tinklenberg received nothing. Unlike Sheldon, she perhaps thought Jesus might seek a patent, and she did eventually win a trademark for *WWJD* from the U.S. Patent and Trademark Office in 2000. But by then the market was flooded.

EVE

THE FIRST WOMAN WAS MADE

In Genesis, there are two versions of how the first woman was created. The most familiar is that Eve was fashioned from Adam's rib, though contemporary researchers think the true Hebrew word for "rib" actually meant "penis bone" (citing this as a "logical" reason why men do not have a bone in their penis, while men and women both have twenty-four ribs). Nevertheless, after she was made from one of Adam's body parts, God immediately introduced Eve to her new mate with this: "Here is your husband, and he shall rule over you." While some consider this patriarchal portrayal of Eve's origin a reason why women have been treated unequally for more than three thousand years, there's another translated account that states that God fashioned man and woman simultaneously, with the blessing "Fill the Earth and master it." In most artistic renderings, Eve appears to be in her early to mid-twenties, always quite beautiful, with hair long enough to conceal nudity no matter what angle she is turned at. She's also fingered as the one who listened to an evil serpent and took a bite from the one fruit tree in all of Eden that God had listed as off-limits. The fruit she tempted Adam to nibble contained all the knowledge of good and evil. Although both Adam and Eve thought this a good thing to know, they had no idea God would be so disturbed and wrathful for disobeying his warning not to touch it, or that he would punish them and all mankind with banishment from the Garden—described as a heaven on earth—for all eternity. They were survivors, though, and it seemed they didn't harp over who was to blame for their harsher life. They had no in-laws, with Eve being the only woman in all of humanity who didn't have a mother. Even though, as a couple, their parental skills were questionable (*See also*: Cain), Eve apparently kept herself in excellent shape, even after giving birth to three boys. We will never know exactly how long she lived, though it seems she died sometime after Adam, who lived for 930 years.

ADAM'S WIFE BEFORE EVE

According to references in the Bible, the Dead Sea Scrolls, and folklore, there was an original version of Genesis that had God creating a woman named Lilith before Eve was formed. He made Lilith and Adam from the same clay, but from the start they were bickering with each other, despite the paradisiacal environs. Legend has it that Lilith didn't like Adam's earthy smell and refused to have sex, or at least have it as often as Adam wished. In addition, Lilith refused to assume the missionary position, and claimed that she had the right to be on top as much as he did since God created both equal. After

a short time, Lilith was disgusted with the relationship and ran off. She began to hang out with the serpent in the Garden of Eden and she found she could have more fun with the demons.

In the Sumerian version dated to 4000 B.C., noting a similar creation story, and in later Hebrew accounts, Lilith became the spirit of dark storms, who was responsible for plaguing man with erotic dreams. Adam desperately wanted Lilith to return and sent angels to bring her back, though she refused. When Adam told the Creator that the wife He made ran off, God then fashioned Eve from Adam's rib, or bone. Eve knew, though, that as faithful as Adam was to her, there was another woman captivating his imagination, and so she said, "But though Adam's lips [said one thing] his soul always echoed Lilith." She is still blamed for inflicting the subconscious of even happily married men, who dream of the originally perfect woman they cannot have. In Hebrew tradition it was believed Lilith took revenge on newborn males during the period before circumcision, and so amulets were placed on infants to protect from Lilith's curse. For centuries it was a custom not to cut the hair of a male child for three years, as an attempt to deceive Lilith into believing the child a girl, so she would not cast her spell on Adam's new heir.

EVIL EYE

THE LOOK OF DEATH

Throughout ancient times, it was believed that a certain glance had the power to cast a spell and cause harm. Biblical proverbs mention this phenomena, warning not to get "killed by a look." Jesus also mentioned an "evil eye" and noted it could be sensed by observing envy, false flattery, and hypocrisy. In sorcery legends, wizards regularly perfected formulas to make a certain look, or raised-eyebrow configuration, or intense stare (also called "biting eye") powerful enough to cause a sundry list of damages. In modern times, the belief in the power of the evil eye has not waned. Regardless of religion, many Hispanic cultures place amulets on children, namely a plastic eyeball on a chain, so that the evil eye, or *mal de ojo*, cast by envious neighbors, secretly wishing failure and harm, is repelled. Many Italians wear a small red pepper for the same purpose. Jewish parents tie a red ribbon to an infant's crib to keep away the evil eye. And Americans, in the 1950s at least, thought a shiny new penny placed in specially designed slots in loafers kept their schoolchildren skipping along, unharmed by the many evil eyes cast their way each day.

ST. EXPEDITE

PATRON SAINT OF SPEEDY RESULTS

Expeditus, later called St. Expedite for short, earned a reputation for getting things done. He's portrayed in art as dressed as a Roman soldier, holding a crucifix and a palm frond, and under his heavy army boot there is a crow stomped to death, apparently stopped in mid-flight. The flattened bird has a one-word note on a fluttering banner dangling from its beak—"tomorrow." It seems Expeditus was a centurion stationed in Armenia during the third century, and while guarding an articulate Christian prisoner, he decided to convert. Expeditus jumped up so fast to tell of his inspiration that he trapped a crow flying by. He then immediately informed his commander of the decision. Although he knew it would lead to an execution, he asked his commander not to waste time on trials and such and to just go ahead and behead him, a request that was speedily obliged.

Some say no such saint ever existed and came to be from a labeling mistake. In the eighteenth century, when a French convent pleaded with the Vatican to send holy relics, they finally got word that an entire skeleton had been shipped, to arrive in a few weeks time. However, when a crate arrived the next day, the nuns suspected a miracle was afoot, and they wondered whose saint's bones were in the box. When finding no papers to identify the relic, the nuns read the letters painted on the outside of the box: EXPEDITIO, Latin for priority mail, and took that as a message from the saint himself and so came to believe the bones belonged to St. Expedite.

Within no time, statues and prayer cards were made and the saint's popularity spread rapidly throughout Europe. He soon became the patron to beseech in order to get things done in a hurry. In modern times, he came to be the patron saint of computer programmers, those in e-commerce, and even hackers and junk mail senders, though above all the patron of people seeking a high-speed solution.

In India and parts of Haiti today, a cult to St. Expedite still erects altars, says prayers, and makes offerings for the saint's rapid firewall-busting intercession. Small shrines made of hastily mortared native stones, usually painted bright red, dot the countryside. Though it's not good to be seen openly at one, many come in the night to leave flowers and notes of things to be done, and to light candles. Strangest of all is the fact that among the objects most often left as offerings are women's panties.

Third-century St. Martin of Tours was also a Roman soldier and came from a military family; even his name, Martin, was meant to honor Mars, the Roman god of war. When he converted to Christianity, he didn't announce it, but he refused to fight. He said he would continue to go into battle, and be on the front line, but would not carry a sword. He was put in prison and then lived as a hermit for many years. His miracles included raising the dead, changing the course of falling trees, and shifting the direction of fires. He's officially the patron of the U.S. Army Quartermaster Corps, where traditionally many conscientious objectors can serve.

F

FALUN GONG

REPRESSED CHINESE SLOW-MOTION-EXERCISE SECT

In 1992, Li Hongzhi began to travel throughout China teaching grammar school children "body-mind" exercises. Li had studied with a number of Buddhist and Taoist masters and was said to have acquired supernatural powers from the age of eight, even if he later made a living as a trumpet player and a grain-purchasing agent. By 1999, there were seventy million people in China practicing Li's program, known as Falun Gong.

His select choreography of slow movements was similar to the ancient qigong exercise regime, which dates back to a thirteenth-century group called the White Lotus Society. Falun Gong (the practice), or Falun Dafa (the teachings), means literally "the Great Wheel of Buddha's Law." Falun Gong advocates the use of certain movements as a means of achieving harmony of the mind and body for purposes of enlightenment and universal understanding. The group's bible, of sorts, *Zhuan Falun*, offered metaphysical explanations concerning human nature and the soul, as well as moral codes.

The Chinese communist government considered the massive number of followers to be a dangerous force—potentially revolutionary, as were the White Lotus Society of seven hundred years ago. The government banned the group in 1999. Li, then in the United States, was sought for crimes of "public disturbance," although attempts at extradition failed. According to the Center for Human Rights, more than six hundred Falun Gong followers were sent to mental hospitals, and as many as ten thousand were sent to "re-education through labor without trial" during the initial crackdown. Prior to the 2008 Olympics in Beijing, another eight thousand Falun Gong practitioners were arrested and sent to labor camps.

In a *Time* interview, Li said: "Since the beginning of this century, aliens have begun to invade the human mind and its ideology and culture." He firmly believes his practice is for the good of humankind, helps rid the body of disease, and if they are ready, allows followers to access higher levels of consciousness.

Although Li supposedly has many incredible powers and can levitate, he still thinks it unwise to do so in front of people who might misunderstand. (See also: Levitating Saints)

FATWA

ISLAMIC HIT LIST

Most official declarations issued by Muslim leaders (known as imams) are given on a variety of subjects that are intended to help interpret Islamic views and ethics. (*See also*: Imams) Imams can also declare more serious proclamations, as when Spanish imams issued a fatwa, or death sentence, against Osama bin Laden, for preaching ideas they considered outside Islamic tradition. Issuing death fatwas often makes the news, none more than in 1989 when Indian-British writer Salman Rushdie was condemned for writing *The Satanic Verses*, a novel that gave a sacrilegious rendering of Prophet Muhammad. Two of Rushdie's translators were beaten, one killed, a publisher shot; thirty-seven people died in a hotel fire when it was learned that Rushdie's Turkish translator had stayed there. Another fatwa proclamation was issued for Evangelist preacher Jerry Falwell in 2002 after a *60 Minutes* interview in which he offered an unsatisfactory explanation of his understanding of the Islamic faith. The fatwa, issued by Iran's Ayatollah Khamenei, called Falwell "a mercenary and he must be killed; the death of that man is a religious duty, but his case should not be tied to the Christian community." The FBI keeps a file on American citizens under fatwa.

FAUSTUS

WIZARD, DIABOLICAL DEAL MAKER

Johann Georg Faust was a widely known German magician during the 1500s. He claimed his skills came from studying alchemy and astrology, but the few scattered and fragmented descriptions of his feats show he had perfected illusion and sleight of hand. He was considered a wizard, and a powerful one at that, though it was suspected that his skills had a demonic origin. He was not so interested in potions or turning metal into gold, but rather won awe and fear from his audience by attempting to duplicate many of Jesus' miracles. He would often use a smoke bomb of sorts to distract, such as his famous trick of walking into a town with his dog, which, after an incantation and a puff of smoke, became his manservant. He apparently made money by reading fortunes as well, and even charged for physician services, though with less success. In one instance he prescribed an arsenic paste to cure a nobleman's rash but instead it caused the skin to shrivel to the bone. Faust was very persuasive, and although he was run out of towns and repeatedly accused of fraud, he was once given a teaching position, but there again was forced to flee after he was found sodomizing a number of the students. He died in the 1540s from an explosion while working with chemicals in his room. He apparently didn't die immediately, but was mutilated and burnt when discovered. This convinced people he had made a deal with the devil to learn his magic, and that hell itself had risen up to bring him downward in a tongue of fire. Witnesses claimed that Faust flopped on the ground, and when bystanders turned him over to administer care, he kept flipping back to face the ground, which again confirmed to many that Faust had made a deal with the devil and was destined for hell.

SIGN THE DOTTED LINE

In literature, the first book about Dr. Faustus appeared in the 1580s, *The Historie of the Damnable Life, and Deserved Death of Doctor John Faustus,* by P. F. Gent. He was also the subject of a famous play, *The Tragical History of Doctor Faustus,* written by Christopher Marlowe and published in 1604. Faust's life was the inspiration for numerous other works, including by composers Hector Berlioz and Franz Liszt and notable authors such as Johann Goethe and Thomas Mann. Even Charlie Daniels's country song "The Devil Went Down to Georgia" taps the fascination with a Faust figure and the still-tempting idea of trading your soul for a charmed life.

ST. FIACRE

PATRON SAINT OF HEMORRHOIDS

St. Fiacre died in 670 A.D., and a cult soon grew around the two miraculous gifts he displayed while an abbot of a monastery in France. It was said that when he was given land to build a hospice on, he walked and surveyed the virgin timbers with his wooden staff. Whatever tree he pointed to immediately fell, and when he dragged his staff on the ground, plowed furrows appeared. But it seems St. Fiacre's divine staff worked on people, too. Each morning he was greeted by a line of village folk bent over and exposing their naked behinds. With a point of his staff, and for the most serious cases a good poke, Fiacre immediately cured hemorrhoids. St. Fiacre is depicted in art carrying a big, soft cushion. However, since he is also the patron saint of gardeners he might be portrayed more respectfully toting a shovel.

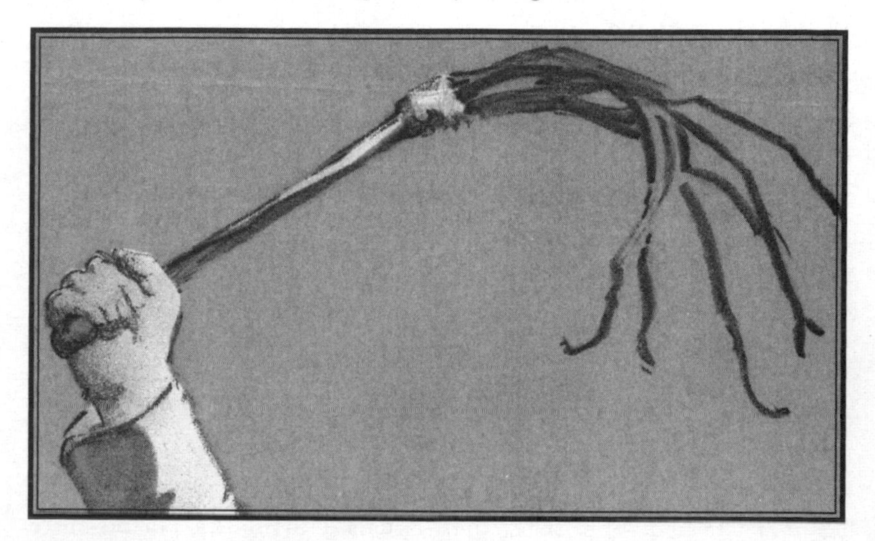

FLAGELLANTS

WHIPS AND CHAINS FOR GOD

For Christians, getting whipped as a means to attain a religious experience began within years of Jesus' lashing. Even though it was a punishment typically administered by Romans prior to all crucifixions, for the earliest Christians it was seen as a self-sacrificing act, and a sign of commitment to imitate Jesus to the extreme. The drastic shock the body undergoes during flagellation from leather straps, stiff rods, or chains has been medically proven to alter perception. Depending on the intensity of the blow, each lash can trigger the body to release massive dosages of endorphins, a hormone commonly associated with numbing pain and capable of inducing a euphoria-like state.

Until the thirteenth century, self-flagellation was mostly practiced by a few mystics and saints in the privacy of monasteries and convent cells. At the time it was considered a respected—if not officially approved path—to spiritual purity, and for penance. (*See also:* St. Dominic Loricatus) It became a mass movement, however, starting in Perugia, Italy, in 1259. An epidemic was killing people in record numbers, and it seemed that prayers alone were not strong enough to stop it. Thousands gathered in the town to commit ceremonial flagellation, parading with banners and wooden crosses and inflicting pain on themselves and on each other until the streets turned red from blood. All surviving citizens were expected to offer this severe penance in order to appease God, and as a divine plea to halt the illness and dying. In the frenzy, those who wouldn't join were considered devils. Disapproving priests and many Jews were dragged from hiding and whipped until dead by the mob. The flagellation craze spread throughout Italy, and eventually to the rest of Europe, until it resembled a loosely controlled militant army of self-beating pilgrims, with numbers reaching ten thousand.

In 1261 the pope ordered flagellant practitioners to cease. The movement was forced underground until 1349, when after two years of suffering through the Great Death, or bubonic plague—and 75 million corpses were rotting everywhere—the flagellants surfaced again and rallied support for their rituals. During this period more than eight hundred thousand people practiced it. Small bats with a tangle of four-inch leather straps and tipped with iron spikes were used to scourge participants. Another flagellant fad took place in 1399, when fifteen thousand marched to Rome, but it was quashed when the pope enticed the leader to a meeting and then burned him at the stake.

Today in Spain, Portugal, Italy, and the Philippines, certain sects take to the streets for public flagellation during Lent. In the Philippines, the penitents dress in white robes and use supple bamboo whips tipped with broken glass to beat themselves hard enough to draw blood. Self-crucifixion is also practiced, though only among the more dedicated.

MORTIFICATION OF THE FLESH

These loosely translated words found in Aquinas-era Bibles, namely in the Colossian chapters, set off a tradition in Catholicism noted as corporal mortification: "Put to death what is earthly in you: fornication, impurity, passion, evil desire, and covetousness." For the reverend to get on the fast track to bliss, and as added insurance to pass through the gates of heaven, thirteenth-century devotees believed harming oneself was an insightful translation into the true meaning of those words. Persons who whipped themselves with cat-o'-nine-tails, sticks, bats, or leather straps, or who held the palms of their hands on a candle flame without flinching, or who bashed their heads against stone walls for hours were seen as individuals seriously dedicated to holiness. However, the interpretation changed during the Age of Reason, when flagellation took on a new meaning.

NICOLAS FLAMEL

IMMORTAL ALCHEMIST

Nicolas Flamel was employed as a scribe during a time when most people were illiterate: He scratched out a living as a notary of sorts for the University of Paris and sometimes dabbled in buying archaic manuscripts in the hope of selling translations to the wealthy with private libraries. In 1378, he purchased a twenty-one-page booklet from relic sellers, only to be mystified by what he discovered. He and his wife spent the rest of their lives trying to decode the text and symbols of this slim volume. What Flamel believed he had stumbled upon was nothing less than the secret guidebook passed down through antiquity on how to turn lead into gold. The legendary "Philosopher's Stone" that so many contemporary alchemists sought, Flamel discovered, was actually a chemical procedure that somehow managed to produce a substance similar to either silver or gold. It was apparently pure enough to pass for the real thing at a time when

testing usually involved holding the metal up to light, or biting it to determine its worth. Nevertheless, by 1392 the once-poor couple became lavish spenders, and eventually philanthropists, especially noted for helping the local poor. Flamel and his wife dug further into deciphering the symbols found in the booklet and learned how to apply the process on humans. It reportedly worked for his wife, who became immortal and not only stopped aging but was seen to be rejuvenated to the splendor of her younger womanhood. Subsequently, Flamel ordered his own tombstone, and died shortly after. A year later, a local priest had a nagging suspicion that Flamel had faked his own death. He employed a known grave robber and went at night to dig up Flamel's crypt. As suspected, Flamel's coffin was empty. It was assumed the wizard likewise took the potion for immortality and went off to be with his wife. Some believe they are still alive, appearing normal, though affluently comfortable; most likely are still philanthropists funding low-key charities somewhere. The house where Flamel lived is said to be the oldest still standing in all of Paris, located in the Third Arrondissement, somehow equally unaffected by the passage of time.

GEORGE FOX

QUAKER FOUNDER

The group originally known as the Religious Society of Friends was soon nicknamed the Quakers after its founder George Fox was arrested and brought to court for his radical views. While at trial, Fox boldly admonished the judge that all good people should "quake" at the word of God. As a young man, George was set to be a shoemaker in Leicestershire, England, though from the start he was more interested in all things religious. Once convinced that all established religions were flawed, he formed his own ideology and favored audacious tactics to spread his message. One technique Fox used effectively was to charge into packed churches while congregations sat silent during sermons, and he shocked all by calling the priests and preachers at their pulpits liars. Subsequently, he was ganged up on by old ladies, pelted with psalm books, and arrested for disturbing the peace on multiple occasions. However, the more noise he made the more popular he became, and he trained his followers in guerrilla recruiting tactics. Many filed into churches with faces blackened with coal soot, or were instructed to go about town wearing yokes to protest current religious beliefs. His primary grievances with traditional Christian views stemmed from the importance many denominations placed on Christ's suffering during crucifixion, and for exclusively extracting biblical scriptures for guidance. Instead, Fox simply advised to look at Jesus' life as a role model, and to find the "Inner light within." Fox remained a tireless preacher, traveling throughout Europe and North America, though he fatally collapsed after a two-day preaching tour while in England, at age sixty-six, in 1691. Throughout the 1700s, it seemed that Quakerism would be the majority religion in the American colonies, though their strict pacifism and advocacy for the abolishment of slavery came in conflict with political notions. (*See also*: Shakers)

OATMEAL AND GOD

William Penn's father, a distinguished admiral, had connections with royalty. After young William caused enough disturbances and embarrassments for advocating Fox and the Quakers, he was given the opportunity to get out of town by way of a governorship of the entire state of Pennsylvania. Along with him, he took many English religious dissenters. He made the new American colony a place that welcomed any religion, and attracted Protestant reform groups of every stripe. Friends with Fox, Penn had ideas of establishing a utopia by promoting individual freedom, including trials by jury, free elections, and other ideas, a number of which were incorporated in the U.S. Constitution. Many mistakenly believe it is Penn's portrait on boxes of Quaker Oats, though it was a composite of a genteel Quaker man, who, in 1877, when the cereal was first trademarked, was supposed to represent good old-fashioned honesty and integrity. Some contemporary members of the Society of Friends don't like it, since it makes the public think Quakers still dress that way.

MARGARET AND KATE FOX

SÉANCE SISTERS

Twelve-year-old Kate and her fifteen-year-old sister Margaret together pulled off one of the greatest spiritualist pranks in history. It came back to bite them bitterly in the end, however.

In 1848, in their upstate New York house—which had long been rumored to be haunted—the sisters began to hear tapping sounds within the walls. Through a system of finger snaps they could get the spirits responsible to answer back. One for no, two for yes, and so forth, until the girls eventually devised an alphabet with rapping sounds for the spirits to communicate. News of this phenomenon spread across the country, and within a year the siblings were displaying their gift of talking to the dead before sold-out auditoriums. When they went to New York City to conduct more séances, the famed editor of the *New York Weekly Tribune*, Horace Greeley, defended the girls and their special skills. Subsequently, séances became the rage across the country, with even Mary Todd Lincoln bringing one to the White House so she could converse with her dead children (the first lady had a history of irrational behavior and was later committed to an insane asylum). The Fox sisters became celebrities for years to come, mingling in high society and earning a small fortune for their appearances. However, from their teenage years, while under Greeley's wing, the girls showed an exceptional fondness for drink, and both became serious alcoholics as they aged. Desperate and nearly destitute in 1889, they agreed to sell their secret to a tabloid, and explained they had made the rapping sounds by snapping the joints of their toes and fingers. Believing this would somehow make them news items once again, the scheme ultimately backfired, and both were shunned by former supporters and friends. Within two years, Margaret and Kate died penniless and were buried in a potter's field.

MR. SPLITFOOT

The Fox sisters claimed contact with a friendly ghost, one Mr. Splitfoot. The spirit was supposedly a thirty-one-year-old husband and father of five children who had been murdered and buried in the basement of the Fox home in Hydesville, New York. He was also apparently a traveling ghost, since he made

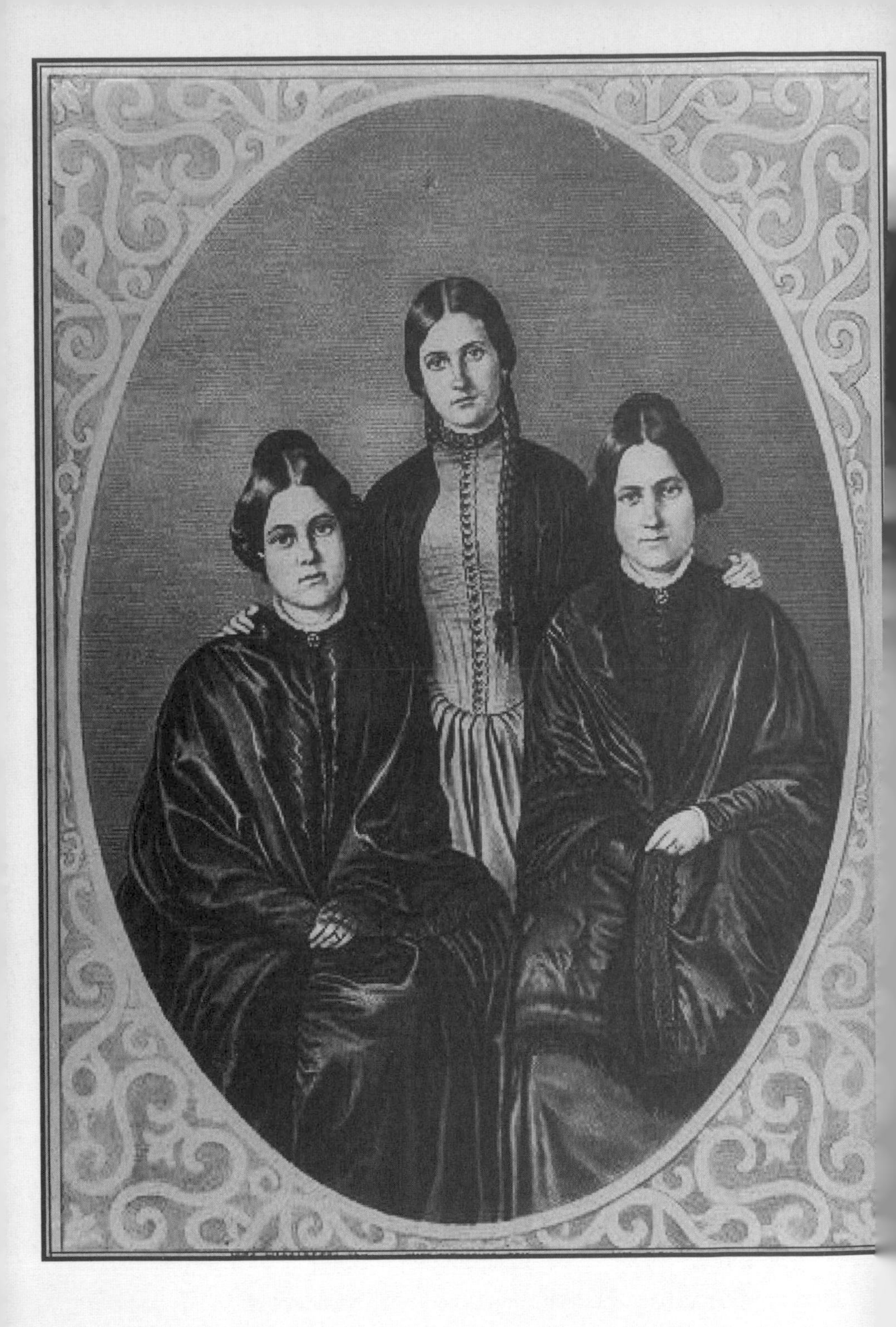

appearances to the girls when P. T. Barnum sponsored the sisters in a sideshow attraction. Despite the sisters' undoing, they were credited as founders of the modern spiritualist phenomenon, with more than forty thousand spiritualists working in America by 1860. Today, the United Kingdom's Spiritualists National Union, founded in 1901, claims not to be a religion but "helps the work of Spiritualists to be carried out in this physical world," representing "350 affiliated churches & centres." Splitfoot was a colloquial name used in colonial times to refer to the devil.

FREEMASONRY

SECRET ORDER OF AMERICA'S FOUNDING FATHERS

The origins of Freemasonry have been so cloaked in secrecy that hard facts on how or who actually formed these fraternal organizations—with their overt mystical overtones—remain a mystery. It seems logical that its gestation began during feudal times, when lords were given power over various tradesmen who wished to labor in their region. Everyone from minstrels to masons was required to answer to the nobles in their charge. These feudal overseers settled disputes, granted permission to work, and also levied taxes. However, those who held control of the stonemasons had the most power, for nothing of substance could be built without the ancient knowledge of the mason trade. The desire for the skilled artisans to break free from this system, which kept them impoverished, was the catalyst behind the secret societies that came to be known as Freemasons. Even if Freemasonry origin mythologies are varied, one prevalent legend traces their roots to biblical times and claims the group's ritual hand signals and symbolism were first used during the construction of the Tower of Babel, when in the middle of the project workers found they could no longer speak each other's language. (*See also*: Tower of Babel) Others say the crusaders, or the builders of Egypt, or perhaps the Druids passed the secret knowledge of the craft down through the centuries. It never became the union or strong trade guild its feudal organizers had hoped for. Instead it developed into one of the most popular secret societies of the next three centuries, beginning in the 1700s. In Britain, France, and later America, membership came to confer a sense of distinction among all class of tradesmen and eventually aristocrats. Many Freemason lodges

splintered and altered ceremonies, though all employed ancient symbolism. New members were initiated to secrecy, promising to uphold certain moral standards and maintain a dedication to self-improvement, under penalty of no less than death. In the earliest days, Masonic groups met with persecution from the church primarily because of its secrecy and for referring to God as "The All-Seeing Eye," or the "Great Architect." Additional bad publicity came when some Mason lodges began practicing elements of the occult, especially adhering to the writings of Hermes Trismegistus. (*See also*: Hermes)

All-male Masonic societies had more than 1.5 million members in the United States in 2009, with most sponsoring charitable works. Only Masonic members of the Ancient Arabic Order of the Nobles of the Mystic Shrine, commonly known as Shriners, still wear the Turkish-style red fez. The first Shriners group started as a drinking club; the only requirement to become a member was that one first belong to another Masonic lodge, acknowledge a "Supreme Being," and have an ability to hold one's liquor. Now the group raises funds for children's hospitals.

BY INVITATION ONLY

During a Masonic initiation ceremony in 2004, a seventy-six-year-old man accidentally killed a new member. Part of this secretive group's ritual ceremony requires pledges to symbolically place their lives in the hands of their brethren. The initiate, a forty-seven-year-old Long Island man, knelt with his back to the altar, facing the assembled. The Masonic leader, dressed in ceremonial vestments, was to fire a gun loaded with blanks above the younger man's head, and into the ritualistically stacked cans arranged on the altar. However, the senior Masonic man had two guns on him, and unintentionally used the real one instead of the ceremonial weapon. In addition, he not only missed the cans but hit the new man in the forehead, killing him instantly.

Many founding fathers of the United States were also Freemasons. (*See also*: Deism) Other Freemasons included Francis Scott Key, composer of "The Star-Spangled Banner," and Francis Bellamy, author of the Pledge of Allegiance. A popular Freemason slogan, "In God We Trust," wasn't used on U.S. currency until the Civil War, appearing on a two-cent coin in 1864. Once only a Freemason slogan, it was made an official U.S. motto in 1956 and appeared on all paper money soon after.

JOHN FRUM

THE CARGO-CRATE CULT

Near the end of World War I, in 1918, a black American soldier washed ashore on the small South Pacific island of Tanna. He was found unconscious and barely alive, though still wearing portions of his torn U.S. Navy uniform. The natives of the island had few regular contacts with civilization, though they were aware that certain other cultures seemed to enjoy prosperity and wealth. Who John Frum actually was remains a mystery to this day. "Frum" seems likely to be a shortening of "John from America." Regardless, because of the islanders' belief that a god who lived in a volcano was expected to reemerge to spread wealth, "John Frum" assumed the role of this long-awaited messiah. Subsequently, Uncle Sam, Santa Claus, and John the Baptist soon became revered gods based on the "teachings" of John Frum. By the 1930s, many on Tanna gave away their possessions, abandoned their villages, and traveled inland near the volcano to live in more traditional communes as John Frum advised. During World War II, when American forces stationed three hundred thousand troops on the island, the Frummers believed the prophesies of John Frum were about to be fulfilled. They hoarded discarded U.S. cargo crates and took to religiously wearing parts of American military uniforms. They made imitation rifles out of bamboo and marched around fires and performed rituals based on what they believed were American religious activities, namely passing cigarettes and chocolate bars among themselves with a sacred reverence. Elders of Tanna remember another part of the legend of the beached American soldier, and say John Frum became deranged, literally believing he was a god. John Frum either died by being pushed or by willfully leaping into the volcano. Nevertheless, there remain more than 2,500 members of the John Frum or Cargo Cult who still believe heaven is a great manufacturing plant where God provides unlimited supplies. The return of John Frum is awaited and foretold to rise out of the active volcano in the year 2015, to usher in permanent prosperity for his followers.

THE CURIOUS CASE OF TOLSTOY AND THE CULT OF ABRAHAM LINCOLN

In 1908, Leo Tolstoy visited a nomadic tribe living in an isolated region of the North Caucasus. During his stay, the great Russian novelist told stories about the world's greatest leaders, but the tribal chief seemed most fascinated by Abraham Lincoln. When Tolstoy presented the chief with a picture of the martyred American president, "He gazed for several minutes silently, like one in a reverent prayer, his eyes filled with tears." Afterward, for a short time, stovepipe hats and mustache-less beards became a symbol of enlightenment among the mountain tribes.

G

MOHANDAS GANDHI

PEACE MAKER, URINE DRINKER

Considered one of the great political figures, as well as a religious icon of the twentieth century, Gandhi studied in England to become a lawyer, but instead eventually came to favor a spiritual path. He formulated his ideas of nonviolent protest—for which he is most remembered—after reading the Hindu Bhagavad Gita, Madame Blavatsky's *Key to Theosophy*, and the Bible. Through advocacy of mass noncooperation and civil disobedience, he managed to facilitate the liberation of India from British control. He used fasting, to the point of death, to bring attention to issues. Ultimately heralded as a Mahatma, or "Great Soul," and unofficial leader of 400 million Indians, Gandhi had a number of strange personal habits that puzzled many. When it was discovered that he regularly drank his own urine, and that of cows, and enjoyed receiving two enemas a day, Gandhi explained it was part of his ideas on spiritual cleansing. However, it was harder for the public to accept Gandhi's rationalization when paparazzi caught him in action with an entourage of beautiful women. Although he had refused to have sex with his wife for most of their sixty-plus years of marriage, he had frequently taken many other women to bed. Gandhi was a self-declared *brahmachari*, vowed to celibacy, but he insisted he merely liked to lay naked next to naked women to test his religious promise, and insisted when questioned by the press about his seemingly rampant sexuality, "I have done nothing wrong." Gandhi was assassinated in 1948, at the age of seventy-nine.

WHY A RED DOT ON THE FOREHEAD?

Hindus, Buddhists, and Jainists wear a red dot on the forehead for varying reasons. Legend has it that it marked the location of the "third eye," where a fiery sensation is said to manifest itself during deep meditation. It was also intended as a reminder to look inward to find God. For some Hindus it became a sign of marriage, with the one on women called a *bindi* and on married men the *tilaka*. A U-shaped mark represents a devotion to the god Vishnu; a horizontal line indicates worship of Shiva. However, some people in India consider it to have no religious significance, and apply the *bindi* as a cosmetic mark of beauty. Additionally, the red dot can be worn to bring good kismet, or luck.

ST. GENESIUS

PATRON SAINT OF COMEDIANS

They didn't do stand-up comedy back in Roman times, but actors specializing in satirical plays were often the most popular. One such actor, Genesius, had considerable talent and comedic timing; at age seventeen he was cast in the lead for a production to be attended by the Emperor Diocletian, the principal force behind Rome's last and cruelest campaign against Christians in 303 A.D. The emperor's favorite plays were dramas that mocked Christian customs, and one in which Genesius was to star, which featured a priest and exorcist acting as clowns baptizing everyone. While Genesius was playing a stricken man calling to be baptized, he recited his verse but then suddenly paused. The other thespians tried to mouth him the next line, but instead Genesius witnessed two angels float down to him, with one showing him a list of sins that he had already committed in his short life. Genesius jumped up and waved the other actors away, and in front of the emperor he affirmed that he had just experienced a vision. He declared that the angels wanted all of Rome to be baptized in earnest. Diocletian had Genesius immediately chained and taken to torture. When Genesius refused to recant his newfound Christian notions and balked at saying a prayer to Roman gods, he was beheaded. Clowns, comedians, and other theatrical organizations still claim St. Genesius as their patron—despite his assuredly unfunny ending and questionable comedic timing.

GENEVIEVE

PATRON SAINT OF PARIS, FRANCE

Genevieve was born in Nanterre, France, in 422 A.D. to a family of modest means, but when her parents died she was sent to live with a relative in Paris. By age fifteen she became a nun, vowed to poverty and austere fast-

ing, eating only twice a week. From the onset, Genevieve announced her communication with angels and offered prophesies. These revelations upset some, to the point that she narrowly escaped an attempt to silence her by drowning. However, during a crisis, namely a threat of a barbarian invasion, she found a wider audience and claimed the angels told her that the only way for Paris to be spared was for citizens to stay in their homes and erect a divine "force field" that would protect the city. She had convinced the entire population to pray, fast, and visualize a thick stone wall around the city. When the barbarian Attila the Hun changed battle plans and concentrated on attacking Orléans instead, she was considered a miracle worker for the power of her prayers. In 1129, six hundred years after she died, it was thought that her bones exhumed from her grave had cured the city of a plague. Nothing stopped the infectious spread until her remains were gathered, placed in a coffin, and marched with a solemn parade to a cathedral. She is also the patron saint of high fevers.

COUNT OF ST. GERMAIN

DISCOVERER OF THE ELIXIR OF LIFE

A mysteriously odd man appeared in Paris during the 1700s claiming to be the Count of St. Germain. Today, skeptical historians suspect he was

born in 1651 as Harold, the son of Francis Rakoczi II, prince of Transylvania. For a while he became the talk of Europe and was sought by all the heads of state for his fascinatingly stunning tales and vast store of knowledge. He claimed to have traveled the world and knew the intimate details of famous persons living in ages past. He never ate in public, and conversed fluently in French, English, German, Dutch, Spanish, Portuguese, and Russian, besides displaying literacy in Chinese, Latin, Arabic, and Sanskrit. In addition, the count seemed to be extremely well-off, although when investigated he had no traceable means of support or bank accounts. He was noted as being musically gifted, playing the violin masterfully, and claimed to have attained this virtuosity over the many centuries he had been alive.

At each town he visited, Germain immediately set up a lab for alchemical practices. Though he refused observation of his techniques, it was widely believed he had discovered the Elixir of Life, a formula that had been the quest of the best minds of civilizations from antiquity. According to his account, the count had attained immortality after ingesting "white drops of liquid gold." Eyewitnesses attested to the fact that he never seemed to age; documented sightings of Germain—he was once detained in England as a suspected spy—consistently described him as forever appearing no more than forty-five years old. He said he had first been a high priest of the lost city of Atlantis. And then he claimed to have been St. Joseph, husband of Mary: He had been present when Jesus performed his first miracle at a wedding in Cana, and in fact he had showed Jesus how to turn water into wine.

From dining with Cleopatra, to attendance at the Council of Nicaea in 325 A.D., St. Germain seemed to have had his hand in every major event in history. He was still around when Napoleon came to power, and the emperor ordered a complete investigation, though all the documents were destroyed in a suspicious fire. Germain also made an appearance while the founding fathers of the United States prepared to write the Constitution, and he is credited for inspiring the U.S. seal, designing the American flag, and somehow being involved in the formation of the Boy Scouts. He influenced the creation of the Theosophical Society (*See also*: Madame Blavatsky), the members of which believed Germain was a resident of a subterranean colony of rarely seen immortals located deep in the Himalayas, sort of like spiritual Bigfoots. From 1651 to 1896, more than 245 years, many declared to have had personal encounters with St. Germain, and these claims continued throughout the twentieth century, the last being in 1971. Spiritualists and clairvoyants seeking to converse with the dead have reported interfacing with St. Germain more than any other figure from history, while others support the notion that the count was actually a time traveler, or an extraterrestrial. A manuscript composed somewhere between 1710 and 1822, "La Tres Sainte Trinosophie" (The Most Holy Three-fold Wisdom), is attributed to St. Germain, preserved at a library in Troyes, France, and contains still undeciphered text and codes.

We moved through space at a speed that can only be compared with nothing but itself. Within a fraction of a second the plains below us were out of sight and the Earth had become a faint nebula.

—from "La Tres Sainte Trinosophie"

GIDEON

FATHER OF 50 MILLION BIBLES

According to the book of Judges, at one point a number of Israelites were back to worshiping idols when it fell upon a commoner named Gideon to bring change. Before he took the job, Gideon called on God to give a sign; he wanted Yahweh to perform a test concerning early morning condensation. God was to make the wool mat upon which a pile of wheat sat to be wet; the next morning he asked God to make the wool dry and the wheat moist. Apparently a divine hand accomplished this feat and it convinced Gideon to do God's will even if he didn't feel up to it. Subsequently Gideon assembled a huge army and ravaged a neighboring city, whose people were considered promoters of idolism. After great victories, the Israelites wanted to make Gideon king, and although he refused, he kept much of the looted gold from smelted idols and went on to have numerous wives and concubines. Another translation for Gideon's name means "Feller of trees," supposedly connoting his rapid deployment of supplies needed to keep his troops in ready. This same moniker might apply to internationally known suppliers of free Bibles.

In 2008, the Gideons International organization printed more than one hundred Bibles a minute, and distributed these to hotels, prisons, and hospitals, equaling more than fifty million copies in circulation. The Bible that Rocky Raccoon, the character in the Beatles song, and countless others found in their hotel rooms was the result of a chance meeting of two strangers. In 1898, in a small Wisconsin hotel, a pair of salesmen were asked to bunk together due to a room shortage, and although in the midst of a wild lumber convention and sleeping in a cramped suite—above a saloon, no less—they kept each other from temptation by sharing Bible stories to pass the time. Before parting ways they talked about the need for a Bible-reading businessmen's fellowship and a year later formed a company called Gideons with the ambitious mission to comfort the lonely traveler by placing a Bible in every hotel room around the world. It took nearly a decade before the rudiments were worked out, though one hundred years later the group still supplies Bibles free of charge to hotels in 180 countries, from the Plaza in New York, with its Louis XV nightstands, to any number of the no-tell-motels. A Gideon volunteer makes a ceremony of it, presenting a copy to the manager when a new hotel opens, and then gives a complimentary supply to the housekeeping staff to put in night table drawers. If a Bible is stolen, the Gideons consider this a good thing and gladly replace it, hoping the absconder will get to the "thou shalt not steal" part eventually. However, Gideons say each of their Bibles have a six-year life span and during that period are read by more than two thousand people, or 25 percent of a typical room's occupancy, before a fresh copy is required.

GNOSTICISM

BACK TO THE COSMIC WOMB

In Gnosticism, the divine essence of man, or soul, is trapped in a disagreeable human container and merits no attention from a disinterested and aloof Supreme Being. However, salvation for the Gnostic can be attained through knowledge of one's self, even if the exact means remains open to interpretation. If enlightened understanding is achieved, one's soul will rejoin this great oneness from which we were separated at the moment of creation. While many religions believe God created the world in a perfect manner, here the serious Gnostic disagrees, instead believing that nearly everything we experience in reality is flawed, purposely so. God, for the Gnostic, didn't actually create anything, since the act of "making" is a human activity and ultimately an imperfect characteristic. Instead human existence and the world as we know it merely "emanated" from this one transcendent force, the way we might produce an exhaled breath.

Gnostics assert that humanity's long lineage of suffering has occurred because most never become aware of the unifying spark of God within them. There are, however, intermediate beings, existing in another dimension, that many Gnostic sects trust are there to guide us to knowledge. One tenet of early Gnostic scriptures contends that an arrogant and inferior god, called Demiurge (from a Greek word for common craftsman), set out to handcraft a purposely cruel world in order to keep us trapped in pain and suffering. In other words, physical reality was created to distract humanity from trying to reach reunification with the "Source." In other Gnostic writings, mankind was offered help by various "Messengers of Light," which often included Seth (Adam and Eve's third son), Jesus, and the Prophet Mani. The purpose of life for the Gnostic is to attain an understanding of our inherent separation and transcend it. If this is not achieved the soul will return again and again to work out the details. (*See also*: Manichaeism)

By the fifth century, most of the ancient Gnostic writings and their followers had been destroyed and scattered because many of their beliefs were seen as heresy to the rising authority of Christianity. In the 1940s, psychologist Carl Jung brought new attention to a few remaining documents found at the Nag Hammadi library in Egypt, which contained some authentic ancient Gnostic texts.

SWEET DADDY GRACE

FOUNDER OF THE HOUSE OF PRAYER; LOVER OF CADILLACS

In 1883, Marceline DaGraca was born a Catholic on Cape Verde, the Portuguese islands off Africa. He immigrated to Massachusetts at twenty and made a living as a dishwasher and itinerant laborer, until he had an awakening, he said, and tried his hand at the pulpit at the evangelical Church of the Nazarene. When shunned, Charles Grace, as he began calling himself, decided to spread his own version of evangelical preaching by traveling in a "Gospel Car" throughout the South, though he found few followers. Undeterred, in 1919 Grace built a shack-size church, which cost forty dollars to construct, and called it the United House of Prayer. He also declared himself bishop.

A long-haired, flamboyant preacher with painted fingernails and bedecked in flashy jewelry, Grace attracted converts with his enthusiasm

but also drew media criticism. Nevertheless, to his mostly black and poorer constituents he was affectionately known as "Sweet Daddy"; in him they found hope through his promises of upward mobility—as he said, "Anything God offers you, I got it." With a genius for taking the nickels and dimes gathered at collections, and with abundant free labor, Grace built a financial empire that included lucrative real estate investments stretching from Harlem to Los Angeles, and even into Havana. In addition, he opened a diverse array of businesses—including healing medicinal powders and creams, as well as a line of spiritual soap. By the 1950s he reportedly owned the largest personal fleet of Cadillacs in the United States and had a mansion in every city where he had a church. In addition, Bishop Sweet Daddy Grace traveled with bodyguards, lawyers, chauffeurs, and an entourage of personal aides. His followers thought he was the messiah, which he didn't fully deny, saying, "If you sin against God, Sweet Daddy Grace can save you, but if you sin against Grace,

God cannot save you." Sweet Daddy's brass-band parades, clapping, singing, speaking in tongues, and massive baptism by fire hose made for public spectacles, all of which he called "getting to the mountain." He left a multimillion-dollar corporation to the church at his death in 1960; Sweet Daddy Grace died from heart disease at seventy-eight, in his eighty-five-room Los Angeles mansion. The church continued after some turmoil over leadership, though succeeding church bishops opted to add "Daddy" in their official titles as tribute. In 2009, there were approximately fifty thousand followers of the United House of Prayer, though spokespeople for the organization claim the church has nearly two million on the roster.

GURDJIEFF

FEEL THE VIBRATION

George Gurdjieff founded the spiritual organization known as the Fourth Way (or as the Gurdjieff Society) in the 1920s. It was originally called the Institute for the Development of Man. He believed reincarnation has a limit, and that one is given only so many lifetimes to get it right before one's soul is considered nonrecyclable and used for other things. The group also believes that even with all the knowledge available, only 20 percent of humanity is intellectually capable of understanding the process needed to achieve enlightenment, and only 5 percent of those, at most, willing to do the work. The work involves intensive studies of enneagrams, which are complicated puzzles of sorts; Gurdjieff's ten-volume tome, *All and Everything* and the book *The Fourth Way,* by P. D. Ouspensky; as well as quantum physics. Ritual dances are also encouraged as a way to connect with universal "vibrations." It is also recommended to say "I AM" one time each hour. Group meetings are intensive; humiliation issued as a tool to coax the lazier disciple into working to be among the 5 percent. If unprepared, one could be ridiculed with a ceremony called "Toast to the Idiots." There are approximately twenty thousand Gurdjieff devotees. (*See also*: Enneagrams)

ST. GUY

PATRON SAINT OF OUTHOUSES

Guy of Anderlecht was born to poverty in 950 A.D. and didn't like being poor, so he decided to become an entrepreneur. He invested in a ship and a cargo deal, but when it sank, he lost his own and investors' money. Guy took it as punishment for the sin of greed. He then became a professional pilgrim and eventually a tour guide in Jerusalem. It seems he had a special knowledge of where the best unlocked bathrooms could be found. He also knew how to handle mad dogs attacking his tour groups, and carried certain smelling salts for the occasional epileptic in his charge. He died walking back home to Belgium. Because of his many skills, starting with his early years working on his parents' farm, St. Guy became a patron of stable boys and those working with horned animals. He is also prayed to by those who need help finding a bathroom, as well as by people prone to convulsions, and as a protector from dogs with rabies.

HAMATSA

SECRET CANNIBAL SOCIETY

An indigenous people of British Columbia, the Kwakwaka'wakw believed all mankind was created on the very spot where they lived—their Garden of Eden—when, during a great flood, whales, seabirds, and bears transformed into humans by putting on masks. By the late 1700s, when the first English settlers arrived, the Kwakwaka'wakw had a structured class society, currency, and a flourishing artistic and specialized craft culture. Nevertheless, the English threatened to exterminate them if they didn't cease what the queen's forces considered primitive and pagan worship practices to their ancestral animal deities. Before contact with the outside world, there were seventeen tribes, totaling a million people, but their population dwindled to near extinction by epidemic—the common cold—a disease that their spiritual medicine had previously kept away for eons. Today, there are only five thousand remaining, with less than 250 having some knowledge of the original language.

Hamatsa was a fringe sect within this society that carried out shamanistic rituals in strict secrecy. To be a member of the Hamatsa was considered prestigious, and although treated with reference afforded the royal class, they were also feared for the magic they possessed. No one could ask to be a Hamatsa; instead candidates were observed and those chosen were kidnapped and then indoctrinated. These candidates were kept in hidden locations until taught all the secrets, though if for some reason they were found not to be worthy of membership, they were usually roasted, or sometimes boiled and eaten. The photo included with this entry was taken by famed Western photographer Edward S. Curtis in 1914; it depicts a Hamatsa initiate emerging successfully from the ceremony "possessed by supernatural power."

In winter the Hamatsas traveled through the region's villages performing the year's most sacred dance ceremony. During the frenzy and euphoria, certain participants were bitten on the neck. If this happened, it was considered good fate, with the Hamatsa shaman then bestowing gifts and other spiritual blessings on that person. The ceremony ended with a feast and included eating various parts of human flesh. Some reports claimed Pacific Northwest tribes' cannibalism was merely an allegoric part of their rituals, while other anthropological findings indicate otherwise. The remaining guardians of the Hamatsa shamans' mysteries still adhere to their code of secrecy.

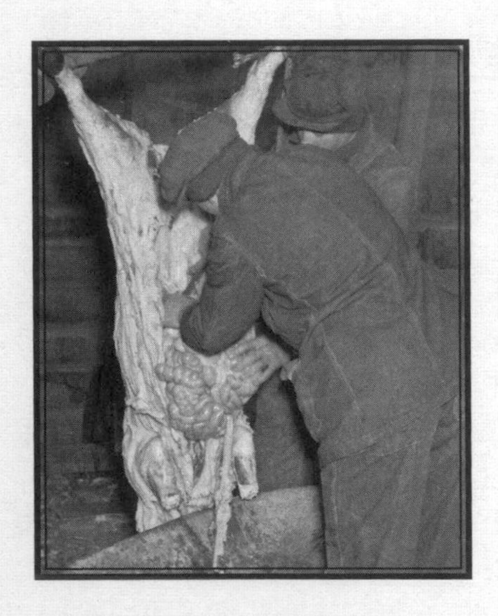

HARUSPICY

READING DIVINE MESSAGES IN ANIMAL GUTS

Since earliest records, animals have been sacrificed for religious purposes. What to do with the dead carcasses became a problem, however. The Babylonians created an entire class of specialized readers believed capable of determining whether the gods were pleased with the offering. After a ritual killing, the animals were opened and the guts were literally handled, smelled, and pressed against the forehead to search for omens. Charts were made to interpret sheep livers to see what gods had to say about the sacrifice, with discoloration or certain burst blood vessels giving clues. Colon examination of livestock was another skill for weather forecasters, and some European city streets that seem to have been designed by happenstance were often actually constructed according to how sheep entrails fell from the carcass after sacrifice. Animal omens remain common to the present day, with many thinking, for example, that a black cat crossing your path when the moon is full brings bad luck. From such superstitions, and throughout the Middle Ages, cats in particular were brought to the point of extinction. In France, more than a thousand cats of all colors were routinely burned each month, until King Louis XIII outlawed the practice in the 1630s.

THE MANY FORMS OF SUPERSTITION

Superstitions are beliefs that certain actions, events, or signs that seem illogically unrelated somehow have the ability to foretell future events; it ultimately implies a certain faith in magic. There are numerous superstitions popular today for good or bad luck. For some baseball players, it's believed that spitting on a new bat is assurance of a fruitful hitting season. If you blow out all the candles on a birthday cake with the first puff, you get the wish granted, though if it takes two tries, you'll likely get the opposite. In addition, if someone else helps to extinguish the candles, they might steal the wish entirely. If, when getting out of bed, you place the left foot first to the floor, know the day will go badly. Likewise, if you wear a hat in bed, or lean a broom against it, no good will follow. If you start to make the bed in the morning and stop, plan on having nightmares that evening. For smokers, don't light three cigarettes from the same match or all are doomed. If a spoon falls to the floor, expect visitors, but if a dish breaks, someone you know is about to die. Driving past cemeteries has three rules: hold your breath, tuck your thumbs inside your fingers, or if inclined, never fail to make the sign of the cross. As for the black cats that King Louis XIII spared, they bring good luck if walking toward you, and bad if seen only moving away. Remedies to nullify the effects of bad omens range from throwing a pinch of salt over your shoulder to knocking on wood, to turning around counterclockwise three times.

HHH

"HEALTHY HAPPY HOLY" SPIRITUAL MOVEMENT

Yogi Bhajan announced "happiness is a birthright" and started a movement in 1969 dedicated to a cheerful approach toward enlightenment. HHH believes there is one God, but he never takes human form. Rather he spreads out through every living thing. Bhajan's writings are the primary text; certain practices, such as always bathing before sunrise, are recommended, as is keeping a constant smile on your face. Manual labor is a badge of honor. You must also share everything you have, since it is best not to be attached to anything. Most Healthy Happy Holy followers don't cut their hair, and use only a wooden comb, as well as donning nothing but cotton underwear. These are some of the many rules established by Yogi Bhajan to make one happy. There were more than eight thousand followers in 2009.

HEAVEN'S GATE

UFO CULT

Hale-Bopp was a comet that became the main focus of a religious group known as Heaven's Gate. The first flying saucers were spotted in the United States by a pilot flying over Washington state in 1947. By 1950, the sudden rash of alien craft sightings led the U.S. government to place them in a category listed as unidentified flying objects, or UFOs. Interest in science fiction bloomed during this period in part from the demonstration of a real doomsday invention, the atom bomb. When the top-secret Manhattan Project came to light, many wondered what other strange technology remained concealed. Some suspected the government had captured one or more of these alien craft and held them in secret locations in the southwestern desert. By 1970, Marshall Applewhite, the son of a Presbyterian minister, became convinced extraterrestrials were numerous and were waiting to show enlightened minds

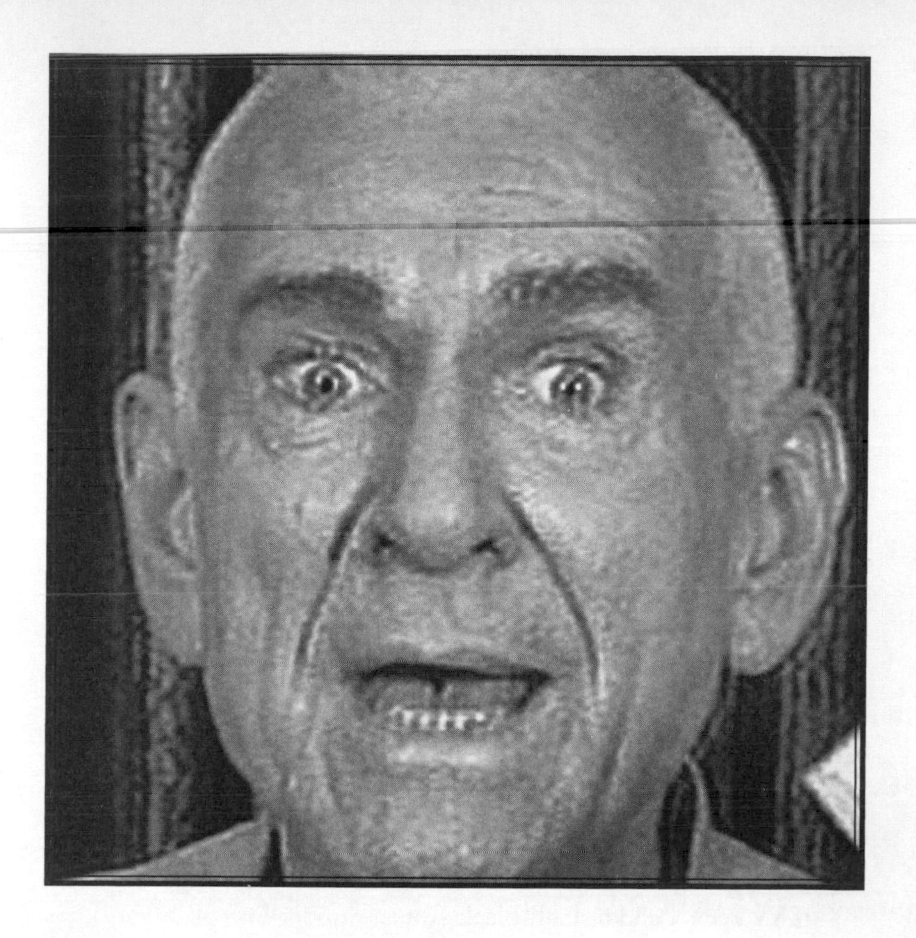

the way to redemption. Once he began to share these notions, his career as a music teacher was over, and he was fired from the University of St. Thomas due to "health problems of an emotional nature."

Two years later, he was in a psychiatric ward when he met nurse Bonnie Nettles, a dabbler in astrology. They became close friends, and with her encouragement Applewhite concluded that he was actually the reincarnation of Jesus Christ. Together they gave lectures, seeking to recruit others to join their group. He claimed that as Jesus, he was in communication with many outer-world beings, which imparted to him great wisdom. For long stretches, Applewhite and Nettles, along with their followers, hid in wilderness areas, believing people would murder Applewhite if they found out who he really was.

After Nettles died in 1985, Applewhite castrated himself and amended his original tenets, saying instead that Jesus was on a spacecraft visiting various planets in the galaxy. To go undetected, the spacecraft trailed the fumes of a comet. In 1997 there was great urgency when Applewhite identified the comet they were seeking to be the fast-approaching Hale-Bopp comet, which was rightly predicted by astronomers as one of the brightest comets visible to the naked eye in a thousand years. Applewhite saw the comet's close passing as an opportunity for him and his group to join Jesus and the enlightened aliens. On March 26, 1997, in a multimillion-dollar San Diego mansion, Applewhite and thirty-eight cult members—twenty-one women and eighteen men, ranging in age from twenty-six to seventy-two—shaved their heads and dressed in black pants, oversized shirts, and wore brand-new black Nikes. They lay on their backs in cots and bunk beds throughout the mansion, their hands at their sides ready to "shed their containers," a process Applewhite compared to the metamorphosis of a butterfly from a cocoon. They were given a choice of either pudding or applesauce, both laced with a deadly dose of phenobarbital, and had plastic bags tied over their heads. One shot of imported vodka was the chaser offered as their last earthly dessert. When the bodies were discovered, the rented mansion was found to be immaculate: trash bags taken to the street, toilet bowls disinfected, the lids down. Each dead cult member had a new five-dollar bill in his pocket and a small black suitcase under his cot. Eight other members had also been found to have been previously castrated. Only one member changed his mind and did not commit suicide.

HERMES

ANCIENT EGYPTIAN ALCHEMIST

A mysterious figure from antiquity, Hermes Trismegistus attempted to achieve unity with the divine by harnessing and altering the elemental forces of nature through the use of alchemy and magic. Hermetic origins are so old that oral traditions believed that the person responsible for founding the religion was a god turned man. In some accounts, Hermes was an Egyptian priest and a possible contemporary of Moses, though others claim he was actually a human incarnation of the god Thoth, an Egyptian deity often depicted with a man's body and the head of an ibis. Although scant archeological evidence can be found of his tangible exis-

tence, others place him in Greece during Plato's time, where he was a counterpart of Hermes, the Olympian associated with trickery and alchemy.

The ancients treasured the writings of whoever Hermes really was, and held great significance to his words, such that every ancient library kept many of his voluminous works in collections. Over twenty thousand books made of papyrus scrolls are attributed to his hand, and they circulated widely through the civilized world until the fall of the Roman Empire. Many of his treaties were philosophical and had a poetic mysticism, such as his guide on how to find God: "Be everything at once; be not yet born, be in the womb, be young, old, dead, beyond death. And when you have understood all these things at once—then you can understand God." Other Hermetic texts were manuals to teach one how to trap souls into statues, for example, or what gems, odors, and herbs are best used to cast incantations. During the dawn of the Renaissance, many scholars who had found it difficult to distinguish between science and magic rediscovered him. (*See also*: Crystal Balls) Advocates of Hermes, such as Roger Bacon and Dr. John Dee, believed he received his knowledge from an emerald that contained encyclopedic information; it was thought that the blueprints to build pyramids were accessed through Hermes' magic gem.

The power of his alchemy and legacy remains today, since one of his spells worked so well to protect objects from inhabitation by evil spirits that the phrase "hermetically sealed" signifies an imperviously sterilized and airtight container, in honor, of sorts, to Hermes Trismegistus's magic.

PSYCHEDELIC ROMANS

From 1600 B.C. to 300 A.D., one of the most popular ancient cults centered on worship of the gods Demeter and Persephone, who were considered bearers of Eleusinian mysteries that supposedly contained the answers to the unknown facts of life and the afterlife. This secret sect conducted an annual celebration attended by most ancient Greeks, and later, Roman citizens. It was so secret that if anyone divulged exactly what these mysteries entailed, the punishment was death, thus leaving few records of the true nature of what was revealed to followers. Phallic symbols and various grains were part of the rites, but they also included hallucinogenic drinks made from fungus or mushrooms, which produced an LSD-like component. It seems that one mystery recommended by the cult was to take an acid trip at least once a year, in order to see and experience what the gods only knew.

HERMETIC ORDER OF THE GOLDEN DAWN

ROGUE FREEMASON SECT

In 1888 a branch of British Freemasons broke away to devote themselves to the study of ancient mysteries by examining Egyptian, Greek, Hebrew, and various treatises on alchemy. Astrological charts, tarot cards, secret ceremonies, and alchemical experiments were the primary tools to seek enlightenment. There were more than fifty thousand members by the mid-1920s. It is now mostly dissolved, but many of its principles were incorporated by certain Wiccan and neo-pagan religions. (*See also:* Aleister Crowley)

L. RON HUBBARD

FOUNDER OF SCIENTOLOGY

Lafayette Ronald Hubbard, a science fiction and self-help author, founded the Scientology religion in 1953. He claimed that he knew how mankind came to be, and why we have so many conflicts. It all began more than seventy-five million years ago, when divine beings called thetas decided to experience human form. They lived on one hundred inhabited planets, orbiting around nearly a dozen different stars as part of a "Galactic Confederation."

At one point in theta history, a cruel leader, Emperor Xenu, noted unrest among the planets and reasoned that overpopulation was the root of the problem. To curb dissent and reduce sheer numbers, many thetas were tricked into reporting to a tax office. There they were paralyzed with a mixture of alcohol and glycol. These unfortunate thetas were then transported to planet earth, where they gathered around a series of volcanoes into which H-bombs had been dropped, down the craters. Their souls, or "indestructible spiritual essences" (weighing approximately 1.5 ounces), were gathered from the atmosphere by a vacuum and then made to watch movies designed by psychiatrists specializing in brainwashing techniques. This was done in order to feed the trapped spirits lies about their past lives, in the hopes they would remain forever confused and never learn of their superior status—and thus stay imprisoned on earth.

There were a few thetas, however, that survived physically, and they became our human ancestors. The spirit-thetas released from the vacuum became the nemesis of the physical people left on earth. Since those spirits could not be fully destroyed, and harbored bad energy, they remained, hovering about and ultimately invaded the bodies of our early ancestors. For that reason, an average person currently has nearly 2,500 negative thinking and acting thetas within them. Humans, therefore, consist of three parts: theta spirits, mind, and body.

Emperor Xenu, by the way, was captured and held in seclusion by a force field that is still powered by an "eternal battery," believed to be somewhere on earth. It's said to be anywhere from a secret spot in the Pyrenees Mountains to a secured basement in a Beverly Hills mansion.

In 1950 Hubbard published *Dianetics*, which made him internationally famous and has since sold more than twenty million copies. In it, Hubbard reveals a program to attain a new personality and a prosperous life. He developed an "e-meter," a device that sends electrical currents through the body to locate and "audit" the harmful spirits preventing a person from becoming "clear." When his clinics were shut down—they were accused of falsely practicing medicine—he made his system and ideas into a religion. When further problems arose, he then called the meters religious artifacts; they are still legally required to carry a disclaimer and Scientologists are forbidden to refer to the device's healing powers.

In 1948, prior to his book's success, Hubbard had attempted to have his theories accepted by professional psychiatrists. In his original thesis he stated: "an adult has three minds: reactive (engramic), analytical (conscious), and somatic (motor-effector)." When not taken seriously, Hubbard openly condemned psychiatry. He compared psychiatrists to the forces who had captured the early thetas, and accused them of keeping people "unclear," especially through their reliance on prescription medications. His many critics, however, claim that Hubbard often came up with his most imaginative mythologies and methods under the influence. He admitted as much, saying that while writing he had been "drinking lots of rum and popping pinks and greys." Hubbard was noted to be an extremely skilled hypnotist as well.

Hubbard stayed involved and helped grow his church, through which he acquired great wealth. By the mid-1960s, Hubbard was awash in legal problems, accused by the IRS of skimming millions of dollars from the organization. Thereafter he spent many years writing new books and theological documents while living on a three-hundred-foot boat in exotic locales. In the end he became reclusive, and was only seen by a few caregivers. Hubbard officially died of a cerebral hemorrhage in 1986 at age seventy-four. His Church of Scientology survived, and the main headquarters franchises its program to many Scientology centers, with membership estimates ranging widely between 1 million and 8 million worldwide.

As for recruitment, Hubbard (just as Mary Baker Eddy had done when she grew her Christian Science church) focused on convincing celebrities to join. Conversion of a celebrity not only brought great numbers to their fold, but also provided large cash infusions, which allowed the organization to silence critics with relentless and costly litigation. Celebrities listed on Scientology rosters have included Tom Cruise, Priscilla Presley, Kirstie Alley, Isaac Hayes, Chaka Khan, Sonny Bono, and John Travolta. Although Hubbard and Baker used similar means to expand their religions, Scientologists do not deny the use of medicines, as does Christian Science. Scientologists also embrace most technologies and advances in medicine, though they still vehemently oppose—and march in protest against—psychiatry.

Even though the symbol of Scientology is similar to a crucifix, with an added eight points, Hubbard questioned the existence of Jesus and thought the brainwashers under Xenu had invented Catholicism as the best tool to confuse man from ever discovering his higher self. Hubbard claimed he had achieved the thetas phase, while Jesus and Buddha, if they did exist, were a mere step above "clear." Scientologists believe in reincarnation, and that many gods are about in the universe, with layers of gods above them. The goal is not to reach heaven but to survive prosperously through each reincarnation, until the divine level is attained. There is no hell for a Scientologist.

I

I AM

"THE ASCENDED MASTERS OF THE GREAT WHITE BROTHERHOOD"

In 1930, a fifty-two-year-old mining engineer named Guy W. Ballard was hiking in the forests near Shasta, California, when he met a peculiar man along the trail. According to Ballard, he had crossed paths with no ordinary fellow nature enthusiast, but with the spirit of an Ascended Master. The stranger identified himself as the long dead though revered Count St. Germain. During repeated reunions at the special site, Guy learned of his own many reincarnations—they included George Washington—and that he was chosen to usher the world into a new age by the ranks of "Ascended Masters of the Great White Brotherhood," which included Germain and Jesus, among others. Guy said the spirits likewise permitted his wife, Edna, and later their son, Donald Ballard, to be "the sole accredited messengers." Ballard believed that the "Mighty I AM," or a divine presence, was within everyone. He then devised an ideology to gain great knowledge and personal prosperity, based on the hidden truths revealed to him by the Ascended Masters—even if his experience was nearly identical to one described in a 1905 book *A Dweller on Two Planets*, by Phylos the Tibetan.

Nevertheless, by actively seeking converts through lectures and seminars, he claimed more than three million followers by the late 1930s, labeled by the press as "Ballardists," especially after his nationally successful book, *Unveiled Mysteries*, described his mystical experiences in greater detail. More conservative estimates believe I AM actually had around fifty thousand adherents—a number that both amazed and alarmed others, since Ballard managed this massive recruitment feat in a relatively short time.

To those who read this work, I wish to say that these experiences are as Real and True as mankind's existence on this Earth today.

—from *Unveiled Mysteries*, Godfré Ray King (Guy W. Ballard)

Ballard, from the start, was intolerant of disagreement and demanded that his members be "100 percent students." As more messages came from the Masters, he modified and further defined how one should live and connect to the Mighty I AM force assigned to each person. Accordingly, the energy of this divine I AM tags along throughout one's life. Ballard thought this hovering luminescence was attached to everyone by an "invisible silver cord," described as an aura about the size of a party balloon, tethered by an invisible string. To keep his faithful unpolluted by contrary ideas, he urged cutting ties with family and friends who were in opposition to his program. He also advocated that sex only be performed for procreation, and that members forgo alcohol, drugs, and meat. Another command, which some thought more unusual, was based on his notion that black magic had created cats and dogs: Ballard urged followers to kill their pets so that their spirits could evolve to a better species after reincarnating. As Ballard's prominence grew, he became more authoritarian, both in means and in methods of controlling followers. Eventually, as World War II approached, he said his was no longer a religion but rather a patriotic movement created to rid the United States of "vicious forces" caused by a secret group practicing black magic on politicians.

After Ballard died in 1939, at age sixty-one, of a sudden heart attack, his cohorts were stunned, since he had previously claimed he had been granted an indestructible body. Edna took charge and revealed that she remained in constant communication with Guy, who while on earth was only a "messenger," though in death he had become an Ascended Master, supplying fresh advice and wisdom. Edna Ballard—incidentally, she was previously reincarnated as Benjamin Franklin—continued the movement, even after the U.S. government charged the group with mail fraud. Ballard literature expressly claimed that by buying and using methods found in their study packages, a person could "heal the sick" besides learning the techniques to communicate with spirits. When they offered portraits of Jesus, which were said to be drawn when Jesus personally appeared for a sitting, the I AMs were banned from using the U.S. postal service from 1940 until 1954. Edna Ballard died in 1971, in her mid-eighties.

I AM centers remain in operation in "Temples," or reading rooms throughout the United States, holding valuable real estate they purchased inexpensively at the height of their popularity during the Great Depression. The twelve-story I AM Center in downtown Chicago still opens its doors on Sundays, staffed by middle-aged and elderly women dressed in lacy white angle-length dresses. They guide visitors to the reading room filled with thick green leather-bound holy texts of I AM theology. In 2009 the group was estimated to have less than a thousand full-time members. (*See also*: C.U.T.)

MEET THE BROTHERHOOD

Although today it sounds like the name of a racist fraternity, the Great White Brotherhood was originally intended to define a celestial government of mystical and enlightened beings residing in the fourth dimension. There's a committee of "watchers" in this realm who periodically keep tabs on mankind's spiritual development. When needed, they reincarnate to effect change. The list of various divine beings that willingly took human shape include Jesus, Buddha, the Virgin Mary, Muhammad, assorted Dalai Lamas, Michael the Archangel, and other notable religious figures throughout history. Christians thought along these lines as well, as indicated by their belief in the "communion of saints." In 1795, Catholic mystic and philosopher Karl von Eckartshausen identified these groups of enlightened mystics as "Secret Chiefs." In the late 1800s, Madame Blavatsky and Theosophy literature elaborated, referring to them as Ascended Masters, members of Brotherhoods, the Great White Lodge, until the name "the Great White Brotherhood" was coined by Anglican priest and occult mystic C. W. Leadbeater in his 1925 book, *The Masters and the Path*.

IDOLATRY

CHEATING ON GOD

The sun, the moon, and the stars were among the first objects worshiped. Meteorite remnants were also highly sought and venerated as idols, and considered as everything from a sort of divine bowel movement to a mysterious message from up above. (*See also:* Kaaba) The sacred stone of Delphi in ancient Greece was such a rock anointed with oils, dressed in fine wool, and fashioned with a crown. It was thought that Zeus himself had thrown this rock down to mark the center of the world. Generally speaking, homage paid to anything other than a religion's designated god is considered idolatry. It was less a reason for punishment among polytheism, and only became a crime after the rise of monotheism. The Israelites made idolatry number one and two of the Ten Commandments, and they considered it a capital offense, or an act of treason. Idolatry was punished by stoning.

STONING

It can be assumed that both the first tool and the first weapon was a stone. To throw a stone is a primitive instinct. Since antiquity, using stones, or stoning, has been a preferred method of execution. Most countries, regardless of

religious beliefs, had outlawed stoning by the twenty-first century; however, it is still practiced by Muslims in Iran, Iraq, Pakistan, Afghanistan, Nigeria, and Somalia. In these countries, stoning is a punishment for adultery, premarital sex, theft, drinking alcohol, perjury, blasphemy, and a number of other offenses. As recently as 2008, a teenage Somali girl was buried up to her neck in a soccer field and stoned to death. She had been accused of adultery, even though she had been raped. One thousand spectators came to see fifty men throw stones at her head.

IFA

AFRICAN GODDESS OF DESTINY

Among African religions, Ifa was one of the spirit deities thought to be in charge of love, fertility, childbirth, and the diseases transmitted by intercourse. More commonly, however, she was considered the general dispenser of destiny. Healers and diviners interpreted geometric lines made in the sand as a form of fortune-telling, or as a morning-after sex test of sorts, to know what Ifa might have in store.

When Pope Benedict XVI visited Africa in 2009, he rejected Ifa's authority as well as the medical community's recommendation that condoms be used to help stop the spread of sexually transmitted HIV. Instead of placing it into the hands of destiny, the pope cited church doctrine, namely that fidelity and abstinence are the way to end the epidemic in the region, where as many as one thousand people die every day of AIDS. Even among many African Catholics, Ifa and other protector deities are called upon to safeguard villages. For example, in 2005, a Zambian woman lost her twenty-three-year-old husband to HIV. Instead of mourning at his grave site and greeting mourners, the young widow hid from her in-laws. When finally found, and as Ifa beliefs recommend, she was forced to have sex with her husband's cousin. This was done to break the destiny the woman had with her husband's bad spirit—and thereby protect the village from being further plagued by his disease. The use of condoms for this exorcising or "cleansing" sex is also forbidden.

IMAM

WHERE'S THE TWELFTH?

In Muslim religions, an imam is a community's spiritual leader, like a bishop or an influential rabbi. Many imams are considered to have a divine connection to Allah. The original imams were blood relations to the Prophet Muhammad, with an established lineage spanning from the sixth century to 872 A.D., when the last one disappeared. It is thought

that since most of the preceding imams had been assassinated, usually by poison, God hid the last imam, known as al-Mahdi, or the "Twelfth," in a cave. Al-Mahdi is believed to be waiting and will reemerge at the end of the world.

Subsequently, there have been rumors of the Twelfth Imam appearing at various times in history, though none has been conclusively identified. Islamic texts say that the Twelfth will have a large forehead and a pointed nose. However, preoccupation with al-Mahdi often involved doomsday forecasts and apocalyptic prophesies. It is predicted that three years of worldwide chaos will precede his return, whereafter Muslims will rule the world in an era of peace and harmony. Recently, there has been a popular revival of interest among Muslims regarding Twelfth Imam mythology, and many believe he will reappear soon.

The Shiite Muslim president of Iran, Mahmoud Ahmadinejad, openly admitted to strong beliefs in al-Mahdi, publicly stating that "time is running out." Al-Mahdi scriptures motivate Ahmadinejad's policies, including his comments about annihilating other nations. Supporters believe he is dedicated to facilitating the Twelfth Imam's return by instigating mass chaos and thereby ushering in a cycle of peace. When Ahmadinejad stated that Iran's nuclear program was for peaceful purposes, he meant it, though he omitted that it was also the catalyst needed first to bring chaos in order to fulfill al-Mahdi prophesies. In 2008, he convinced the Iranian parliament to pass a law that made turning away from the Muslim faith a capital offense, punishable by stoning.

IMMACULATE CONCEPTION

DIVINE REPRODUCTION

The notion that a woman can be impregnated by a heavenly seed is hardly unusual in ancient writings, for numerous characters from mythology bore children in this manner by any number of gods. The mother of Alexander the Great, for one, got pregnant by a godsent lightning bolt. In Catholic dogma the Immaculate Conception specifically refers to Mary—not Jesus—being conceived in the womb of her mother by normal sexual intercourse without the burden of "original sin."(*See also*: Eve) By the second century A.D., most Christian doctrines held that Mary conceived Jesus after being impregnated by the Holy Spirit. Mary therefore remained a virgin, although Jesus apparently followed the normal birthing process at the end of term and was delivered vaginally. Equally miraculous, Mary remained a virgin even after suffering labor, with dogma (namely "the Perpetual Virginity of Mary") explicitly claiming that she remained *virgo intacta.*

In March 2008, a rumor spread through India's Kottayam district that the Virgin Mary's image would appear on the face of the sun. It was said that she would manifest herself to help couples who had been unable to bear children. Subsequently, hordes of the faithful, and even those not wanting to conceive, stared for days into the sun's burning rays, waiting to witness the momentous image. However, no rendition of the Blessed Mother appeared, and more than fifty people suffered permanent eye damage or even blindness.

IMP

PESKY LITTLE DEVILS

Like angels, demons also come in many shapes and sizes. An imp is considered to be one of the smallest and easily trapped inside a bottle, in a ring, or a sword handle. They are believed to be the least evil of the demon class, and are seen more as devilish little jokesters. For example, if you are about to walk onstage to receive an award but trip on the way to the podium, that is the doing of an imp. Stub your toe, bang your head, or misplace car keys just long enough until you've missed your appointment—again, the work of an imp.

"Imp Bottles" are for sale on the Internet. The prankster is sealed within the bottle, but it comes with this disclaimer: "The seller takes NO RESPONSIBILITY for flood, fire, catastrophe, or harm engendered if the buyer opens the bottle."

In Germanic folklore, an imp could be subdued momentarily with music. These creatures were markedly different from full-grown incubi, demonic creatures that appeared to women during the night desiring sexual intercourse. In medieval times, incest was rarely punished if the accused swore on a Bible that an incubus had done it, and not a family member. In satanic circles, an ipsissimus is the highest rank a human can achieve after death, after mastering the skill of both the imps and incubi, though an ipsissimus is never as powerful as Satan or the original devils.

INCUBATION, CAVE OF

MUHAMMAD'S FORTRESS OF SOLITUDE

In 610 A.D., at age forty, the founder of Islam went into a cave on Mount Hira near Mecca to meditate. Muhammad fasted and went into trances at this site for six months. Prior to this period, he was not allowed such indulgences and earned a sporadic living in the camel trade. But after he had married a rich widow fifteen years his senior, he had the time to follow his calling. It was common in that era to go on "vision quests" in search of the particular spirit that might be employed to help protect tribal families. It was thought spirits, or *jinns*, were everywhere, and could be harnessed, similar to the genie in Aladdin's lamp. Muhammad, however, had a visit from a different sort of *jinni*, later said to be the Angel Gabriel. Muhammad was told to forget the other *jinns* and focus on one God, to be known as al-Ilah. He was called upon to be the prophet designated to tell the world. At first Muhammad thought he might be going insane, given all the visions, but his wife urged him to continue. He went back to the Cave of Incubation many times to converse with angels as he formed the tenets of Islam.

INCORRUPTIBLE SAINTS

HOLY WAX MUSEUM

Numerous holy persons left behind human bodies that are impervious to decay. Before John Paul II died in 2005, he wished not to be embalmed, and when his corpse was recently examined, he seemed to be heading to incorruptible status. In 1944, although Pope Pius X had been

dead for more than thirty years, his corpse had yet to undergo rigor mortis and was found to have retained flexibility in its arms and legs. In 2000, the crypt of Pope Pius IX, dead more than 120 years, was opened. Without mummification or embalmment, he was still the same, wearing the exact smile on his face as he had the day he died. One nun, St. Sperandia, died in 1276 and had her body exhumed a record eight times.

She remains incorruptible, still giving off a sweet perfume known as the "odor of sanctity." St. Rita, the patron saint of female lost causes (not to be confused with St. Jude, the patron saint of all lost causes), was found with wounds on her head similar to what a crown of thorns might inflict. After she died, her body was still performing miracles, and more than five hundred years after her death the saint's body appears in fairly good shape, with reports that her eyes still open and close periodically. St. Clare died in 1253 and remained intact until the mid-1900s. The good news was that her skeleton stayed unbroken and strong. It was then encased in a wax mask, and the bones were dressed in the habit of a nun. The most startling example of this phenomenon is St. Catherine of Bologna, who died in 1463 and was originally buried without a casket. Her uncorrupted remains were not placed in a glass coffin for display, as with most incorruptible saints; instead her corpse has been arranged in a seated position for more than five hundred years, stationed next to a bank of votive candles. Although her skin is a little darkened from candle smoke, it's no wonder she was named the patron saint of artists, who wish that their body of work will remain similarly immortalized. The devout believe incorruptible saints are prevented from the process of natural decomposition by the intercession of a supernatural power. There are more than one hundred Catholic uncorrupted saintly bodies displayed around the world.

Science explains that some corpses can decompose slower than normal due to environmental conditions, such as high concentrations of salt in the soil where the body was first buried, and humidity. (*See also:* Mummy) However, there are some cases of well-preserved cadavers where there is no reasonable explanation and so they remain a mystery. An example is St. Silvan, who is the oldest of the incorruptible saints still on display. His corpse is dressed in Roman garb, and it is believed he was martyred around 350 A.D. It's not known who he actually was, though a "Silvanus" is mentioned in the letters of Paul; perhaps he was a priest serving in Rome. He was beheaded, as his uncorrupted remains clearly show: His waxy-looking head is severed and pressed against the neck and body, with his mouth still slightly agape. The Silvan relic was crated in 1847 and moved from Rome to the Church of St. Blaise, in Dubrovnik, Croatia, where it continues to be a popular attraction. Scientific tests on the body are not permitted.

INDULGENCES

SCALPING TICKETS TO HEAVEN

According to Christian belief, sin is either an action or a thought that opposes a standardized moral code of belief, and is considered an offense to God's will. These transgressions are kept on an individual's tally sheet, like marks on a blackboard. Some can be erased; others cannot. A mortal sin is so grievous it can never be forgiven. For example, the Ten Commandments offered a list of possible mortal sins, including murder, idolatry, and adultery. Lesser offenses, called venial sins, can be washed away through various forms of penance, doled out by priests or clergy after the person confesses to the flaw.

In the early Catholic Church, penance was harsh. The Council of Nicaea in 325 A.D. determined that a venial sin could be forgiven only after twelve years of continuous repentance—seven of which included daily public prostration, followed by two years of speaking and answering questions only in the form of prayers. The most heinous sinners were made to become *flentes*, or "weepers." These unfortunates were not allowed to enter a church but rather were commanded to stand at the doors covered in ashes. They were to cry continuously and beg the faithful to say prayers on their behalf. The second rung of penitence cast the sinner into an *audientes* class, a term meaning "hearers." These were permitted to stand inside a special "sinner" section of the church, where they could listen to the sermon and readings; however, they were kicked out to the street before any sacraments took place. The third group of sinners were given a penance that made them *genuflectentes*, which meant they could stay through the entire ceremony but had to lay prostrate or kneel for the entire mass and a minimum of two hours afterward.

These harsh punishments led to the popularity of indulgences, which offered absolution of one's sins for a cost. Prayers, fasts, pilgrimages, and paying a certain sum of money—a fixed rate per sin—allowed the wrongdoing to be wiped clean. By the Middle Ages a person could buy a "certified" document attesting to their forgiveness, marketed by professional pardoners. To get a relative out of purgatory, pardoners hawked their wares with catchy jingles: "As soon as the coin in the coffer rings, the soul from purgatory springs." The system worked better than taxation or tithing and became a lucrative means for churches to collect incredible wealth. During the Crusades, more than six hundred thousand men were recruited to fight by offering a carte blanche indulgence and forgiveness of any sin.

Martin Luther, originally a Catholic monk, was more upset by the sale of indulgences than any other issue and in 1517 he threw a ratchet into the scheme by posting a document called *Ninety-five Theses* upon a church door. Number 86 was the breaking point that later led Luther to be considered the father of Protestants: "Why does the pope, whose wealth today is greater than the wealth of the richest Crassus, build the basilica of St. Peter with the money of poor believers rather than with his own money?"

Today many Catholics confess their sins anonymously in a darkened confessional booth, for which they receive penances—usually a set number of prayers to recite, and depending on the sin, a recommendation to make restitution.

JEWISH SEASON OF PENANCE

Yom Kippur, meaning "Day of Atonement," remains the most observed of Jewish religious holidays. For ten days prior, a penitent must think of the wrongs done through the previous year, both in thought and in deed, and rectify them, as well as ask for forgiveness. If this is done in conjunction with the observation of various rituals, then the slate is wiped clean and absolution is granted. During this period, especially from sundown on the eve of Yom Kippur to the next afternoon, Jews are commanded to fast and abstain from wearing leather footwear, bathing, and sexual activity. The final breaking of the fast is celebrated with a feast, at which everyone preferably is dressed in white.

INQUISITION, THE

RELIGIOUS TORTURE CAMPAIGN

Inquisitions, or queries on the resoluteness of a person's faith, have occurred since the third century, primarily after Roman emperor

Constantine organized various Christian ideas into a "universal" creed. According to the Catholic Church's official historians, "On the whole, the Inquisition was humanely conducted." Originally, excommunication was the punishment recommended for Christian heretics, though during the Middle Ages the church considered those opposed to any aspect of Catholic dogma as traitors to both God and state. In 1252, Pope Innocent IV decreed: "[heretics] are murderers of souls as well as robbers of God's sacraments." He thought they should be treated as criminals and coerced to admit guilt. However, he ordered that "life and limb" should be spared, and that torture should be applied only once per question. Interrogators interpreted this rule liberally, and so persecution of a suspected heretic could go on for a few days. Although priests—usually from the Dominican order—were most often the interrogators, capital punishment was administered by secular authorities. Church law allowed the heretic's property to be confiscated.

The exact number of deaths was not fully recorded, and depending on the source, the injustices range from the impossibly high nine million deaths to the ridiculously lower estimate of a mere two thousand executions. There were at least one hundred thousand known "trials," 4 percent of which resulted in a death sentence. Needless to say, the simple fear of being called to an Inquisition served better to keep people faithful. The notorious Bernard Gui, a French Dominican inquisitor, conducted 930 trials and reportedly killed as many as forty-two people. Another famous inquisitor, Jacques Fournier, conducted 114 cases, and sent five to death. However, it should be said the torture methods often caused permanent disabilities.

The devices for capital punishment were already in place, such as the rack, which stretched and pulled body parts and limbs until they popped loose. "The wheel" was another, where a person was bound naked to the spokes of a large wheel, bludgeoned, then hoisted and left to the elements and scavengers. Most convicted heretics during the inquisitions were burned. The instruments used to extract confessions had creative religious names. For example, the "heretic's fork" was a miniature pitchfork fastened under the soft spot of the chin and to the breastplate, such that a mere twitch would cause it to pierce the skin. "Judas's chair" had an ingenious lever and pulley system that suspended the accused above a pyramid with a very sharp pinnacle. The person was positioned so that the anus would sit on the pyramid tip, raised and lowered a little at a time until confession was given. "The maiden" was a metal sarcophagus shaped like a woman and featuring a carved likeness of the Virgin Mary. The contraption stood upright with a hinged door fastened with spikes that pierced the body, though it was designed not to hit vital organs, so it could instead prolong suffering. It became popular because it was so tight that the moans and cries of the inflicted were inaudible, giving the interrogators much-needed quiet time for further prayer.

The last inquisition officially took place in Rome in 1870, though by then the bloodiest methods had ended. In Portugal, for example, it was officially banned in 1831, but not until after 1,175 were burned, and nearly thirty thousand crippled by penances. The Catholic Church still has an office of inquisition, though its name has been changed from Supreme Sacred Congregation of the Universal Inquisition to the Congregation for the Doctrine of the Faith.

WILLIAM IRVINE

LEADER OF THE "CHURCH WITHOUT A NAME"

Founders of new religions often have trouble deciding what to call their new faiths, as it was for Scottish minister William Irvine. At the turn of the twentieth century, Irvine was an articulate and convincing preacher who claimed the end of the world was going to happen in 1914. He borrowed from Jehovah's Witnesses the concept that only 144,000 were going to be saved, but he went further, demanding absolute asceticism as the single most important requirement for salvation. More than thirty thousand gave away every possession and voluntarily became homeless. All that Irvine kept was his dog. At first he called his movement the "Two by Twos," because pairs of his followers were sent about preaching and securing donations from dawn to dusk. Later he instructed them to say they belonged to the No Name Church, or the Church Without a Name.

Even though 1914 turned out not to bring doomsday, it did see the start of World War I, which prompted many to believe Irvine had some prophetic skills. Nevertheless, in 1928 Irvine was excommunicated from the very movement he started, though by that time he was living in Jerusalem. He believed he was called to be the last of the great prophets and wanted to die in Palestine, at which time he thought he would be risen in three and a half days—twelve hours longer than it took Jesus. When he died of throat cancer at age eighty-four, in 1947, few remembered that Irvine was still living in "no-man's-land" outside Jerusalem. It seems he was not resurrected. Followers of Irvine's ideology persist, and although the church currently owns no headquarters, members do meet at annual conventions. Some groups had to finally pick a name to file for tax-exempt status; they often go by "Christian Assemblies." Since Irvine's days, the movement has required full-time members to be homeless, while "saints" or "workers" can own some things, though no TVs or any other electronics.

ISAIAH

THE NAKED PROPHET

During the eighth century B.C., the learned prophet Isaiah was noted as one of the most outspoken activists in biblical times. Once he protested Judaic policies by walking about the city naked, a practice he persisted in doing on and off for more than three years. Before he met his death by being "sawn in two" with a wooden blade, he made twenty-one prophesies, many against Judah's enemies but mainly about signs that would help people to recognize the true messiah. He said the messiah would be spat upon, rejected, sit silently before accusers, and be buried in a rich man's tomb. Some of his other prophesies included the prediction that the Nile River would dry up, and that the city of Damascus would disappear from the map, though it currently has a population of 1.6 million.

Naked protesting remains a showstopper. More than a thousand nude bicyclists pedaled through Europe in June 2008 to highlight environmental concerns—and stop the end-of-days, which they believe will be brought about by global warming.

ST. ISSA

THE BUDDHIST JESUS

In 8 A.D., a fourteen-year-old Jewish boy from Nazareth ran away from home by joining a merchant caravan headed to Asia. He changed his name from Yeshua (Jesus), a name as common as "Bob" in ancient times, to Issa. The young man was on a quest to seek answers to the doubts he harbored about his own religion. He believed in one God and had listened at his mother's knee, from his earliest memories, as she had told him stories from the Jewish scriptures. Yet during the previous year, while he had been initiated for bar mitzvah, Yeshua had begun to have misgivings about Judaism. The Greeks and the Romans had brought news about other cultures and strange-sounding beliefs, and he became determined to learn of them.

When he returned to his native land eighteen years later, he brought back a credential of having traveled the world, and it had been common knowledge that Issa had been among prophets while a wandering ascetic. While abroad Issa studied Hinduism, Buddhism, and Mozi philosophies, such that many of his most poignant sayings can be traced to lines found in each of these three religions' writings. (*See also:* Mozi) Nevertheless, it was enough to draw the attention of John the Baptist, who was the most popular mystic of that time. Issa was also closely affiliated, or at least had knowledge of, the radical Jewish Essenes sect and their view of an impending apocalypse. They preached that a messiah was about to manifest and destroy Israel's enemies. However, the Vatican has suppressed the facts and documents detailing Issa's, or what are considered Jesus', "Lost Years."

The aforementioned is the opinion of Russian journalist Nicholas Notovitch, who published the bestselling *The Unknown Life of Jesus Christ* in 1894. He claimed he was shown Brahmin manuscripts at the Himis Tibetan monastery, located in an isolated part of the Himalayas, which described the life of Jesus, known as Issa. The ancient manuscript explained how Jesus learned to perform miracles, such as walking on water, called *siddhis*, and other techniques of prayer and achieved enlightenment by becoming "god incarnate" from within. It noted that Brahmins eventually drove him out since he preferred working with the lowest castes and condemned idolatry. Notovitch was immediately ridiculed when Buddhist monks refused to show the proof of his source to other Westerners seeking confirmation. The story wasn't considered probable until the ancient manuscript in question was produced by another head monk in 1922, for Swami Abhedananda, who verified its existence. Lost gospels and documents unearthed in 1945 and at the Nag Hammadi Library in Egypt also relate many of the same details as found in Buddha folklore. Maps drawn by the Brahmin monks indicated where Jesus had been and how Issa was considered a Buddha and enlightened by the time he left at age twenty-six. They traced his route back to Jerusalem by way of Persepolis, Athens, and Alexandria. Subsequently, Jesus learned how to gather a crowd long before he returned home to Palestine.

APOLLONIUS OF TYANA

Some think the Issa that these texts describe was the Greek Apollonius, a contemporary of Jesus. He was a Pythagorean and an ascetic wanderer as well, teaching philosophy and performing miracles. Records indicate he traveled to Mesopotamia, Arabia, and India studying Hinduism and Buddhism, and may have met Jesus at some point in his travels. Apollonius also concluded there was only one god, who didn't want animal sacrifices or in fact any kind of worship whatsoever by humans. God, instead, could be known by tapping into each person's *nous*, the gut feeling within, which might be roughly translated as intuitive knowings. According to legend, Apollonius didn't die but rather ascended into heaven. Incidentally, Apollonius is not to be confused with St. Apollonia: She achieved martyrdom by having all of her teeth pulled after they were hammered loose by chisel. She is the current patron saint of dentists.

JAINISTS
GENTLE GIANTS

Considered one of the oldest surviving religions, Jainism traces its roots to ninth-century B.C. India, when a thirty-year-old nobleman named Lord Parshvanath left his wife and riches to take up an austere life of meditation. After eighty-four days of fasting and reflection, Parshvanath supposedly achieved enlightenment. He claimed that the wisdom he found had been first introduced by a spiritually advanced group of humans called "Jina." These beings were believed to have first appeared on earth eight million years ago and made a settlement in the Indus Valley. When ordinary tribal people began to migrate to that area about forty millennia ago and eventually established an agricultural based civilization, the Jina—or Jainists—began sharing their knowledge about the workings of the universe. Early art depicts them as relative giants, a foot and a half taller than humans at a time when the average height of an ordinary adult male was four feet, eight inches. A number of the objects often depicted with the renderings of the Jainists seem to be spaceship-like, leading some current practicing Jainists to believe that the origins of their religion were perhaps extraterrestrial. Jainist mythology describes these first oversized instructors as calm and nonviolent, and that they either left the planet or became extinct by the fifth century B.C.

Jainists believe that everything has an immortal soul and that everyone must go through various stages in order to become conscious of the god within. There is no Supreme Being for a Jainist, and the universe was never created but always was. Moreover, God is equal to the totality of the world. For one's soul to advance spiritually, a person must live a life of austerity and be devoted to seeking knowledge; eventually those who are dedicated will achieve a liberated state, or *siddha*. The trials one endures in life are the result of past lives and "karma." The reincarnation process never ceases, and all people, civilizations, and even global eras cycle over and over again. The cycles are very long, however; the one we live in now is called the Avsarpini half cycle, lasting nineteen thousand years. In this period, humanity is scheduled to go from its best to worst. One of the last of the repeating cycles is Kalchakra, in which humanity will encounter a landscape of wish-granting trees. In addition, during the Kalchakra cycle all people will be born in sets of male and female twins who will stay together their entire lives.

The last known Jina was Vardhamana, though he left his body in 420 B.C. by *salekhana*, or fasting to death. Nearly ten million people practice some form of Jainism and go to great lengths not to cause harm to any other living thing. Some wear surgical masks so as not to breathe in microorganisms that might be killed by the immune system. But literacy is a chief virtue among followers, stemming from their commitment to constant self-improvement and study; Jainist libraries are credited as being repositories for some of the world's oldest known writings and books. In India, Jainists form a demographic odd to Western sensibilities, for they are considered well-off economically yet pursue a modest, even ascetic lifestyle.

JEHOVAH'S WITNESSES

INFAMOUS HOUSEGUESTS

In the 1870s, Charles Taz Russell held Bible classes in his Pennsylvania house that eventually led to the formation of Zion's Watch Tower Bible and Tract Society, later to be known as Jehovah's Witnesses. Interested in the Millerite and apocalypse religions during his time, Russell looked for a new date for the end of the world and predicted it to be 1914. By studying the Bible and transforming the inches on the height of a pyramid to years, he concluded he had the true date of the Second Coming. When the day passed, it was rescheduled for 1918, then 1920, 1925, 1941, and another in 1975—but it was now predicted to merely end "soon." Russell advocated spreading his message by knocking door to door. And it's still done, with followers going about neighborhoods to spread the message, or "witnessing," dressed in "church clothes." They may leave a copy of the Watch Tower magazine, *Awake!* (twenty million copies printed each month), and despite being frequently ignored and harassed, they pray that someone who unknowingly happens to answer their doorbell will convert. The considerable effort placed on "cold-call" religious recruitment is done not so much out of charity, but because if a Jehovah's Witness does not convert at least a dozen people in his or her lifetime they will probably cease to exist at death—or worse, be shunned from the religion and

called an apostate, with oblivion of their soul nearly assured. Charles Russell supported his early efforts to form a sect through funds he earned as a men's clothing store shopkeeper, and that is one reason dressing fine is still considered a good thing. Over the years, Russell as the religion's founder has been downplayed, especially after questions arose about sexual impropriety with a secretary and his interest in the occult was revealed.

Followers say their religion was founded by Jesus, a special man, and not part of the Trinity, though he has ruled in heaven since 1914. On earth, Satan took over from that date (and note this was the beginning of World War I), and since that time all churches and world governments have been ruled by the devil. Only 144,000 spirits will get to live with Jesus in heaven, with about eight thousand of the souls previously selected thought to be still alive. The rest of the human race has souls that die with the body, but if one lives according to Tract Society principles, it's possible to be resurrected when Jesus returns: You will regain a perfect body and reside in a paradise on earth ruled by Jesus and the 144,000-person sect. Some of the things one must do to be eligible for resurrection: For one thing, never get a blood transfusion, since at resurrection you might be part of the person whose blood is in your veins and subsequently rise from the dead as a mutant. Also, a Witness must not accept a birthday gift or a Christmas present, get a tattoo, serve in the military or government, become a police officer, wear a cross, salute the flag, or stand during the national anthem; nor should they join clubs or school sports; read anti-Witness material, say "good luck," play chess, or own wind chimes. A member in good standing must go to religious services five times a week and use family vacation time to attend society conventions, preach, and avoid using aluminum pots and pans, which were incidentally created by Satan to kill millions. There are currently six million going door-to-door and considering themselves as Witnesses.

Charles Russell died while traveling on a train in 1916 and had requested a pyramid be his tombstone. His successor, Joseph F. Rutherford, modified Russell's doctrines and dubbed the religion Jehovah's Witnesses in the 1930s. Many of Rutherford's rules and much of his ideology remain central.

JESUS OF NAZARETH

GOD'S ONE AND ONLY SON

Jesus never wrote a single document. The question remains whether Jesus could read or write fluently, during a time when 95 percent of his contemporaries were illiterate. In those days, twelve-year-old male Jews were trained to memorize certain Hebrew passages from the scriptures to pass the quiz of sorts conducted by rabbis before bar mitzvah, and that often remained the extent of the education offered to his class. In one reference, Jesus was given a passage from the book of Isaiah to read aloud, which he recited, though adding his own comments and edits. Everything known about the man's biographical details are hearsay, in the scholarly sense that no physical records can be found to corroborate them. New Testament writings about the early life of Jesus are sparse and conflicting, with few verifiable details known concerning his intimate habits or exact familial history. It seems that according to mentions of feast days and planetary alignment, Jesus of Nazareth was born between 6 and 4 B.C.; however, the precise day and year of his nativity remain in question. He was supposedly circumcised eight days after birth, and by divine instruction given the name Yeshua, meaning "God saves." That his mother, Mary, gave birth in a

Bethlehem stable is related in two accounts, even if the particulars, such as when the Magi came to visit or whether barnyard animals actually spoke at his birth, are related briefly. From birth to twelve, there is nothing known. Of his youth, he was noted to accompany Joseph and Mary on a pilgrimage to Jerusalem—a seventy-five-mile trip by foot from Nazareth—for Passover. There he ditched his parents for a nerve-racking three days, until he was found at the temple asking questions of learned religious scholars.

According to the Gospel of Matthew, Mary and Joseph had six other children: James, Joses, Simon, Judas, and two unnamed sisters. Some think there were children from Joseph's previous marriage, with James mentioned more than once as Jesus' brother. However, despite all those siblings, Jesus expressly asked his apostle John to look after his mother after it seemed certain his end was near.

In Nazareth—a town with a population of about two thousand at the time—Aramaic was the primary language. In general, only rabbis and the upper class spoke Hebrew, and Latin was the official Roman language for all government affairs, though few commoners understood more than a handful of words. There were a large number of Greeks inhabiting the area as well, and their language was spoken for purposes of daily commerce; it appears Jesus knew enough of these dialects to use phrases from each, though he preached almost exclusively in Aramaic. He never wrote a single thought, nor commanded that anything he spoke of should be put to paper. He had learned to preach in a two-beat, semi-rhyming scheme to help people remember, a standard technique for ancient orators, as exemplified by the Beatitudes: "Blessed are they who mourn: for they shall be comforted." The *ipsissima vox Jesu*, or Jesus' actual voice, was not recorded verbatim, since the earliest conveyances of his ideas were originally translated from Aramaic into Greek by the first gospel writers long after he had died. When his disciples asked how they were to remember all he had to say, according to the Gospel of John (written between 50 and 85 A.D.), Jesus remained untroubled about the language issue and said, "The Helper, the Holy Ghost, whom the Father will send in my name, he will teach you all things and bring to your remembrance all that I have said to you."

Jesus' stepfather, Joseph, is remembered as a carpenter, but he was actually what we would consider a handyman or semiskilled laborer, noted for making plowshares, digging out foundations, and on occasion hewing lumber used for crucifixions. Historians delving into clues concerning Jesus' lost years (from age twelve to thirty) suggest that young Jesus' disobedience to his parents at twelve was more than typical pubescent angst, and that he seemed uninterested in learning a trade. Instead Jesus was allowed to join a group of merchants that eventually traveled through Persia, India, and Tibet. (*See also*: St. Issa) Most Catholic scholars say Jesus stayed put in Nazareth, or studied in Egypt, and remained out of trouble, though no corroborative references to what he did during that period have been preserved. There's at least an eighteen-year gap before he's noted to be baptized by John at Bethany, a village across the Jordan River from his hometown. For a little over two years, Jesus then appears at various locations in a hundred-or-so-mile radius before his execution in 29 A.D. on Calvary Hill in Jerusalem. The one time during his active years when he returned to his familial base, the citizens of Nazareth nearly stoned him. It also seems odd that Mary, by all accounts a good Jewish mother, wouldn't have shared more boyhood facts about her eldest son with his disciples. The question of Jesus' marital status likewise finds no conclusive answers, though it would have been an anomaly for a Jewish male not to take a wife by the age Jesus had been known to begin his public ministry. The only local group to believe in celibacy at the time was the ascetic Essenes, of which John the Baptist seemed to belong. Those who find the need for documentation irrelevant say that since Jesus was God and knew the future and his own earthly fate, he chose to forgo marriage, not wishing to leave a young widow or children behind.

JESUS ON SEX

Jesus gave nearly three hundred instructions on how to live and in what to believe. On the subject of sex he was nearly mute. Jesus never commented on celibacy, though he did mention that if a person was born or made a eunuch, then "accept it." On homosexuality, or even masturbation, there is no specific guidance, and he deferred to what was said on the subject in Hebrew scriptures. (*See also:* Ruth and Naomi) Divorce was not permissible, except if the woman was discovered cheating. Forget pornography; Jesus believed that for those who were already married, even looking at another with a sexual thought was committing adultery in the heart, a mortal sin. On marriage Jesus quoted a Genesis passage: "A man shall leave his father and mother and be joined to his wife, and the two shall become one flesh." Curiously, however, when the apostle Peter, a married man, asked what the reward was for following Jesus, divorce and even abandonment of children is given credence: "Everyone who has left houses or brothers or sisters or father or mother or wife or children or lands, for My name's sake, shall receive a hundredfold, and inherit eternal life."

Even though the first preserved records of his life and preaching weren't authored until decades after he was gone, it seems unlikely that Jesus would have neglected to get his message put into some tangible form. Every other aspect of his short period of public life was orchestrated masterfully, and it seemed from the start he set out to create a legacy. Josephus, the most reliable of the Jewish historians, and who lived through those times, noted that an archive building in Jerusalem where concrete facts about Jesus were most probably held was destroyed in 66 A.D. The depository stored birth and marriage records, property deeds, and such, as well as collected papers concerning outstanding events. During an attempted Jewish uprising against the Romans, the building was torched by the Jews to destroy census documents that might be used for further retaliation. In 70 A.D. the Romans decimated Jerusalem, demolishing temples and any other likely place where original records of Jesus might have been concealed. In the end the particulars of his personal affairs remain unknown, leading some to claim that he never existed, or that the known life of Jesus is a composite of various messianic preachers, who were in abundance during those times.

THE "Q DOCUMENTS"

The four gospels of the New Testament have many similarities, with some lines appearing word for word. It appears the authors were citing an earlier record of Jesus' sayings, now lost, called the Q-Gospel. These firsthand, handwritten notes were removed from Jerusalem as trouble brewed, and were circulated among the faithful. The documents were used as the primary source and served as the basis, though either abbreviated or elaborated upon, when John, Mark, Luke, and Matthew composed their Greek language gospels, and offered similar details concerning the life of Jesus. It's suspected that Jesus had one in his group who knew how to write and made notes and transcribed sayings soon after he spoke. His cousin Matthew was a tax collector and needed literacy as a job requirement, and some think he may have taken notes of what Jesus said.

Writing something on the spot was not easy. There were "notepads" in that era, small wooden tablets coated in wax that could fit into a robe pocket, but they were considered a luxury item and primarily used for business. A stick or stylus made impressions on the wax, which could either be erased or transferred by making impressions to leather scrolls or papyrus. However, scrolls were very expensive and papyrus had to be imported from Egypt. Not a single contemporary record relating to Jesus' miracles remains, even though such miracles would ordinarily get the equivalent of newspaper headlines. There is not a letter to a friend describing heavenly happenings in town, or even a passing mention of Jesus' doings that was definitively composed in his lifetime. (*See also:* Scientology)

JESUS PEOPLE

EVANGELICAL HIPPIES

The so-called Jesus freaks of the 1960s—then considered hippies with an Evangelical slant—organized as an official church in 1972, headquartered in Chicago. Communal living is recommended as well as imitating the lifestyle of Jesus and the early Christians. Although robes and sandals are optional, they support their communes by owning a Christian rock music label and manufacturing religious T-shirts. All the money that members earn is put back into the commune. There are currently about five hundred Jesus People.

JIHAD

MURDER IN THE NAME OF ALLAH

Jihad means "struggle" and in Islamic doctrine it is defined as the holy call to join an armed conflict against non-Muslims. Beginning with the Prophet Muhammad and continuing throughout the centuries, jihads were issued to fight off an array of enemies, from Mongolians to the Crusaders, with the goal of not only winning but also causing the humiliation of its enemies. In 1998 five imams issued a jihad "requiring the killing of Americans, both civilian and military."As a recruitment device, jihadists who died in battle were promised guaranteed comfort in paradise, with all wishes granted, even a prize of seventy virgins. One Arabic scholar noted, however, that the word for "raisin" and "virgin" are very similar in ancient texts, and it is thought that a suicide bomber, for example, might receive a handful of raisins instead. Female jihadists get their sins forgiven, and likewise whatever they want in paradise, except seventy virgins. (*See also*: Fatwa)

JÍVARO

SHAMANS OF THE SHRUNKEN HEADS

The shamans, or *wishinu*, were the most powerful leaders of the Jívaro Indians. They are perhaps the most famous of the Amazon rain forest tribes because of their custom of shrinking an enemy's head to the size of an orange. These macabre trophies were worn on belts or hung as decoration in their huts; the heads harnessed the spirit of the enemy, trapped forever as the shaman's slave. Jívaro shamans were also responsible for concocting poison-laced blow darts made from piranha teeth that paralyzed enemies and prey instantly.

For the Jívaro the physical world was an illusion that could only be pierced by shamans. The only way to venture into the lower and upper worlds of true reality was through their guidance—and a potent hallucinogenic tea called *natema*, made from a recipe of jungle vines and leaves. Anthropologist Michael Harner lived among the tribe and described his first experience with *natema* in 1961: "I met bird-headed people, as well as dragon-like creatures who explained that they were the true gods of this world. I enlisted the services of other spirit helpers in attempting to fly through the far reaches of the Galaxy."

Today the Jívaro, and offshoots of their tribe, can only sell imitation shrunken heads to tourists, unlike the heyday of human skull trade during the 1930s, when an authentic head cost less than twenty-five dollars. In the 1950s, an American hot rod was truly souped up if a shrunken head hung from the rearview mirror. Hallucinogenic drinks are still in use by the Jívaro, and many new cults advocate a variation of its use to achieve higher consciousness. In Amazonia it is still legal to possess a human shrunken head, though it must have been harvested prior to 1976. (*See also*: Santo Daime)

JOB

BIBLICAL WHIPPING BOY

In the Old Testament, Job's faith was famously tested by having to endure a series of misfortunes. One included body growths: "Job was smote with sore boils from the sole of his foot unto his crown." Job's tale of woe was more than bad luck; rather, it seems he was subjected to a type of divine game, between God and Satan.

Previously, Job had a Midas touch. He was prosperous and well liked, and he followed all religious principles as required. Just as he was nearing a comfortable retirement at the age of 148, God and Satan took notice of Job. Satan remarked that it was easy for a man with so much wealth to be righteous. To prove the devil wrong, God then let Satan have at him. First, Job's herds and servants were killed by marauders, and then the house where his daughters were eating lunch was blown down, killing them all. When Job still didn't blame God for the wrongs, Satan was given permission to afflict Job's health, covering him with boils, which Job tried to scrape away with a broken piece of pottery. Eventually God had the devil temporarily vanquished as Job proved to keep his faith. God then rewarded the battered old man with even greater riches, and Job died happy at age 248.

ST. JOHN THE SILENT

THE QUIETEST SAINT

Young John Encratius was a go-getter in the sixth-century church and was made a bishop of Colonia, in Armenia, by the age of twenty-eight, partly because he was a rich noble who inherited a fortune that he donated to the pope. Although given a comfortable appointment, by his mid-thirties John had had enough of all the sermons and speeches required of his position, and so he resigned his post to make a pilgrimage to Jerusalem. At a small monastery there he asked for a private room, but when people kept seeking out his opinions he again grew weary and decided to keep his mouth permanently shut. Eventually he had his door removed and walled over with bricks. How he got food to survive remains a mystery, but for the next seventy-plus years St. John the Silent never spoke a single word. Even on his deathbed in 558, at age 104, John did not utter a sound.

JOHN THE BAPTIST

BEHEADED COUSIN OF JESUS

There were hundreds if not thousands of rough-looking, wild-haired and bearded ascetics during the time of John the Baptist, all proclaiming to be prophets or messiahs. Many warned of the approaching apocalypse. Some wandered, gathering a handful of followers as they moved from town to town, but John claimed as his primary preaching spot a bend in the Jordan River, the way a panhandler might commandeer a street corner as his "franchise." For a while John was the most popular, dressed in a camelhair sack, carrying a staff, wading knee-deep in the

"living water" and dunking pinched-nosed converts under the surface. Most believe Jesus was one of his early followers, while other sources say they were cousins.

However, despite his success, around 30 A.D. John got in trouble by passing judgment against King Herod; the prophet had condemned Herod's taking of his brother's wife as his own. Herod was a Jew, and had publicly declared himself "King of the Jews." In fact he was the highest ruler of the Jews, although he was an extremely paranoid one whose power depended on pleasing Rome. Herod imprisoned John for being a troublemaker with a following that was getting a bit too noisy; the king even thought John might spearhead a rebellion. During a royal party, discussion arose as to what to do with John, and Herod's stepdaughter, Salome, suggested that it would be entertaining if John's head were served up on a platter. After an exotic dance performed by Salome, Herod granted her request. While awaiting his fate, John still harbored doubts about Jesus being the true messiah and sent his disciples, hoping to get a straight answer. Jesus didn't visit John in prison, and replied: "Go tell John what you hear and see. The blind regain sight, the lame walk again, and lepers are cleansed," though it seems the desperate John was hoping that Jesus would perform a miracle and get him freed. John's head, or one claimed to be, is currently on display in Rome.

JONAH

THE PROPHET WHO WAS VOMITED UP BY A WHALE

Jonah was a minor prophet who lived around 800 B.C. He made his fame by living in the stomach of a sea creature for three days. Jonah had some clear prophetic credentials even before the event: He was sent from Judea as a delegate to the neighboring city of Nineveh, then a prosperous force that was poised to destroy the Israelites. Jonah was reluctant to go, though he finally went the long route by boat. When a storm raged, the sailors fingered Jonah as the cause and tossed him overboard. Soon after, the seas calmed. Whether Jonah was a good swimmer we never learn, but soon after his immersion a huge sea creature, a miraculous species of whale, swallowed him whole. The anatomy inside the creature's stomach was the size of a studio apartment, furnished with a

table, chair, and a glowing pearl that served as a lamp. Jonah constructed a ladder from waterlogged timber and climbed up to the creature's eyes, which had the transparency of glass. After Jonah prayed for forgiveness for his less-than-enthusiastic effort to carry out his mission to the enemy city, Jonah was vomited back onto dry land and he resumed his quest toward Nineveh. Upon getting an audience with the king, Jonah warned that if Nineveh's idolatry weren't halted, his stronger god would annihilate them all. The king said he could appease any god by commanding all inhabitants to merely fast and don sack clothes for a short time; it seemed true, since God did nothing and the city went on about its business. Jonah was extremely annoyed by this lack of divine power punch, especially after his ordeal inside the whale. Angry at God, Jonah decided to live in a hut as a hermit. Still, the story of Jonah is remembered today and held as a lesson on the power of repentance.

JIM JONES

CULT LEADER; MASS MURDERER

His father was a Klansman. When Jim Jones reached maturity, he declared himself a blended reincarnation of Jesus and Lenin, and he formed a cult known as the People's Temple, which attracted a diverse membership, including a large number of African-Americans. He drew thousands of followers by promising to lead them into salvation and save them from an impending nuclear holocaust. In 1977 he was forced to move his church out of San Francisco to Guyana, South America, where he planned to create a utopia at "Jonestown."

Within a year, the families of members had grown worried and implored Congressman Leo Ryan to go to Guyana to free anyone who wished to leave. Officially, Ryan went to investigate alleged human rights abuses perpetrated by Jim Jones against American citizens at

Jonestown. Shortly after Ryan arrived, a mob of cult members tried to hack him with machetes. Only slightly injured, Congressman Ryan quickly decided he was ready to leave, taking with him eighteen cult members who wanted to return to the States. Jones sent a brigade of loyal followers to the jungle airstrip. They gunned down Leo Ryan, three journalists, and one of the cult turncoats.

That evening, the cult leader ordered his followers to drink a purifying elixir—grape Kool-Aid, spiked with cyanide and tranquilizers—from a bathtub. Most of the nearly one thousand residents of Jonestown did what they were told; those who didn't were forced to drink or were shot. Infants received an oral dose by syringe. As the mass suicide progressed, one cult member, who spit out her drink, grabbed a gun from a fallen guard and shot Jim Jones in the head. Two months later, Jones's killer allegedly slit the throats of her three children and finally committed suicide. A year later, in 1979, three other survivors of the People's Temple, Jeanne and Al Mills and their daughter Linda, were to start out on the lecture circuit, speaking about their cult experience, when they were shot to death in Berkeley, California, during a home invasion. The Mills had devised a theory that Jim Jones was linked to the CIA, insinuating that the cult was allowed to flourish as part of a mind-control experiment.

JUCHE

OFFICIAL RELIGION OF NORTH KOREA

Although not actually a religion, as per communism doctrine, Juche is instead referred to as an "ideology." Invented in the 1950s by North Korean leader, Kim Il-Sung, Juche combines elements of Stalin's philosophies, some Confucian ideas, and the teachings of Mao. In short, Kim and his family line were deemed to be an all-knowing dynasty. In the Juche calendar, New Year begins on the day Kim Il-Sung was born, and stories about his boyhood miracles, such as his ability to turn pine nuts into bullets and his skill at walking on water, are used as lessons to inspire Korean children. Other rituals include bowing to his picture in homes upon departure or arrival; attendance at every government-held parade and rally is mandatory. There is no heaven or hell, only the

"State." But Juche followers do believe that North Koreans have been selected to exist for all eternity and that they will ultimately rise to world domination. Hard work, self-sacrifice, and stoicism are characteristics of a follower obedient to Juche teachings. Approximately twenty-three million people practice Juche ideology.

JUDAS

INFAMOUS APOSTLE

For the last 1,700 years, most people have believed that Judas Iscariot, one of the original twelve apostles of Jesus, traded his master over to Roman soldiers for thirty pieces of silver. That he did indeed alert the arresting authorities to Jesus' whereabouts that led to crucifixion is not in contention. The motivation, however, behind Judas's actions has recently come under debate. Since the discovery of a document written on papyrus, dating to the second century, called the "Gospel of Judas," an entirely different story has emerged, one that does not portray Judas as the vile traitor he's commonly believed to be. The scrolls were found in an Egyptian cave in the 1970s and were finally deciphered in 2000. All but a few scholars vouch for its authenticity, even if the true author is unknown. Judas, in this version, is given instructions by Jesus: "The star that leads the way is your star. You will exceed all of them for you will have sacrificed the man that clothes me." Here Judas is not a traitor, but rather Jesus' best friend. Jesus asks Judas to help release his soul from his body so he can go back to heaven. Judas, as is known, then informed the authorities with the signal of a kiss. If Judas was in fact Jesus' closest confidant, as well as the apostles' treasurer, as this "Lost Gospel" attests, then the suicide of Judas shortly after Jesus' ordeal can be considered in another context. This concept of Jesus setting up his own final torment was not in line with Christian thought around the second century, with some saying that this gospel and about thirty other documents found and once considered divinely inspired were omitted. Only four official gospels made it to the New Testament as we know it, and it's unlikely the Lost Gospel of Judas will ever be added.

KAABA

SACRED METEORITE OF ISLAM

Muhammad was driven out of Mecca in 622 A.D. because he preached about believing in one god, which was considered a serious problem primarily because of the city's thriving pilgrim business.

Long before Islam was established, a multitude came annually to worship a black stone and a cubic-shaped building in Mecca, a popular city in the central portion of what is now Saudi Arabia. The rock was revered as an object delivered by the gods. The cubic structure was on the exact spot where Adam, the first man, had built his house. Later, the Prophet Abraham and his son Ishmael made a small shrine there, incorporating the black stone and following the foundation of Adam's original floor plan. In Muhammad's time pilgrims came to worship

the stone, which was thought to be a portal between heaven and earth. Around the shrine there were temples that featured statues of 365 other gods from many religions, which pilgrims of all faiths came to make offerings to: even Christians did. When Muhammad returned eight years later and converted Mecca to his religion by conquest, he destroyed all other icons, but kept the stone, calling it and the houselike structure Kaaba. It was made even more hallowed when Muhammad kneeled to kiss it. The Kaaba has since become the most holy and sacred site of Islam. When Muslims pray five times a day, they do so in the direction of the stone. One of the Five Pillars of Islam—the rules left by Muhammad to guide his followers—recommends that if one is able to make a pilgrimage to the stone they should do so at least once in their lifetime. At the Kaaba, it's best to walk around it counterclockwise three times. Currently, during the annual feast called Hajj, or the pilgrimage made to Mecca, as many as two million people at one time stroll about the rock chip from outer space, and hope to gain the Prophet's blessing.

The stone, when one kisses it, has a softness and freshness which delights the mouth, so much so that he who places his lips upon it wishes never to remove them.

—Ibn Jubayr (1145–1217)

KABBALAH

MYSTICAL JUDAISM

The Hebrew word *Qblh*, meaning oral tradition, is at the root of the mystical Judaic religion called Kabbalah. When Adam and Eve were clueless on what to do after their expulsion from the Garden of Eden, a lesser-known angel named Raziel gave them a handbook full of divine codes to make them "whole" with God again. The other angels thought this a bad idea and snatched the booklet back, scattering the pages across the tides. What was remembered of these codes had been passed orally. However, in the first century B.C., it was a rage in Israel to interpret the most revered Torah, and turn Hebrew letters into numbers, looking for Raziel's divine clues. The preoccupation faded in popularity until the Middle Ages, when Spanish rabbi Moses de Leon wrote

The Book of Zohar, considered the Bible of Kabbalah. Told in allegory, the Zohar (meaning "radiance") describes the 125 stages a soul must overcome before achieving the last, or "the end of correction." This is done in part by heightening sensory perceptions. De Leon humbly claimed all his ideas were from others and that he was merely a scribe of a celestial muse: In his life he remained a wanderer, though considered extremely holy until dying of a constricted trachea at age fifty-five, in 1305. The sixteenth-century ascetic Isaac Luria, after spending six years studying the Zohar, attracted a huge following and added the notion of transmigration to the belief system (the soul moving into another body after death). Before dying from the plague at age thirty-eight, in 1572, Luria supposed that individuals, through right actions, could redeem the world. Today, Kabbalah has been adopted by celebrities, some of whom have been photographed wearing T-shirts saying "Kabbalists Do It Better." Men and women, usually wearing white clothing, are often separated during services at many Kabbalah centers, referred to as "un-synagogues." One ceremony requires celebrants to form their arms into a circle around their heads until the Torah is brought out, and then to hold out their hands, with palms up to "receive the Light."

KAMA SUTRA

ENLIGHTENMENT THROUGH SEX

The Kama Sutra is often thought to be holy Hindu text, though it was originally composed as a supplemental guide (*sutra*) on how to get the most enjoyment (*kama*) out of life. It gave ideas on how to find a wife, and have the best sex with one's wife and courtesans. It also provided tips on genital piercing, aphrodisiacs, and the most pleasurable sexual positions. With the purpose to get the soul to the next level and achieve ultimate liberation, or the Moksha phase, the booklet described sixty-four different sexual acts, including assorted positions, "marking with nails," biting as foreplay, and oral sex techniques. It's believed some notions had been compiled from folktales, though organized first into a format that somewhat remains written by a second-century Indian philosopher, Mallanaga Vātsyāyana. He was considered a learned scholar and was said to have spent his early years raised in a brothel, where he apparently learned many techniques firsthand. Others claim he was simply a celibate monk with a wonderful imagination. The original work had no drawings; those were apparently added when an updated and revised version was commissioned in the fifteenth century. The many acrobatic positions that seemed to require double-jointed partners were inspired by original textual descriptions and from examples of erotic statues and frescoes preserved on Hindu temples.

KARMA

YOU GET WHAT YOU DESERVE

In Hinduism and Buddhism, our actions directly determine our status in future reincarnations. Thus our deeds are indelibly imprinted on the soul, for better or worse. For example, if a person lived guiltlessly, his or her station in the next go-round could change, such that a person once of a lower caste could come back and live the next life in the Brahmin class. Under this plan, bad deeds can conversely cause rebirth of the soul into nonhuman form and force one to return to live as anything from a fly to a rat. However, coming back as an animal is not always appalling, since some species are revered—this is one reason why in India, cows, pigs, and monkeys are seen roaming the streets unbothered and frequently cause traffic jams. Many consider these animals as embodiments of divine reincarnated spirits.

The karmic accounting of actions done twenty lives ago can hinder or help in the present. Karma is considered a universal system of cause and effect on the immortal scorecard of the soul. In other words, you get what you deserve—good or bad—even if you have no memory of what it was you had done to earn it. In recent times, karma is seen to have more immediate results, such that what one does today will come back as reward or punishment as quickly as tomorrow.

SACRED COWS

From ancient times, Indians considered cows as a symbol of wealth and providers of life-sustaining milk. In Hinduism, the cow's sacred status is tied to the religion's story of creation: Lord Krishna, an important figure in Hindu mythology, was reincarnated five thousand years ago as a cowherd. Coming back through reincarnation as a cow in India would be a positive turn, since the animals are treated with the rank of the highest Brahmin priests. Feeding a cow in India is considered good luck, but injuring or killing one is still a criminal offense. The cow remains a representation of generosity and motherhood; in 2008, a population of more than 200 million Indian holy bovine roamed the countryside and city streets. If you step in cow dung it's still thought of as a blessed omen, and Indian cowpies are believed to have antiseptic qualities as effective as any industrial-strength disinfectant.

MRS. KEECH

"OUTER-SPACE SUBORDINATE"

In the early 1950s, a middle-aged woman using the pseudonym Mrs. Marion Keech of Lake City, Missouri (aka Dorothy Martin) began to receive messages from the inhabitants of the planet Clarion. The aliens

communicated only with Mrs. Keech, and through automatic writing she dutifully transcribed the extraterrestrials' prophesies. The aliens had made contact with her to warn of an impending worldwide flood that would take place on December 21, 1954. The Great Lakes would swell to destroy the Midwest; the entire Pacific Coast would be flooded, from Seattle to Chile in South America. The good news in this bleak pronouncement was that the planet Clarion had dispatched a spacecraft to arrive at exactly 11:59 P.M. on December 20. It would land on Mrs. Keech's front lawn to take her and select followers to safety. She referred to herself as an "Outer-Space Subordinate" and became an impromptu UFO cult leader of a group she dubbed "the Seekers." Only true believers would be given passports, and this required the followers to quit their jobs, leave spouses, and give away all money and possessions. In addition, they had to learn the alien greeting "I am my own porter." They spent many hours repeating it in a robotic monotone.

Minutes before the spaceship was to arrive, Mrs. Keech remembered that no one could wear metal. Women removed bras with metal clasps and men cut the zippers from their trousers with razors in a frenzy so as not to be left behind. However, when the designated time came and went, the followers cramped inside Mrs. Keech's house grew severely depressed, since they still believed a great flood would begin at dawn. Mrs. Keech began to weep, although a few hours later, at what eyewitnesses reported to be 4:31 A.M., she began to tremble. She then announced that a new message was coming to her: The planet Clarion advisers informed her that God had intervened and saved the planet at the last second, and their spaceship subsequently got orders to return to home base. While the group had previously held their plans in tight secrecy, they now began to call media sources to alert the world of the narrow escape and say that indeed everyone would be fine. The once-shy Mrs. Keech did numerous interviews and said the "lessons" the aliens had given her generated so much energy in her small house that God was alerted to the plight. She encouraged others to follow her path, as she had become convinced that channeling positive energy into the universe was a good thing. Dorothy Martin became "Sister Thedra" and continued to attract followers by channeling intergalactic beings, serving as leader of the Association of Sananda and Sanat Kumara. She died in 1988. (*See also*: Raëlism)

KEMETIC RECONSTRUCTION

MODERN PHARAOH WORSHIPERS

"Em hotep" is the greeting to offer a fellow Kemetic Reconstructionist. Meaning "Welcome in peace," this phrase and many others used in Kemetic Orthodoxy are from documents believed to be part of early Egyptian religious practices.

"Kemet" is what Egyptians called their native land and the region of "black soil" surrounding the Nile. Certain hieroglyphics, and the pronunciation of ancient Egyptian words, are used in this Kemetic religion to help one now and in the afterlife. They believe in one self-created god, called Netjar, from which all the other gods manifested. The goal is to practice *ma'at*, or virtue, and a moral code of forty-two rules, known as "the Declaration of Innocence." Some are very specific, as in advising not to eat a piece of cake taken from a child. Among the rituals, it's recommended to get tattoos that look like bracelets, put notes to ancestors in special clay vases, and make altars consisting of a candle and two bowls, one with water and one empty, in secret spots such as kitchen cabinets. To become a Kemetic Reconstructionist, one must be purified in a bath of natron, the ingredient used to prepare mummies, though baking soda and salt water are acceptable today. Kemetics favor all things Egyptian and often don the *khat*, the head garb of pharaohs. There are approximately 2,500 practicing Kemetic Reconstructionists in the United States.

KHLYSTS

PROPONANTS OF GROUP SIN

This Russian sect, dating from the 1630s, interpreted the Christian notion of the Second Coming differently. They believed Jesus and the Holy Ghost came individually to each person at some point during their life, and not to all en masse through an apocalypse. However, exactly when one was to receive this visitation was unknown, which encouraged the sect to find ways to accelerate the process. The group and its various offshoots were all disparagingly referred to as Khlysts, closely associated with the Russian word for self-whipping. In fact, many belonging to the group indulged in the practice. They also believed in frenetic dancing and group song fests, and in some instances the ceremonies turned into orgies.

Originally, the sect reasoned that Christ would pay more attention to a mass of sinners than to the sole sinner, and that his return would be accelerated if more people sinned together simultaneously. The idea of achieving divine insights through sex and song was further expanded, such that if people consciously sinned together, then the power of the sin was nullified. The leaders were called Christ once they experienced the internalized Second Coming through the "Holy Ghost"; women at the higher levels were referred to as "Mother of God." At the turn of the twentieth century as many as six million people considered themselves adherents to certain aspects of the sect, even if they likewise attended other mainstream congregations. After the Soviets came to power, all but twenty thousand members remained and went into hiding. The others were reeducated or killed.

KKK

THE INVISIBLE SOUTHERN CHRISTIAN EMPIRE; CROSS BURNERS

Originally formed as a social fraternity by Confederate army veterans, the Ku Klux Klan never claimed to be a religion, though from its inception it has relied on quoting scripture and fire-and-brimstone ideology as the best way to raise funds and acquire new members. They claimed to be defenders of the "Invisible Empire of the South" as early as 1867—and to have been divinely appointed by God to do so. Fraternities frequently use Greek letters and mystical symbols, and "Ku Klux" was the phonetic spelling of the Greek word *kyklos*, meaning "circle"; and by changing *c* to *k* in the word *clan*, they meant to connote that their ties were stronger than even blood-bound family clans.

The first leader who was bestowed the highest title of Grand Wizard was former Confederate General Nathan Bedford Forrest. He was opposed, on both moral and religious grounds, to the federal government's economic plan to revitalize the southern economy without slavery. He believed those "radical" reconstruction efforts had to stop, and saw the KKK as a means to do just that. As ruler of the Invisible Empire, Forrest renamed the southern states as "realms" and divided territories into eight districts that would be under the authority of Grand Dragons, called Hydras. When asked about the Klan's murdering of blacks as well as whites in opposition to his beliefs, and its spread of fear and intimidation, Forrest replied: "There were some foolish young men who put masks on their faces and rode over the country, frightening negroes, but orders have been issued to stop that, and it has ceased." However, from 1867 to 1929, an average of two African-Americans, including men, women, and children, were lynched each week.

THE IRISH WHITEBOYS

A secret fraternity called the Whiteboys formed in Ireland in 1761 to protest Protestant controlled land use and the paying of exorbitant tithes. At a time when southern Ireland's population was more than 90 percent Catholic, it was created primarily to combat entrenched Protestant control. Members donned white shirts and after reading Bible passages, set out during the nights to destroy fences, intimidate, and on occasion kill Protestant landowners and lawmakers. Their methods of justice were extremely harsh; during one incident, a tithe collector was dragged from his home, stripped, and buried up to his neck in a grave filled with briars. However, they eventually turned their methods against Catholic landlords and religious authorities as well. Whiteboy ceremonies were not elaborate, though an initiate was sworn in for life, and murdered if he quit. The Ku Klux Klan, most of which were of Irish and Scottish descent, had firsthand knowledge of Whiteboy tactics and costumes.

In 1915, a part-time Methodist preacher and insurance salesman from Stone Mountain, Georgia, William J. Simmons, became inspired after watching D. W. Griffith's smash hit movie, *The Birth of a Nation*. The film included scenes in which the early Klan was portrayed as heroes of the South. Though the movement had been nearly forgotten, Simmons rewrote Klan doctrines, added new texts, and enlisted the aid of fundamentalist preachers. Forty thousand congregations from across the country soon joined the secret order—the Klan revived. Simmons, calling himself "the Imperial Wizard of the Knights of the Ku Klux Klan," extended the group's policy of intolerance to immigrants, Catholics, and Jews, and pronounced that all its members were God-fearing and "100 percent American." Subsequently, membership reached over two million by the mid-1920s. Wearing white robes and hoods, and burning crosses was seen as consistent with fundamentalist Christian theology. They didn't view the destruction of the most sacred symbol of Christianity, the cross, as burning it, but referred to the practice as "cross lighting," a statement of their strong and fiery belief in Jesus. It was alleged to be no different in symbolism than lighting a votive candle in church. They claimed the Bible actually indicated that Nordic races were the "Lost Tribe," God's true chosen people, and must be preserved, and among many of its beliefs they condemned "mixing," deeming interracial marriage as sinful. Due in part to the Klan's efforts, interracial marriage was illegal in sixteen states as late as 1967, and today members still post on websites: "Interracial marriage is a violation of God's Law and a communist ploy to weaken America."

Before the Civil War, KKK founder Nathan Bedford Forrest was a slave trader, and during the conflict he rose from the rank of private to general. He was heralded as a genius of cavalry warfare, though later he was tried for war crimes stemming from the massacre of black Union army prisoners at the Battle

of Fort Pillow in 1864. At six foot two, and two hundred pounds, Forrest presented a formidable figure with or without his hooded robe. In addition to his Grand Wizard duties, Forrest spent his last years running a prison work farm, until dying from the complication of diabetes at age fifty-six, in 1877. According to a contemporary, Dr. J. B. Cowan, "[Forrest] had the most profound respect for religion." Twelve Tennessee parks are named in his honor, and more than thirty historical plaques have been installed in remembrance throughout the state.

DAVID KORESH

BRANCH DAVIDIAN CULT LEADER

Born Vernon Howell, David Koresh at first wanted to be a rock musician and hoped to achieve celebrity status and the adoration of fans. However, religion seemed a surer route and instead of rock chords, he studied the work of Jim Jones, the infamous cult leader. In 1990 Koresh became the charismatic head of the Branch Davidians, a radical offshoot of the Davidian Seventh-Day Adventists, the Christian sect formed in 1935 in Waco, Texas. Koresh had not always been the main leader, but a follower under Ben Roden, who racked up massive debt. Koresh was elevated after paying some of it down, and was the only one people could turn to when Roden was arrested for an ax murder and then institutionalized. Koresh armed the compound originally so they could protect themselves from Roden, who had escaped lockdown. Nevertheless, Koresh, like Roden, focused on an impending doomsday prophecy. Once in power, Koresh demanded unquestioning loyalty from his followers as a requirement for salvation. Among other things, this included his right to have sex with every follower's wife. He also liked weapons, explosives, night-vision equipment, and all the latest in warfare technology. In 1993, when federal agents went to Koresh's compound to ask about the cache of heavy artillery he kept, a cult sniper killed four agents and wounded sixteen others. The Davidians refused to surrender and held out for fifty-one days. After attempts at negotiation, and the government's psychological warfare techniques failed—which included blasting Billy Ray Cyrus's song "Achy Breaky Heart" around the clock—the FBI decided to force their way in. A fierce shoot-out ensued. Some believe Koresh set the compound on fire, killing eighty-six cult followers, while others blame the government

for firing the first shots. When the ashes settled, the dead inside the compound were found to have expired from burns, smoke inhalation, suffocation, and self-inflicted gunshot wounds.

Gulf War veteran Timothy McVeigh was upset with U.S. government interference, especially when it came to meddling with one's religious beliefs. He saw in Koresh's death a primary example of this policy. In 1995, on the anniversary of the Waco siege, McVeigh detonated a bomb in Oklahoma, killing 168 people at the Alfred Murrah Federal Building. Both Koresh and McVeigh believed the Bible was the literal word of God.

KOSHER CHICKEN

HOW TO TASTEFULLY SLIT A NECK

Regulations concerning dietary issues are found in the Torah, in a section called "Kashrut," the Jewish laws that require certain foods to be handled in a specific manner. The Torah never fully explains why certain customs should be followed, though many are still practiced. If the word *kosher* is part of the label, it signifies a stamp of approval and that it's "fit to eat." Meat processing has some of the more explicit requirements and specifically details what to do with slaughtering, butchery, and animals' blood. A special knife with no nicks in the blade must be used by a pious person while reciting prayers. Slitting a chicken's neck, for example, is called *shechita* and requires a practiced hand that makes only one stroke to slice open the trachea and sever the esophagus without hitting the spinal cord. Since the Torah asserted that an animal's blood contains its "life," it was recommended to bury the drippings in dirt. In kosher butcheries, a symbolic plate covered with soil is situated to catch some drops from the chicken carcass while it drains.

In Islamic tradition, a butcher is asked to say "Allahu Akbar," or "God is great," when killing an animal in order to make it more holy to consume.

As for Christians, there is no right or wrong way to kill a chicken, though one company added a Christian philosophy on how to run a franchise to serve it. When businessman S. Truett Cathy opened the Chick-fil-A restaurant chain, he didn't forget his fifty years as a Baptist Sunday school teacher in his business plan. He stated that the purpose of the company was "to glorify God," and he made Sunday a day of rest for every store, saying, "it's a way of honoring God."

THE HOLY SNACK

The pretzel, a famous snack food, with more than $55 million in annual U.S. sales, can also be baked kosher. However, its unique design was supposedly conceived by a sixth-century Italian monk who had some leftover dough and fashioned a biscuit to symbolize arms folded in prayer. He handed one to children as treats for reciting psalms and called it *pretiola*, Latin for "minor reward." For the next ten centuries the popularity of the pretzel spread throughout Europe. It was eaten particularly during Lent, since it was made without ingredients, namely milk and eggs, that were forbidden during that period.

HARE KRISHNA

CHANTING FOR KRISHNA CONSCIOUSNESS

In the early 1960s, Indian businessman A. C. Bhaktivedanta gave up all his earthly possessions and dedicated his life to the service of the Hindu god Krishna, considered an avatar of Vishnu, the Supreme God. In 1965, at age sixty-nine, Bhaktivedanta came to the United States as Swami Prabhupada, with no more than ten dollars in his pocket. Within a few years he started the International Society of Krishna Consciousness (ISKCON), which now has more than four hundred living centers, sixty agricultural communes, three dozen schools, and owns and operates nearly one hundred restaurants around the world. Chanting and dancing in public with "Hare Krishna, Hare Krishna," as its catchy mantra-cum-jingle attracted many. Members had to live communally, abandon Western names, clothing, and culture, and adopt the wardrobe of saris

and saffron-colored robes. Men shaved their heads, with the exception of one long lock. If unmarried, there was no sex. The purpose of public chanting was to raise funds, spread the message, and encourage worship to Krishna. As strict vegetarians, with an ascetic lifestyle that included avoidance of drugs, alcohol, and gambling as requirements for membership, the movement at first attracted many as a spiritual rehab of sorts. After ex-Beatle George Harrison joined the group and dedicated his song "My Sweet Lord" to the sect, the Hare Krishna ashrams filled to capacity. In 1992 the ISKCONs sued the Port Authority of New York and New Jersey for not allowing them to hand out literature in airports, which hurt recruitment efforts dramatically. In cities where their begging and tambourine jangling became a nuisance, "No Solicitation" signs were specifically designed to keep Hare Krishnas at bay, though the notice is still seen posted at many businesses. Prabhupada used street-collected donations to open bakeries, food stores, and publishing companies, and invested in real estate. Today, more than one million people consider themselves worshiping members of Hare Krishna, believing all material desires and wants of the flesh are evil and need to be suppressed in order to achieve a better level in the next life through reincarnation.

L

LALIBELA

UNDERGROUND HOLY CITY

In the twelfth century, chaos and wars in the Middle East made pilgrimages to Muslim-controlled Jerusalem a deathtrap. Ethiopian emperor Gebre Lalibela (since deemed a saint by the Ethiopian Orthodox Church) had a vision to make a replica of the Holy City in his kingdom. He had more than eleven churches and other structures carved out of volcanic sediment below the surface and connected by tunnels. Through the center of the town he even fashioned flowing water that he called the Jordan River. The buildings contained no wood or beams but were hollowed from single blocks of stones so that the subterranean mini-city was invisible from the horizon. It required tremendous excavation, and although some historians think it took decades to complete with technology as it was then, others believe it was created in record time by divine aid. According to a thirteenth-century account of the emperor's life, Lalibela (a name meaning "even the bees recognized his sovereignty") assembled many workers for the massive project; however, each night when the workers quit, a team of angels came to resume digging out huge piles of stone and dirt. This seemingly superhuman creation has since been deemed the Eighth Wonder of the World. Legend has it that the emperor grew tired of governing this unique pilgrim site and that he transferred rule of his kingdom to his son. St. Lalibela then became a hermit for the remaining twenty years of his life, said to subsist on nothing but roots before dying around 1249.

DALAI LAMA

HE KEEPS GOING, AND GOING, AND GOING . . .

In Buddhism, everyone has the opportunity to turn divine. However, some are *more* divine, and because of reincarnation, they can return. Nevertheless, it's difficult to know whether a newborn has acquired an enlightened one's returning soul. Such is the conundrum of find-

ing the next Dalai Lama, Tibetan Buddhists' highest spiritual leader. A Dalai Lama is considered the incarnation of an ascended master that had achieved freedom from the endless cycle of death and rebirth, yet wishes to come back to help mankind. The 14th Dalai Lama, born Lhamo Thondup, was seventy-four in 2009, but was discovered at age three in 1938 to be *the one*. The recently deceased and embalmed head of Dalai number 13 somehow miraculously became animated and spun around, pointing with its nose toward an obscure village in northeastern Tibet, indicating where his next incarnation was to be found. In addition, a strange, star-shaped fungus appeared on the tomb, and letters were formed from fog to help guide those in search of the next Dalai Lama. The clues led them to a small rural hamlet, and precisely to the doorstep of a poor potato-farming family, with five children. The youngest of the siblings, through a series of tests, was determined to be the reincarnated "High Holiness," and enthroned in 1940.

HOLY LAUGHTER

RELIGIOUS INTOXICATION

Although there's no Breathalyzer test to determine how much "Holy Spirit" one has been inebriated with, there have been numerous reports of people unable to drive after attending certain Pentecostal meetings. "Falling out," "slain in the spirit," and "holy laughter" are phenomena experienced by attendees of charismatic religious gatherings intended

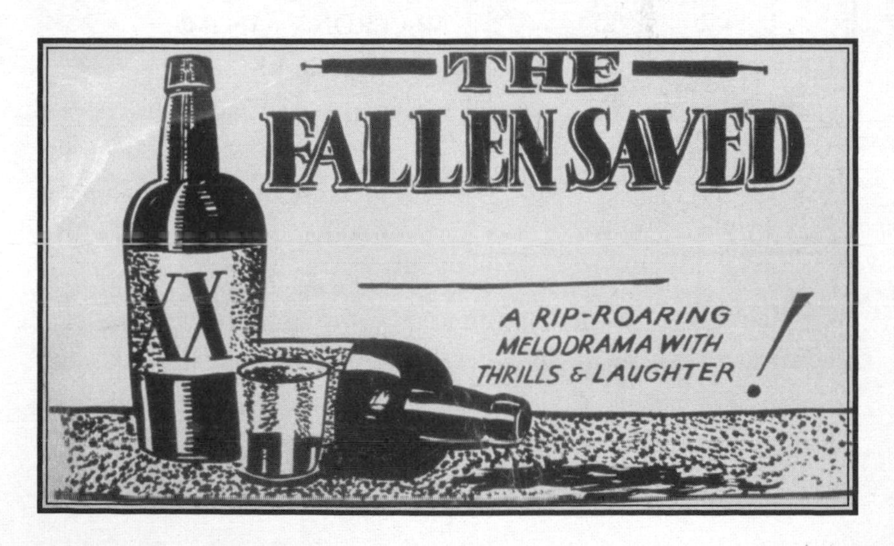

to fill one with the Holy Ghost. Speaking in tongues, collapsing to the floor, and inconsolable weeping are common, though occasionally uncontrollable laughter can seize an entire congregation, even if the preacher is in the midst of describing the agony of crucifixion or some other saintly suffered torture.

What makes something "funny" has yet to be understood by science. "Ha-ha," a typical laugh identified by a series of short vowel-like sounds, takes sixty milliseconds to pronounce and is usually repeated at continuous interims of two hundred milliseconds apart. If the laugh exceeds this speed, the brain will become starved of oxygen. A hyperventilating type of laughter seizes many charismatic congregations and has proven to be contagious, though apparently it also makes motor skill coordination difficult. One Evangelist minister, Reverend Howard Browne, specialized in getting his flock into a hilarious uproar, and he believed it a good thing to be "drunk with (or in) the Holy Spirit." At a "Holy Laughter" meeting in Toronto in 1995, the reverend had more than seventy-five thousand people falling down from hilarity, without a single joke told. It was considered an outpouring of the Holy Spirit. Previously, it was thought that only possession by demons made one laugh hysterically, though it's now believed the Holy Ghost, and all members of the Trinity, in fact, have a good sense of humor.

ST. LAWRENCE

PATRON SAINT OF BARBECUES

Lawrence was an Italian priest during the second century who was caught up in the purge by Romans to stop the spread of Christianity. Officially he is the patron of grill cooks and tanners, though he is often prayed to by barbecuing devotees. He got this distinction by being put on a spit and roasted while alive. In martyrdom he retained a kind of genial attitude and even called out during his torture, "I'm cooked on this side, so turn me over and have a bite." Besides this misfortune, St. Lawrence is also remembered as a church deacon who was assigned the duty of safeguarding the original chalice used by Jesus and his disciples at the Last Supper. He supposedly gave it to a traveler to bring to his family in Spain. This chalice, or the Holy Grail, was misplaced, though one golden cup did arrive and was passed among a cache of monks throughout the centuries. One such version is currently protected under glass at the Church of Valencia in Spain. St. Lawrence is sometimes depicted holding a frying pan.

LEFT–HAND PATH

THE SIGNIFICANCE OF BEING A LEFTY

There are two roads to take toward enlightenment in Hindu and Buddhist Tantric traditions. The right-hand path is self-denial and self-sacrifice, and includes avoidance of meats and alcohol. The left-handers, however, see indulgence as a way to master desire until illumination is achieved. They eat whatever they want, drink intoxicants, take hallucinogens, sometimes offer animal sacrifices, and practice sex rituals. In Eastern thought neither path is condemned, though the left-handed route is thought to be more risky. With the rise of occult practices the left-handed path came to mean black magic, or evil sorcery. From biblical times "left" was bad and "right" good, and accordingly, at Judgment Day the saved will form a line on the right while the condemned will be to the left side of God. In Rome, for example, if you were told to take the door on the left, they would say take the *sinister* one, and it became a word that connoted that left-handedness was not only bad but equally unlucky. Subsequently, due to the biblical and Roman belief being born a lefty was trouble, in medieval times babies seen to favor the use of the left hand were reportedly killed. In Christianity the devil baptized with the left hand, while priests used the right, and it is still forbidden to make the sign of the cross using the left hand. Heretics and sorcerers were routinely condemned merely because they were lefties, as was Joan of Arc, who held her sword with her left hand. It was further thought that the devil sat on the left shoulder, whispering into your ear, while an angel who was perched on your right side countered with better advice. For gypsies, if your right palm itched, money was coming your way, while scratching the left palm meant money would be soon lost. In addition, before the invention of toilet paper, Christian children were taught to wipe only using their left hand. In modern times lefties are considered more creative, while right-handed people are better at math. Eight of the forty-four U.S. presidents were left-handed, including Ronald Reagan, George H. W. Bush, and Barack Obama.

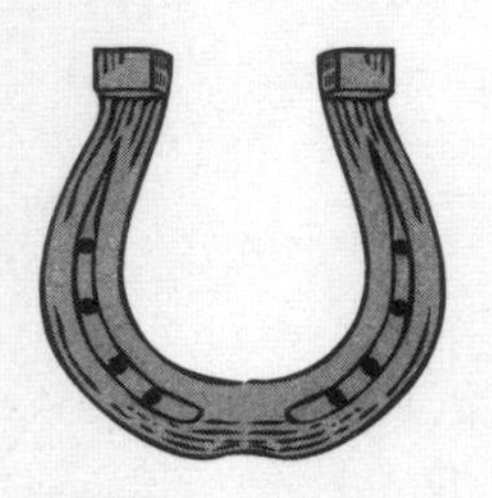

LUCKY HORSESHOE

One of the most popular of amulets to bring good luck is the horseshoe. Even in ancient times it was recommended to place doors only on the right side of buildings or barns, but if an opening was on the left, nailing a horseshoe above the threshold was considered a powerful neutralizer to left-handed evil. Throughout Europe and America the lucky horseshoe remains in wide use. It is believed to have acquired its good-luck symbolism beginning with Nordic mythology. Before the invention of horseshoes it was used frequently as a symbol of the moon goddess, with its curved design meant to represent the shape of the goddess's vulva. Debate continues among lucky horseshoe enthusiasts as to how it should be nailed to the wall. Some think it must point upward in order to catch the luck dispensed from above, while others urge to place it down so that luck pours over you. Both parties are convinced that hanging it the "wrong" way will bring twice as much *bad* luck.

ELIPHAS LEVI

OCCULT MAGICIAN

Alphonse Constant, a nineteenth-century Parisian, had aspirations to be a Catholic priest but left the seminary for love and continued his quest for spirituality through research and writing treatises under the name Eliphas Levi. He occasionally performed as a practicing magician and partook in necromancy, or conjuring up the dead. Toward the end of his life he saw a running thread of truth in many beliefs, from the Vedas to the secrets of the Sphinx, which he believed could be accessed through occult practices. He is credited with introducing the West to tarot cards and many other aspects used by mediums to contact spirits through magic and ancient symbolism. Although his notions were

the foundation for secret societies, such as the Order of the Hermetic Dawn, and Aleister Crowley's cults, Levi was uninterested in seeking followers or forming sects. He believed that physical reality held only a portion of total knowledge and that "astral light" provided passage to other realms, which he defined as a "cosmic fluid" within reach of everyone. He thought that human will, if focused, could cause the fluid to transform physical reality and was powerful enough to create miracles: "To control the astral light was to control all things; a skilled magician's will was limitless in power." He is most remembered for drawing a "Baphomet," a bearded and horned gargoyle with huge female breasts, claws for feet, and demonic wings, and representing male and female duality inherent in the universe. He also used pentagram symbols to indicate good and evil forces that are still in use. He died in 1875 at the age of sixty-five, from blood poisoning and handling alchemy chemicals.

LEVITATING CLERGY

FLYING NUNS AND AIRBORNE SAINTS

More than two hundred saints had abilities to defy the laws of gravity and hover. St. Edmund Rich started floating in England during the early 1200s. His mother was seriously pious, dressing her children in haircloth, a coarse fabric that was tediously itchy. She also encouraged prolonged fasts on bread and water. When Edmund was a teen he had a visitation from the Christ Child, and after that Edmund compulsively traced the letters for "Jesus of Nazareth" on his forehead with his fingertip, a practice he promoted throughout his life. He was known to spend most of his night in prayers, and he never slept lying down. While St.

Edmund served as archbishop of Canterbury, his chancellor, Richard of Wyche testified that he saw Edmund "raised high in the air with knees bent and arms in the air." It took only seven years after his death to be canonized a saint.

St. Ignatius Loyola was another noted to reach heights of one foot off the ground, while St. Dominic floated only six inches while serving mass. Still, St. Dominic was notably higher than St. Philip of Neri, who achieved levitation of a mere "palm" width above the sheets of his deathbed in 1595. St. Alphonsus de Liguori was serving mass in Foggia, Italy, in 1777, at the Church of St. John the Baptist when he levitated a full yard into the air, an act witnessed by the entire congregation. In the mid-1850s, Sister Mary Baoude, a Carmelite nun stationed in Bethlehem, frequently levitated to the treetops in the convent's garden, and remained there for hours in prayer among the birds. Although considered the original "Flying Nun," she has not yet been canonized. In 1911, one Father Suarez, a priest in Southern Argentina, levitated numerous times, though unfortunately never before a camera. In 1973, Jesuit priest Oscar Gonzalez-Quvado could levitate other people, and he seemed to silence skeptics as he allowed witnesses to check for wires with a Hula-Hoop.

MASTER OF GRAVITY

THE FREQUENT-FLIER SAINT

The most prestigious levitator was one seventeenth-century friar, Joseph of Cupertino, born near Naples, Italy, in 1603. From the onset, the young man was absent-minded, losing track of his thoughts halfway through a sentence, and extremely jumpy, such that a church bell or tap on the shoulder flustered him enough to make him weep. In addition, he was a sickly child, catching every flu or illness, which caused his mother to lose patience. She punished him without mercy. At seventeen, he watched a begging friar come to his town, and from that moment he saw this as the only career path to which he was suited. He tried to be admitted to a monastery but was quickly dismissed for reportedly dropping dinner plates, even if the shards were sewn to his habit as a humiliating penance and a reminder to not drop them again. Joseph failed at the most rudimentary tasks and suffered extreme poverty until accepted as a stable boy at a Franciscan monastery. Due to his cheerfulness, and despite the fact that

he was unable to read or write except in the most rudimentary manner, Joseph was reluctantly admitted into the order. His sudden distractions or spiritual blackouts were attributed to what he claimed to be visions of God. Frequently Joseph fell to his knees, mouth agape, lost in the ecstasy of "God's wonder." Talking to birds and animals was one of his gifts, as well as levitation, with over one hundred reports attesting to his spontaneous flights. When brought before Pope Urban VIII, and after kissing the pope's ring, Joseph suddenly soared into the air, until fellow Franciscan officials ordered him down. Wherever Joseph went, people tore his habit as a relic. Eventually he was turned over to priests of the Inquisition, ordered locked up in a remote monastery, and forbidden to fly. More than a hundred years after his death, Joseph was canonized a saint, and he currently serves as the patron of air travelers, astronauts, pilots, and students with learning disabilities.

ST. LIDWINA

PATRON SAINT OF ICE SKATERS

In the winter of 1394, a fourteen-year-old Dutch girl from the town of Schiedam, Holland, was ice-skating home from school when she tripped

and fell, breaking a rib. She did not recover, and instead her parents grew understandably alarmed when blood began to spurt from her ears and nose, and then more so when parts of her body simply "fell off." Her parents kept her discarded organs in jars and the parts were soon claimed to cause miraculous healings. Although eventually deteriorating to total paralysis, except for the use of her left hand, Lidwina was wheeled about when divine intervention was required, especially after it was known she went without sleep or food for months at a time. When she did eat, documents indicate it was no more than a small bite of an apple and that she favored drinking seawater. Although she is revered for claiming frequent religious visions, scholars now believe St. Lidwina was actually suffering from multiple sclerosis, a condition then thought either caused by demons or angels. She died at age fifty-three, and within years her gravesite became a popular pilgrim destination. Many skaters still wear amulets and jewelry of a "Skate Angel," imploring St. Lidwina to keep them from tumbling, or worse.

LIMBO

AFTERLIFE FOR UNBAPTIZED BABIES

Medieval theologians described various levels encountered in the afterlife. While heaven was the state of complete happiness, purgatory was the place where good-natured sinners could be rehabilitated. Hell was the eternal prison for the damned and the completely hopeless. Meanwhile, limbo, situated on the edge of hell, was less painful, even if the souls stuck in limbo had no hope of ever escaping.

From the third century, Christians rushed to baptize their babies, since if they were not cleansed of original sin and washed away by baptism, the infant would be assigned to limbo if they died. St. Augustine

suggested infants were subjected to the "mildest" punishment, though Catholic dogma has yet to clarify limbo, since the notion that innocent babies would be punished in any way seems particularly cruel.

In the 1300s church scholars conceived of two limbos, one for non-Christians of all stripes, while unbaptized babies born to Christians had a special "kid limbo" where no suffering occurred, though there is no hope of ever seeing God. Thomas Aquinas believed infants were cheerful in this massive spiritual day-care center, since they were ignorant of how much happier they could have been if they had been baptized and got to know God. There is no mention of whether babies sent to limbo developed physically or otherwise matured.

LIMBO ROCK

The limbo dance, now used at many Caribbean resorts as a fun activity, and one that requires shimmying under a constantly lowered pole, was originally a ritual of ancient African religions. The dance under the pole symbolized the skill needed to successfully complete the cycle of life and death, and not get stuck in limbo.

ST. DOMINIC LORICATUS

PATRON OF DISCIPLINARIANS

In 995 A.D., Domenico Loricato's father took one look at the newborn and discerned something unusually spiritual about his offspring. He found a local bishop, and for a stiff sum, he had Dominic ordained a priest while still a child. By his teenage years Dominic was already seeking a life of hermitage and penance. He eventually found his way to various monasteries, but made his fame as a Benedictine monk in 1040 at Fonte Avellana, an Italian abbey noted for having many hermits living in the nearby forests. The abbey was run by Peter Damien (later made a saint), who made all visiting hermits practice his doctrine, called "La Disciplina," which required the devout to give themselves three thousand lashes per year. Dominic went the extra mile and in six days he had whipped himself an incredible three thousand times. He continued this feat every Lent for the next two decades. The idea was that each lash would reduce a year in purgatory, and with St. Dominic's arithmetic, known as "One Hundred Years of Penance," the devout could gain a windfall of spiritual credit. Dominic devised a system that required lashes to coincide with the recital of psalms (one hundred lashes per each psalm), as well as a method of whipping that tried to hit a different spot on the body with each stroke.

LOT'S WIFE

HUMAN PILE OF SALT

Lot was a nephew of Abraham, and like many Israelites, he moved around frequently, depending on famines or other opportunities. Lot traveled from Canaan areas, to Egypt, and finally to the city of Sodom, a place noted for luxury and a life of ease. Sodom also had a reputation for a wicked nightlife scene and for being the place to go to for homosexual sex. (Sodom is the city from which the word *sodomy* was derived.) The activities there made God angry and he sent angels to destroy it. However, the men of the metropolis had somehow been tipped off and instead planned to lure the angels in and then rape them—the only case of a premeditated sexual assault intended against angels found anywhere in scriptures. Lot offered the gang his daughters as an alternative, saying they were virgins, but the men had no interest. Since Lot was a favorite nephew of Prophet Abraham, the angels personally warned him of the impending doom and destruction of the city, instructing Lot to flee with his wife and daughters. There wasn't a moment to spare, and two angels physically dragged Lot's family by the hands and warned, "Save yourselves with all haste. Look not behind you. Get as fast as you can to the mountains." Sure enough, Sodom and four other "sinful" cities in the region burst into flames. Understandably, Lot's wife (who was never mentioned by name) wanted one last glance at their hometown, but as soon as she looked, she screamed and was stopped in her tracks. In the next instant, Lot and his daughters watched as Lot's wife turned into a pillar of salt. There was no time to gather her remains, and Lot and his two daughters hid in a cave. According to some biblical interpretations, the girls thought the world had ended and that it was left up to them to repopulate it. They got Lot drunk and both became pregnant by their father. In Islamic texts Lot also appears, but without the incest angle to the story. Today, the area where Lot's wife turned to salt has a number of sodium deposits, and some believe she was frozen in midstep by volcanic eruptions, the way petrified citizens of Pompeii were later discovered.

LADY OF LOURDES

FAMOUS FAITH-HEALING HOAX

Jesus was asked to heal many, but seemed to only use his touch on a fraction of requests, though he was credited with about forty-five "faith-healing" incidents that reportedly cured ailments considered beyond medical understanding at that time. About 20 percent of those healed by Jesus had incurable leprosy. Many saints afterward likewise claimed the gift, but the record holder for the most faith healings belongs to a place, where a special statue stands in an alcove of a cave.

In February 1858, Marie-Bernarde Soubirous, a fourteen-year-old daughter of poor laborers, claimed she had seen a ghost in the caves

a mile outside Lourdes, France, where she had gone to gather firewood. During the next five months the apparition manifested a total of eighteen times to the young girl, though she never truly learned of its identity and referred to what others saw as "a small lady in the rock," or "that thing." Marie's announcement caused immediate interest in her town, especially after she described the vision in detail, saying the spirit wore a "white veil, a blue girdle and had a golden rose at each foot." When Marie said the spirit held a "rosary of pearls," everyone assumed it was the Virgin Mary, and so by mid-March of that year, twenty thousand people had followed the teen into the forest in hope of witnessing the miracle. Encouraged by clergy and onlookers, Marie was told to ask the apparition for identification, and though it initially ignored the girl's request, the "lady in the rock" finally revealed during the penultimate vision that she was "the Immaculate Conception." Marie said that the Holy Mother wanted a church built on the site, where all could come to get her blessing and bathe in a pool of muddy water that suddenly bubbled up from a spring at the foot of the grotto. However, priests wanted tangible proof, and then coaxed Marie to ask the Virgin to perform a miracle, which was achieved when a wild rosebush bloomed in winter.

When the visions stopped, Marie was put under serious scrutiny and interrogated by numerous clergy. They concluded that the young woman was retarded, and in fact found Marie to be nearly illiterate. However, these findings seemed to only further convince them the girl was incapable—short of experiencing a miracle—to make such stuff up. Marie was eventually sent to live at a convent where she worked making embroidery for religious vestments. When she had an asthma attack she went back to the Lourdes spring for a drink that cured her, but when she got mortally ill with tuberculosis at age thirty-five, she refused to go. When she was exhumed thirty years later, her body remained remarkably preserved, and is still on display 150 years later. During this period more than seven thousand people claimed healing after visiting the grotto at Lourdes, while the Lourdes Medical Bureau, a nondenominational group of doctors and scientists who, upon examining X-rays and other medical data, officially recognized sixty-seven miraculous cures attributed to the wellspring at Lourdes.

WHAT QUALIFIES AS A MIRACLE?

Miracle comes from the Latin word *miraculum,* implying an unusual though welcomed occurrence that is both marvelous and wonderful. It is something that happened that does not fit into the normal understanding of the laws of nature. In original biblical text "miracle" was interchangeable for "a sign," meaning that God wished to show his power by superseding physical reality and making something entirely remarkable happen. Miracles are harder to validate in modern times, and often fall into the category of coincidence, requiring time to determine its benefits. For example, in 2009, when a man missed the doomed Air France jet that broke apart in midflight en route from Brazil to Paris, killing 228 people, the serendipitous circumstances that led to his delay were considered a miracle by himself and his loved ones. However, it was a short-lived wonder: He was killed in a car crash two weeks later.

ST. LUCY

THE SAINT WHO SERVED HER EYEBALLS ON A PLATE

Sicilian-born Lucia was raised by her mother after her Roman father died. She was left a considerable dowry to make marriage within the noble class still attractive. However, by her late teens, Lucy was a dedicated Christian, especially after she had a visitation from the recently martyred St. Agatha, who healed her mother's "blood flux." During the visitations, Lucy was promised that she too would one day be as revered as Agatha. However, when Lucy's mother arranged a marriage, she refused and instead wished all her dowry money be given to the poor. The groom-to-be was upset by the loss of the considerable jewels he'd get once betrothed, and reported Lucy's Christian ideas to the authori-

ties. She was sent to a brothel, where she allowed them to do whatever they wished to her body. She claimed that since it was against her will, her soul remained pure: "I will remain guiltless in the sight of the true God." When one client fell in love with her and admired her beautiful eyes, Lucy gouged them out and had them waiting on a plate upon his next visit. Apparently Lucy's behavior was extremely disruptive to brothel business, and in short, guards came to remove her to a place for execution. However, Lucy was so filled with the Holy Spirit that her thin body seemed to weigh a thousand pounds and could not be budged. They had to get a team of oxen to drag her through the streets. They tried to burn her, but she wouldn't ignite. By then Lucy's eyes had grown back, or in some recollections, there were bright glowing orbs in her eye sockets that allowed her to see. When they slit her throat, Lucy still talked. Repeated stabbings eventually did her in at age twenty, in 304 A.D. Depicted as carrying a plate with her eyes, St. Lucy remains the patron saint of the blind.

M

JESÚS MALVERDE

PROTECTOR OF DRUG TRAFFICKERS

Though still waiting official canonization, Jesús Malverde is nonetheless considered a saint in certain parts of Mexico and the United States. Born in the 1870s, Malverde had tried his hand at honest work and watched his mother and father toil tirelessly at the same, but he could do nothing to prevent their deaths by starvation. Afterward, he hid among the hills of Sinaloa, Mexico, and became a thief, though with a Robin Hood–esque reputation for using portions of his ill-gotten gains to help the poor. His career was short-lived and he was captured and hanged before his fortieth birthday, in 1909.

Soon after his death, the mustachioed apparition of Malverde began appearing to poor people in desperate situations, divinely intervening on their behalf. His saintly status was solidified when two smugglers prayed to Malverde to return lost mules packed with stolen gold. When the beasts and loot miraculously reappeared, Malverde was then heralded as a patron for religious criminals. The poor refer to Malverde as "the Generous Bandit," and shrines built in his honor are even now funded in part by narcotics dollars. Many drug traffickers crossing the border from Mexico into the United States carry a Malverde religious photo, or icon, and pray to him to elude capture. He's such a popular saint among the illegal narcotics industry that DEA agents look for Malverde medallions and such as part of profiling before searching certain vehicles or persons suspected of dealing. As one convicted drug seller noted, "St. Jesús Malverde makes it possible to live life on the margins."

NEW SAINTS WORK FASTER

Perhaps because of heavenly distractions, the older and longer-dead saints seem to take more time to answer prayers than the newly venerated. Like Malverde, the Venezuelan doctor José Gregorio Hernández is not officially canonized, though many in South America already consider him a saint. Born in 1864, the demure José Gregorio had tried twice to become a priest, but was rejected for health reasons. Instead he became a brilliant doctor, studying at the leading universities in Europe, only to return to Caracas, where he made a reputation for attending to the poor for free. In 1919 there were probably less than five hundred automobiles in all of Caracas, but one managed to run over the absentminded doctor and kill him at age fifty-six. It happened on a quiet Sunday morning when after attending mass, José Gregorio stopped to buy medicine for a poor patient. Even as his casket was paraded to the cemetery, large crowds gathered and wept, "Our doctor has expired! Our saint has gone." Shortly after, miraculous healing among the poor was attributed to José Gregorio's intervention, and it is reported that today nearly all Venezuelans have at least one statue or portrait of a thin man in a black suit and hat, to which they pray for health. However, to get him in the running for complete canonization, the Vatican requested that new images dress the good doctor in a white suit, which is considered more in line with saintly garb.

MANDAEANISM

THE LAST CHILDREN OF ADAM

Seventy thousand people believe they are direct descendants of the first man, Adam. Known as Mandaeanists, they are considered the last of the ancient Gnostic sects. It is not possible to convert to the religion unless one is born of a Mandaean mother.

In the mayhem following the U.S. invasion of Iraq in 2003, neighbors targeted the secret sect for extinction. Afterward, less than five thousand remained in their traditional homeland, an area south of the Tigris and Euphrates rivers in Iran and Iraq, which is believed to be where Adam was created in the Garden of Eden. The remaining survivors are exiles in Syria and Jordan, with only a few allowed into the United States.

In addition to Adam, the Mandaeanists' bloodlines are traced to Adam's son Seth, to Noah, and to John the Baptist. They do not recognize Abraham or the Prophet Muhammad, and they believe the Torah was written by demonic forces as a spiritual hoax that was unwittingly accepted by the early Hebrew people. They likewise ignore many Islamic traditions. The souls of all Mandaeans are female, with the universe in a struggle between Light and Dark, or Father and Mother. The creator, Mana Rba Kabira, is in charge of all other spirits and uses astrological signs, prayers, and secret rituals that only a few of the remaining high priests know. It is recommended to seek a life centered on family, which will eventually send their souls back to the realm of light. Mandaeans consider men and women equal, even if on the 6th of January, "pious" men are granted any wish they desire. Most Mandaeans wear white and get baptized in a river once a year. When they pray, it must be done facing north, considered the gate of heaven. (*See also*: Gnosticism)

MANICHAEISM

THE ORIGINAL UNITARIANS

Mani was a gifted child, noted for reading and writing fluently in five languages, painting artistically, and preaching by the age of ten. He lived in Babylon, part of the Persian Empire, or modern-day Iraq, during the second century. He claimed that he received spiritual revelations from an entity called Syzygos, or an enlightened twin of himself, similar to a guardian angel. By his mid-twenties, Mani had formed a new religion, called Manichaeism, and wrote seven voluminous sacred texts that were translated into most of the known languages of the day. Manichaeism eventually spread from Great Britain to China.

Mani consciously tried to incorporate the best elements of Buddhism, Zoroastrianism, and Christianity into a one-world religion. Mani prescribed the path to salvation through education, vegetarianism, chastity, and other self-denials. He believed there were two forces, good and evil, forever in battle in the universe, with the soul belonging to the light, or virtuous powers, while the body, and the earth, adhered to the dark side. Sin was the natural by-product of man's true spirit nature forced as it was to occupy a physical body and interact with matter. For Mani, there was no one omnipotent god to help, but he taught that the soul could remain incorruptible by getting to know one's inner self through asceticism, detachment, and avoiding material temptations.

In the end his popularity was seen as a problem for the then most widespread and politically connected religion, Zoroastrianism. Certain priests drummed up charges against Mani and threw him into prison. He asked to be crucified, since he was an admirer of Jesus and likewise viewed himself as a savior. Instead he was sentenced to beheading, though he died waiting for his execution in a Persian prison at the age of sixty-six, in 277 A.D. Most of his original writing was lost during the rise of Christianity, when his views were considered heresy. (*See also*: Zoroastrianism)

Manichaeism was fiercely persecuted, though during certain periods between the 2nd to the 7th centuries it was the most popular world religion. The Roman emperor Diocletian considered Manichaeans as an equal or even greater threat than Christians: "We order that their organizers and leaders be subject to the final penalties and condemned to the fire with their abominable scriptures." St. Augustine of Hippo had been a follower before converting to Christianity in 387 A.D. Augustine reworked Mani's theology in order to mesh with sacramental views, even borrowing and elaborating on the Manichaean notions of the universal struggle between good and evil. However, a medieval-era resurgence of Manichaeism in Europe was brutally crushed by the Inquisition. Currently, there are approximately one thousand followers of Neo-Manichaeism in the United States. They live in small communes and emphasize veganism, fasting, and celibacy. Although original Manichaeans considered water evil, and baptized with oil and bathed in urine, the new sect allows normal hygiene. (See also: *Gnosticism)*

CHARLES MANSON

HIPPIE CULT LEADER

Charles Manson spent most of his youth in detention centers and served more prison time as an adult for crimes including burglary, prostitution, and fraud. He nevertheless was seen as a guru in the Berkeley, California, area at the height of the hippie movement in the late 1960s. While in prison he became acquainted with Scientology, and he claimed to have achieved "theta status," the highest level, though he seemed most interested in learning mind-control techniques. (*See also*: L. Ron Hub-

bard) Manson likewise became interested in satanic cults, particularly the Church of the Final Judgment, whose followers went about the San Francisco streets dressed as Grim Reapers and predicted an impending Armageddon. Manson focused primarily on female recruitment, runaways, and hitchhikers. He had enlisted dozens of women as his sexual servants by advising, "You must live now. There is no past. The past is gone. There's no tomorrow." Eventually he claimed to be God, the reincarnation of Jesus, and finally, Satan.

By the late 1960s his messages grew more apocalyptic, especially after he took numerous acid trips and listened for codes he believed were directed to him in the Beatles' *White Album*. He prophesied that a racial war was about to change America and he saw himself and his "family" as called upon to jump-start the turmoil. He thought the national conflict would be won by blacks, though Manson would emerge from the group's underground bunker to become the new world ruler. Subsequently, his family went on a seven-person brutal killing spree, leaving messages written in blood on the walls, and making international headlines by murdering pregnant actress Sharon Tate. Manson intended for police to blame blacks and planned ultimately as a catalyst for riots. During a sensational trial he carved an X on his forehead, symbolically removing, or "X-ing" himself from the establishment. That particular prophesy turned out to be true, and moreso after his death sentence was commuted to life in prison. Manson replaced the X with a permanently tattooed swastika instead, revealing more than all else his true philosophy. He remains in prison, now in his seventies, and he has stopped showing up for parole hearings, saying he is a "prisoner of the political system."

MARGARET OF ANTIOCH

EATEN BY A DRAGON

Margaret was the daughter of a pagan priest in the early third century A.D. When she converted to Christianity, she was tortured by everything imaginable. She allegedly was even swallowed by a dragon. However, she hacked her way out of the beast's stomach with the crucifix she wore around her neck. This qualified her to become the patron saint of childbirth.

MARGARET OF HUNGARY

SELF-CHASTISING BEAUTY

During the 1200s, King Bela IV of Hungary made an agreement with God that if the Mongols were defeated and his family returned to power, he would give one of his daughters to a convent to become a nun. When his prayers were answered, he donated Margaret to a strict religious order. She was apparently very good-looking, and, in order to stop her own temptations, she wore an iron corset with spikes that dug into her side. Margaret filled her shoes with glass shards and preferred only the dirtiest duties, such as cleaning chamber pots, and attended to the most hopeless of the ill. Six years later her father decided to reverse his decision and marry Margaret to a neighboring king. But Margaret refused, even after her dad threatened to cut off her nose and lips if she didn't marry. More than forty miracles were already attributed to Margaret, including raising the dead. She only intensified the self-inflicting torture devices upon herself, and seldom ate or slept until dying on the cross of her own hard work at age twenty-eight, in 1270. She was canonized in 1943.

A SINGING HEART

Also vowed to celibacy, St. Cecilia was instead forced into marriage by her wealthy Roman parents in the second century. During the wedding ceremony, however, a miracle occurred when religious hymns were heard coming from her heart. It was reportedly loud enough to drown out the festivities. It was such an odd occurrence, that once the guests were gone and Cecilia was alone with her new husband, Valerian, she convinced him to become Christian and likewise take the vow of celibacy—right there on their wedding night. Both were eventually martyred. They tried to execute Cecilia by boiling her alive, but she was taken from the cauldron because she kept on "singing to God." When three blows with the executioner's ax failed to completely sever her head—or to stop her singing—she was left to die; she bled to death three days later.

Naturally, St. Cecilia is the patron saint of music. Her skull remains on display in a church near Venice, Italy.

MARIAN APPARITIONS

VISITATIONS ON GRILLED CHEESE AND ELSEWHERE

The Catholic Church has an entire branch dedicated to the study of Mary, the mother of Jesus of Nazareth. In particular, the Marianum Pontifical Institute in Rome is responsible for official inquiries regarding new sightings of the Virgin Mary. (The Marianum library has more

than eighty-five thousand unique books on the Holy Mother.) Throughout history, the church has investigated 295 cases, though the Vatican has only recognized twelve incidents as supernatural events. However, independent investigators ranging from secular scholars to spiritualists and clairvoyants have noted that the majority of sightings are usually described as an apparition of a lady wearing white. In only a few instances has Mary identified herself. (*See also*: Faith Healing) It has been proposed, in all seriousness, that the spirit people encounter is a Babylonian one that has come to enjoy masquerading as Mary.

In 1997, when Ugly Duckling, a used-car company, was leasing an all-glass building in Clearwater, Florida, a sixty-foot-tall image of the Virgin appeared on the mural-size panels near the entry door. People marveled at it until a teenage boy threw a rock and smashed it. (He was sentenced to ten days in jail.) When the glass was replaced, the image returned, though before long someone else sprayed acid on it. When the rainbow-like silhouette appeared again, a Catholic organization came to take over the lease. The glass panels then alternated between images of Mary and what some say sort of looks like Jesus in a robe.

Although not yet officially approved by the Marianum Pontifical Institute, another apparition of Mary has been appearing in Medjugorje, a small town in Bosnia, regularly since 1981. Here, the Virgin has allowed six people to see her vision—Ivan, Jakov, Marija, Mirjana, Vicka, and Ivanka—and repeatedly returns to tell the group secrets about the future. This select bunch usually remains mum on giving particulars of the future news, but they have quoted Mary: "You have forgotten that with prayer and fasting you can ward off wars, suspend natural laws." She also warned that a supernatural sign will appear on a mountain somewhere, and if the unfaithful aren't converted before that event, it might be too late. Millions have since made a pilgrimage to the small town where for a donation of fifty dollars a "special rock rosary handmade from the stones" gathered from "Apparition Hill" can be obtained, and it comes with a guarantee to be blessed by the Virgin, personally. For one hundred dollars, a holy mass stating specific petitions and requests can be served in a church that has since been constructed near the site of these visions.

The most bizarre sighting of Mary, or at least her image, appeared in Florida on a grilled cheese sandwich in 1994. It eventually sold on eBay for twenty-eight thousand dollars in 2004. It has miraculously remained unmolded and nearly as fresh as the day it was made. The sandwich, made with two slices of Wonder Bread, stayed unrefrigerated, nor was it subjected to extreme preservation methods, other than simply put in a plastic bag, which was placed in a night table drawer. Skeptics, however, point to the high level of artificial preservatives in the bread and the sliced cheese, as the reason it remains preserved, or nearly petrified.

At first, the image frightened the Florida housewife, Diana Duyser. After she had taken a bite and saw a face staring back, she ate no more, though had the foresight to save it. The special grilled cheese sandwich, with its missing bite mark, was purchased by an online gambling site, which shows it occasionally to raise donations for charity. In 2008, a Dallas woman was about to throw away a bunch of rotten grapes when she noticed that one had the Virgin Mary image; the woman is still in debate whether to sell it online or build a shrine around it. There is also a cinnamon bun bearing the image of Mother Teresa on display in a Tennessee café.

People ask me if I have had blessings since she has been in my home, I do feel I have, I have won $70,000 on different occasions at the casino near by my house

—Diana Duyser, sandwich cook.

MARBURG, CONRAD OF

HERETIC HUNTER

Conrad, a German preacher noted for his zeal and knowledge of Catholic dogma, was asked by Pope Innocent III to help muster support for the Crusades. He believed in flagellation as a way to curb his own temptations and advocated it for others. In church records he's considered the confessor of St. Elizabeth of Thuringia, a Hungarian princess. He had ordered debilitating supplication with rods to keep the princess "virtuous" and diligent with her work among the poor.

St. Elizabeth nearly died at the hands of Conrad's "penances" more than a dozen times. She often required hospitalization after his "holy" inflictions. However, through his tutelage, she became canonized in 1235 and remains the patron saint of the wrongly accused and of people mocked for excessive piety.

Conrad hit his fanatical stride in 1227, when by personal request of Pope Gregory IX he became the chief papal inquisitor of Germany and southern France. During his six-year reign of terror, all were considered guilty of heresy before his eyes. He used torture to prove their guilt, usually at the hands of his main men, John the One-Eyed, and a Dominican friar, Conrad Dorso. He eventually employed a loyal but deadly mob of followers that attacked and gathered up everyone suspected of being unfaithful. He particularly went after a sect called the Waldenses, who were deniers of the Catholic doctrine of Transubstantiation. John the One-Eyed liked to insert red-hot iron rods into rectums or vaginas. But scourging, hot oil, or breaking bones were the methods used to first encourage repentance. Once a suspected heretic was in Conrad's court, he would only accept guilt as a plea. Afterward he shaved their heads, usually cut their tongues off, and severed the sinners' hands. Others were chained to walls until dying of starvation and complications of torture. If one maintained innocence, Conrad considered them beyond redemption and they were burned at the stake. What finally made some regard Conrad as insane was his penchant for singing a solemn and beautifully peaceful Gregorian chant, an early version of "Miserere mei, Deus," while inflicting pain. One verse includes the line "I will teach the unjust thy ways: and the wicked shall be converted to thee."

However, when Conrad's accusations of heresy broadened to include priests and nobility, his own days were numbered. After publicly accusing German count Henry II of conducting "satanic orgies," the nobleman went over Conrad's head to appeal his case to bishops and was acquitted. While traveling in 1233 to protest the decision, Conrad was ambushed and murdered by a band of knights; he was fifty-three. A stone placed at the site of his death, near a farm outside Marburg, Germany, is said to be haunted. Evidence of black magic rituals and sadistic ceremonies are still discovered there. (*See also*: Inquisition)

CONRAD'S LICENSE

We have therefore determined to send preaching friars against the heretics of France and adjoining provinces, and we beg, warn, and exhort you to treat them well, that they may fulfill their office.—Gregory IX in a letter to bishops, 1233 A.D.

MASAI

AFRICAN CATTLE RELIGION

When the one true god, Enkai, created the world, he carefully lowered cattle to the earth and chose the Masai people to be the overseers of these divinely provided herds. Masai believe they are the rightful owners of all cattle, no matter where they may graze in the world. What is thought of as stealing, or rustling cattle belonging to others, is not a crime, but rather a correction of the divine order. More than eight hundred thousand people still practice this belief, a religion without texts and transmitted orally, passed on by spiritual leaders and shaman known as Laiboni. Accordingly, if you lived evilly, spirits will snatch you at death and drag your soul to a desertlike hell, while the good find themselves in a pasture-perfect paradise among billions of cows. A Masai cannot put a hoe to soil since it will destroy the grass cattle eat, and grass is considered a sacred herb. They also drink cattle blood, and sometimes mix it with cow milk to make them stronger and closer to Enkai. The semi-nomadic Masai are patriarchal, with women expected to bear as many children as possible. Wife-swapping is common, with a wife "given" for a night or two to a guest upon request as a way to show good manners. If a child is born from it, the offspring belongs to the husband's line and adds to a man's wealth: Having many children and owning cattle are a measure of riches, while celebrity status is acquired by killing a lion. Known as fierce warriors, with the ability to throw a spear and hit a target three hundred feet away, they also like to wear red in honor of cow blood, blue for the sky, and green for the grass, as well as putting hoops, disks, and heavy beads in the lobes of the ears as religious symbolism. There is only one career path for faithful male Masai—cattlemen. (*See also*: Sacred Cows)

KARNI MATA

THE TEMPLE OF RATS

The Hindu goddess Durga had ten arms, rode a tiger, and wielded swords, whips, shields, a bell, a flower, and a smile. She was noted for never losing her sense of humor as she sought out victory and vengeance. In the fifteenth century, an Indian holy woman named Karni Mata was believed to be the reincarnation of Durga. She didn't have ten arms, but she could poke a stick in the ground and minutes later a fully grown shade tree would appear. A temple named in her honor was built six hundred years ago in Rajasthan, India, and still stands, although it is famously infested with rats.

The thousands of rats that make the temple home are considered the reincarnated souls of previous Karni Mata followers. They are treated with reverence. If a visitor happens to step on a rat, the dead rodent must be replaced with one made of gold. Considering that the furry reincarnated souls have the run of the place, it's impossible for visitors not to squash a good share. In fact, the temple monks count on these donations to meet an unusual monastery expense. Plenty of rat food must be on hand to keep the rats relatively docile—at five hundred pounds of grain per month. Otherwise it's assumed the monks of the Temple of Karni Mata would not make it through the night—at least not in human form.

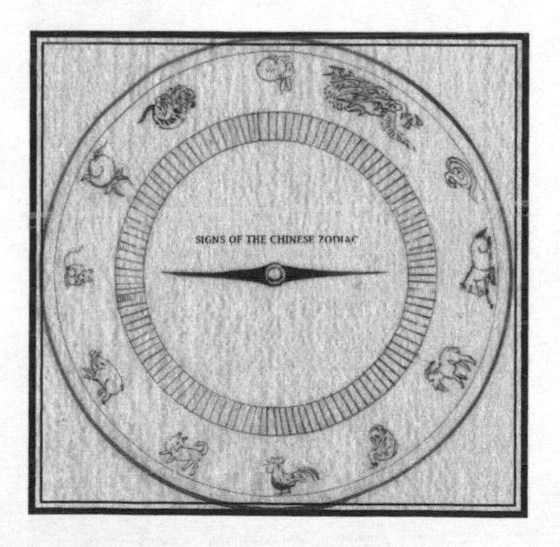

THE YEAR OF THE RAT

Time, according to the Chinese Zodiac, is divided into twelve-year cycles, with each represented by an animal or reptile. Certain creatures are considered to embody various characteristics that become part of a person's lifelong personality traits, particularly if born under its sign. The Chinese ancients thought a rodent the harbinger of prosperity, and admired it for its industry and shrewdness. Those born in the rat years are expected to be very sociable and even eloquent, though with a manipulative tendency. The latest cycle of the rat began in 2008.

MERLIN

WIZARD; ALCHEMIST

Merlin is first noted in *Prophetiae Merlini*, written by Geoffrey of Monmouth, who later elaborated on the famous wizard in his better-known work, *The History of the Kings of Britain*, in 1136. Some believe Merlin was based on the life of a Welsh prophet, Myrddin Wyllt, an historical figure alive around 560 A.D. Myrddin was a hermit and wild-haired madman who emerged from the forests periodically with urgent prophesy. He was eventually driven off a cliff by angry shepherds and impaled below on a fisherman's spear. Others cite a man named Aurelius Ambrosius, a known wizard of some powers, who was brought before the British king Vortigen in the fifth century. The king wanted to know why a tower under construction kept collapsing. Ambrosius said it was obvious: There was an underground lake beneath the site, where two dragons lived, and with their constant clashing it was impossible for the structure to remain intact. Ambrosius and Merlin may have been the same man, since both were said to

have been born of a woman sired by an incubus. (*See also*: Imps) These incubi were creatures or demons that came to lie on women during the night; they were often identified as having a cold penis. Merlin's magical powers grew and in addition to serving as King Arthur's adviser, he was believed to have been responsible for building Stonehenge. Merlin had the ability to see into the past and future and could shape-shift with a snap. His death remains a mystery, with some thinking he is trapped under a stone at the bottom of a lake. Others suppose he is encased in an invisible tower of air in the middle of a forest. This air column is frequently transported to wilderness areas in order to keep him away from people. If you find yourself in quiet woods, listen to the rustling leaves, for the sound is said to be Merlin's garbled voice trying to give instructions to someone who can understand, and concoct the right potion to free him.

MESMER, DR. FRANZ

CHARISMATIC HYPNOTIST

Why are some people so popular, while others can't seem to make a friend or keep one? In the late 1700s a theory called mesmerism was de-

veloped to explain the phenomena and thought it related to an invisible liquid, or to a spiritual energy within, originally described as "animal magnetism." Those who had it in abundance possessed charisma and had a near-hypnotic sway over others. It worked the same way that a metal paper clip, for example, is compelled to move and adhere to a magnet. In 1779, French physician and astrologer Franz Mesmer devised methods to free the internal flow of this animal magnetism, which when blocked caused everything from a sore toe to insanity. His method to stimulate stagnated "fluids" involved staring into a person's eyes, or touching a knee, or pressing his hand to a body part for up to hours at a time. He placed magnets on a person's head, or had patients hold magnetized rods to aid in alignment. He also believed that the position of the stars and the moon affected internal magnetic flows and produced surges or depletion of the energy force.

Dr. Mesmer was driven into exile after his techniques for cures were investigated and deemed "imaginative." However, James Braid, a Scottish physician, explored some of Mesmer's notions and discovered hypnotism in the 1840s. Though Braid rejected the magnetic fluid theory, his research concluded that people were affected and seemed sedated by the repeated suggestions induced by mesmerist practitioners. He revealed that by concentrating solely on a simple visible object and clearing the "mental eye," any person could be hastened to the state that immediately precedes sleep. Braid called it "neuro-hypnotism," or "nervous sleep." He found that it was induced by having a subject stare at a small shiny object, swayed slowly "eighteen inches above and in front of the eyes." Once in this relaxed condition, a person was unable to resist suggestions "planted" in the subconscious. Although some now consider it to be not at all like sleep but rather a heightened consciousness, hypnotism is a technique thought to breach the "social self" and reveal hidden truths. Psychiatrists and even police investigators have employed the practice to reveal suppressed memories. Spiritualists believe it helps one remember past lives or gain mystical insights.

The Magi of Persia and certain yoga practices indicate that similar hypnotic states date to 2500 B.C., and are achievable by staring and concentrating on nothing more than the tip of your nose. A suggestion implanted during a hypnotic trance can remain for years. One hypnotist described how he made a man fall into a deep sleep whenever someone touched the end of his pinky finger. When he met the same person fourteen years later at a party, with one touch, the man instantaneously fell into a sound sleep.

DISCOVER PAST-LIVES THROUGH HYPNOTISM

By the 1950s hypnotism was popular among stage magicians and entertainers. Often they made people cluck like chickens or do equally outrageous things while under a trance. Concurrently, scientists were experimenting with the hypnotic state to make people recall certain childhood events. In 1952, amateur hypnotist Morey Bernstein put a midwestern housewife under trance and then took her step by step back through events in her life until the moment of her birth. He then made a breakthrough and probed further to see if the woman had memory of any thing before birth. In the next moment the woman began speaking in a perfect Irish brogue and came to tell of her previous life as an 1880s Irishwoman named Bridey Murphy. Channeling Bridey, the woman sung Irish songs and described nineteenth-century Irish life in detail. When investigators went to Ireland to debunk the reincarnation theory and to find evidence to prove Bridey Murphy never existed, the results were mixed. Nevertheless, it started a popular interest in past-life regression that has yet to subside. Some unexplainable personality traits and even the cause of phobias are revealed from events experienced in a previous life. For example, one man who was unable to wear a watch or bracelet would even go into a panic with a plastic hospital wrist band. It was "discovered" he had died shackled and chained in Rome in 173 A.D.

MILLERITES

JILTED BY THE APOCALYPSE

William Miller, an upstate New York farmer, liked to read the Bible. During the 1820s, through his casual study of the text, Miller decided he had finally unlocked the secret everyone was waiting for: According to Miller, Jesus would return to earth sometime between March 21, 1843, and March 21, 1844. He had no desire to start a cult or seek converts, and kept the insight to himself for more than ten years. But Miller eventually

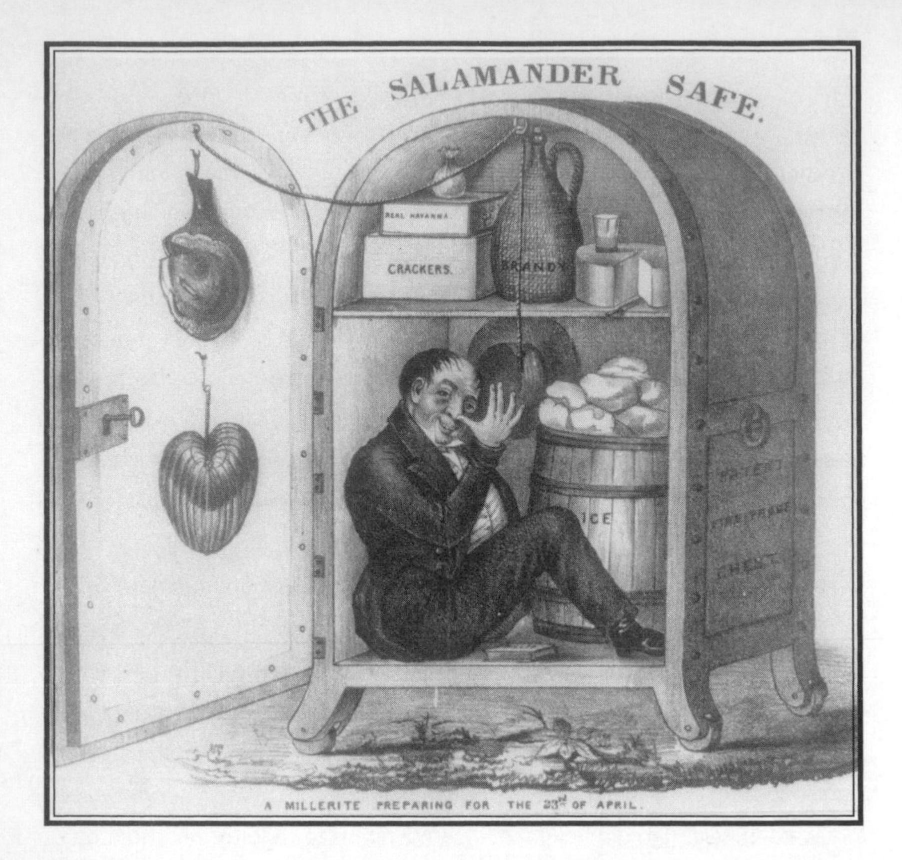

A MILLERITE PREPARING FOR THE 23rd OF APRIL.

thought people might be interested, and in 1833 he wrote a document on his findings. He based his prediction on one line in the book of Daniel: "And he said unto me, in two thousand and three hundred days; then shall the sanctuary be cleansed." By counting "days" as years he calculated that the Second Coming of Jesus was fast approaching.

Miller seemed an unlikely preacher, with terrible stage fright while speaking before his first congregation. But within a few years Miller could elocute with conviction before packed venues wherever he went. When he got the blessing from Joshua Vaughan Himes, an influential Boston clergyman and publisher, Miller's homespun, New England Evangelical phenomenon spread rapidly; Miller captivated a national audience as few other preachers had done before or since. Himes printed pamphlets espousing Miller's view and acted as a booking agent, assembling "great tent" meetings across the country. There were a number of astrological events during this period, including a huge meteorite shower and Halley's Comet (*see also*: Joe

Bullard) that were unusual enough to have the general public discussing its significance. To many, Miller's prediction made it all clear—the end of the world was indeed coming. By New Year's Day 1843, his followers, known as Millerites, numbered more than one hundred thousand, while another million people had attended Miller's lectures and kept a watchful eye toward the predicted date with general fear, as if today waiting for a nuclear holocaust. Millerites took joy in the impending world end, and considered those who scoffed their views to be damned to hell; they alone would be saved. When certain dates came and went without an apocalypse, Miller revised his prediction and stood firmly on a new date, October 23, 1844, as the one and only true finale.

Many sold possessions and gave all their money to church officials, as it was recommended to meet Judgment Day debt-free. In cities and small towns throughout the country, droves of followers donned white robes and went to hilltops with arms outstretched. Others took the words from scriptures that none will enter the kingdom of God unless childlike so literally that adults began to regress, skipping and hopping around while others sucked their thumbs and stayed under the covers in a fetal position. The less religious constructed bunkers, and the Salamander Safe Company had brisk sales when they made modifications in their "Fire Proof Chest," outfitted to accommodate foodstuffs and a small fan as a secure haven to ride out the impending fire purge.

When the day passed uneventfully, Miller tried again to offer a recalculation. However, people were so dumbstruck they abandoned his views, and a nationwide depression settled in, called by newspapers the "Great Disappointment." Millerites were beaten, some tarred and feathered, and spat upon for causing such fear. William Miller believed, to the day he died, five years later, in 1849, that it was all a math error and the Second Coming was imminent: "I confess my error, and acknowledge my disappointment; yet I still believe that the day of the Lord is near, even at the door." Contemporary religions including Latter-Day Saints, Jehovah's Witnesses, and any number of Adventist congregations are traced to Millerites and the aftermath of the "Great Disappointment." (*See also*: Seventh-Day Adventists)

MIRACLES

DIVINE FREE LUNCH

Jesus and his mother, Mary, were attending a wedding in Cana, a small village about twenty miles northeast of Nazareth, just as Jesus was about to begin public preaching in the spring of 27 A.D. Why they were both there and why Mary seemed to be the host is unknown, since it is mentioned only once, namely in the Gospel of John. He called it Jesus' first miracle and one of eight required to accurately identify the new messiah, per clues found in Hebrew scriptures. Accordingly, Mary found Jesus talking among friends and informed him that the wine was about to run out—a situation that would turn the festive affair into a disaster and an embarrassment. Instead of sending a runner with a few coins to the nearest wine merchant, Jesus told the servants to fill huge vases with water. In an instant, simply by stretching his hands over the vessels, and without uttering a prayer, he transformed the atoms that make up water (H_2O) to become wine ($C_6H_{12}O_6$). Some take the entire story to mean that Jesus approved of social gatherings, marriage, and alcohol. Others believe the event actually described Jesus' own wedding, with Mary as host and acting in a traditional role of a mother at her son's celebration. Why Jesus would use such powers to merely keep the bar open at some trivial social gathering in an obscure village is still in dispute. But theologians suggest he was just an obedient son, and that his mother had long known about his special powers and occasionally requested their use. Biblical scholars think the reason the story only appears in John's version is that he was inspired by the Holy Ghost to add it, even though John wasn't an eyewitness, but knew Jesus would have done such a thing if he had been there.

Jesus also had the ability to increase other food supplies, as accounted in multiple sources. On one occasion there was a group of five hundred listening to his message in a remote area near a desert. When Jesus was alerted that there were no provisions to keep the audience from turning into a hungry mob, he gathered five loaves and two fish from a boy who brought lunch along. Jesus subsequently turned these into enough bushels of food to feed everyone, with twelve baskets left over. Jesus did it again for four thousand people when he multiplied the physical mass of seven loaves and two fish into enough food to satisfy the crowd, with seven baskets remaining.

UNEXPLAINABLE PHENOMENA

In Hebrew scripture, miracles are cited frequently, especially when defeating enemies, though they were probably not seen the same way by the losing side. Performing miracles, or causing unexplainable occurrences that defy natural laws, is a criterion still used to prove saints worthy for canonization. In spiritual contexts, a bona fide miracle occurs when God decides to override the laws of mathematics, physics, and all the "workings" of things that he created. As Albert Einstein noted, "There are two ways to live: you can live as if nothing is a miracle; you can live as if everything is a miracle." He implied that the best minds retain a sense of wonder and awe while probing the laws of sciences. Millions of unlikely events and improbable occurrences are considered miracles by many, from winning a lottery ticket purchased with someone's last dollar, to missing a plane that ultimately crashes (but this is certainly no miracle to the person who is waiting on standby and takes the seat). Precisely identifying a divine phenomenon remains tricky, since many of the "laws of nature" are imperfectly understood. Creation itself is currently agreed upon by the scientific community to have originated with the Big Bang, an explosion that produced everything as we know it in the observable and ever-expanding universe. However, what started the Big Bang—or why there is anything at all—has yet to be explained.

Peter the Venerable (circa 1150 A.D.) explained how the Cana miracle worked, attempting to compare it to the process that transforms liquid water to ice, or vapor, yet lets it remain water. He explained that Jesus knew how to harness natural laws, but Peter ultimately concluded that "faith is not the fruit of intelligence, but the reverse is true." Prophet Elisha had been known to perform a similar feat, as described in Hebrew text. In that story an unnamed man in the town of Shalishah (Salim, Israel) offered twenty loaves of bread, which the prophet turned into enough food to feed one hundred people, also hungry after listening to a sermon, eight hundred years before Jesus did it.

In 2008, residents of Marino, Italy, opened their sink faucets and discovered that instead of water their glasses filled with white wine. The miracle was eventually attributed to a plumbing error. During its annual wine festival, the municipality normally filled its fountain with wine, but this time workers had mistakenly opened the wrong valve and channeled the vintage into local homes.

SUN MYUNG MOON

FOUNDER OF THE MOONIES

As a boy in North Korea, Sun Myung Moon had a vision from Jesus. He was told Adam and Eve failed to create the perfect family (*See also*: Cain), nor did Jesus have time to do it himself, since he was crucified so young. Accordingly, it was up to Moon to save the world through marriage. His plan was to focus on making strong families who would produce new generations to populate the world and adhere to his ideas. After World War II, this was not a fashionable approach and so the atheist North Korean communists jailed Moon for more than ten years. It didn't help when he stated he was in a personal, nearly hand-

to-hand battle with Satan. Originally a Presbyterian Evangelist, Moon was excommunicated for his views and started his own sect, the Unification Church, in 1954. Once in America, five years later, his staunch anticommunism and belief in capitalist ventures proved winning, and he was welcomed for a while, until he became suspected of sexual misconduct and fraud. He was eventually convicted of tax evasion. There was also concern regarding Moon's growing financial clout and his ability to amass huge sums of cash from his devoted followers; members pledged to tithe anywhere from 10 to 100 percent of their income—and many of them worked at church-owned enterprises. During the 1970s and '80s his flock grew to two hundred thousand people, through a philosophy he called the Divine Principles, a Taoist interpretation of the Bible. He made news by performing mass weddings, once in 1982, when he married more than two thousand couples at Madison Square Garden in New York, and again in 1988 in Toronto where more than 6,500 brides and grooms were joined by Moon's "Blessing Ceremony"—many of the couples had never met before the wedding event. His followers, who at first referred to themselves as

"Moonies" (they came to think it a slur), not only aim to change the world through marriage, but also believe in Reverend Moon's ultimate political goal—"the natural subjugation of the American government and population." In the 1990s, Moon invested more than $1 billion in the *Washington Times* to further his mission. It wasn't until his eighty-fifth birthday, in 2004, that he admitted outright that he was in fact the new messiah, while Jesus was merely a good man who had left the job half done.

Every people or every organization that goes against the Unification Church will gradually come down or drastically come down and die.
—Sun Myung Moon

Rituals among the Unificationites include waking early on Sunday mornings and bowing at exactly 5 A.M. to an altar that bears the picture of Moon and the first "true family," which consists of Moon's forty children, grandchildren, and other blood relations. Sins are forgiven after fasting. The good go to heaven and the bad to hell, but Moon encourages talking with dead ancestors for extra guidance. The marriage ceremony is the religion's major event and requires five steps for purification, including a forty-day abstinence regime. Moon determines the pairing of partners by looking through a stack of head shots. The brides and grooms are thereafter joined for life, with divorce not permitted. The church claims to have more than one million members worldwide, with fifty thousand in the United States.

Sun Myung Moon divorced his first wife after allegedly getting a university student pregnant. He married his second wife when she was seventeen, and declared she was the "true" mother of the "true family," especially after giving birth to thirteen children. Moon's personal net worth is estimated to be well over $100 million, which he obtained through a network of diverse industries. His church sold more than $200 million in auto parts in 2008 and reportedly has nearly cornered the market for roses sold on Valentine's Day.

MOTHER ANN

SHAKERS FOUNDER

In 1744, eight-year-old Ann Lee was already working full-time, putting in twelve-hour shifts as a velvet cutter at a Manchester, England, mill during the dreariest period of the Industrial Revolution. She was never sent to school, and although she would eventually attract thousands of followers, she could not read or write. When her mother died at a young age, Ann was left as caregiver to a clan of siblings, in addition to her factory work. It's no wonder she forestalled marriage. She claimed she wanted no part of sex, given the misery she saw it cause to women, and she believed it a sinful act.

In 1750 one in four women died of childbirth. Instead of readying a nursery, more women prepared for their own death as the end of term drew near. If they died, a cleansing ritual was performed over the coffin. If the mother survived, she was often thought to be impure for forty days. In Catholicism, the midwife brought the infant to baptism, since the mother wasn't allowed in church. Among Anglicans, a ceremony called "churching" was meant to purify all women, including midwives, present at the birth, and even more so if the mother died.

Nevertheless, Ann's domineering father demanded she marry, and she was dragged to the clerk to betroth Abraham Standley, a blacksmith, like Ann's father: Ann Lee agreed to it and signed the certificate with an X. She would have in rapid succession four miscarriages and give birth to four children, all of whom died before they reached the age of six. At twenty-two Ann had resisted joining her father and husband as most nearly impoverished working-class people did, spending free time in the twenty-four-hour saloons that were open on every corner. Instead she found reprieve and was intrigued by the new charismatic religions preached by enthused ministers searching for converts among the overcrowded urban tenement districts. She was earning wages from a gruesome custodial job at a lunatic asylum when she went to her first Quaker tent meeting. She was enthralled at how the preacher excited the crowd into prayerful frenzy. She learned that wiggling outstretched fingers and gyrating legs, or twitching into a near-convulsive state, was encouraged as a way to release the negative energy of sins, or "stamp out the devil."

When the Quakers stopped the "dance" part of their meeting services in an attempt to appear more respectable, Ann Lee and her mentors, Jane and James Wardley, wanted to keep the quivering revivals going. Lee and the Wardleys formed their own denomination, called the United Society of Believers in Christ's Second Appearing—though from the beginning they were referred to as "the Shaking Quakers." For their persistence at conducting twitching and trembling prayer meetings, and when Ann began speaking in tongues, they were prosecuted, pelted with bottles and stones, and jailed. But there was no stopping Ann once she began having her own visions and claimed frequent contact with divine spirits. None was more startling than when she revealed that she was actually the female Jesus Christ.

In 1774, a vision commanded her to take her own small band of followers, numbering only eight, including her husband, to America, where she was to found the new true Christian church. Once they were settled in upstate New York, her husband deserted her, as she refused to cohabitate with him any longer. Even so, Ann Lee persisted through physical assaults in the United States as well and remained determined to start a utopian community. She traveled through New York and New England on a mission to gather converts. She professed that women were equal to men, and that the true nature of Jesus was both male and female. Eventually, she was accused of breaking apart families by urging women to withhold sex from their husbands. Mother Ann also advocated a frugal, ascetic life summed up in her three C's: celibacy, confession of sin, and communalism. Her insistence on flawlessness in mind and morals made her communes do well financially, and one of their products, Shaker furniture, to this day commands top dollar at auctions.

Do your work as though you had a thousand years to live and as if you were to die tomorrow.

–Mother Ann

Mother Ann never wrote a word of her ideas, though she could recite the Bible from memory. Without signs of debilitating illnesses, she died seemingly and suddenly from exhaustion at age forty-eight, in 1784. Her communes survived and were eventually accepted, such that during the Civil War, President Lincoln allowed the Shakers—due to their pacifist beliefs—to be the nation's first conscientious objectors. Due to stringent celibacy policies, those wishing for children did so through adoption. Others joined the Shakers, it was said, as a way to get a respectable divorce. When laws in the 1970s prevented adoption by religious groups, the religion that had lasted more than two centuries and had in its heyday claimed more than two hundred thousand followers and nineteen communes, teetered near extinction. In 2006 there were only four remaining Shakers, two middle-aged men, and two elderly women, left hoping that the spirit of Mother Ann would intervene and inspire a new generation of converts.

SEX EDUCATION

Celibacy rules at Mother Ann's commune were so strict that men and women were only allowed to meet face-to-face during a weekly "union hour." This was for talking only, with no flirting permitted. Once, two teenage girls were caught watching flies copulate. The girls were stripped and flogged.

MOTHER TERESA

ADMIRER OF SUFFERING

Born in Albania in 1910, Mother Teresa became a Catholic nun at the age of eighteen and went on to start her own order, the Missionaries of Charity, which was noted for its work in the poorest sections of Calcutta, India. She was awarded the Nobel Peace Prize in 1979. When Mother Teresa died in 1997, at age eighty-seven of heart failure and malaria, she was immediately put up for canonization, especially after surveys ranked her as one of the most admired persons of the twentieth century. It usually requires a minimum five-year waiting period before anyone can be considered as a saint, with proof of miracles required. Subsequently, a Bengali woman claimed that a "laser beam" emanated from Mother Teresa's picture and was strong enough to remove a cancerous brain tumor.

However, upon closer scrutiny, it was learned that some of the many millions of dollars in donations collected by Mother Teresa came from known unsavory characters, though she never refused the money or questioned how donors came to their potentially ill-gotten gains. (Charles Keating, of the notorious 1990s savings-and-loan scandal, was a big donor, and was praised by Mother Teresa.) A nun who had worked with her said that the poor coming to her mission clinics were often given a meal instead of medicine, though they had to come on schedule or else would be turned away even for a bowl of rice. Mother Teresa was likewise suspected of finding a bit too much joy in suffering. She did not give out painkillers, and dissuaded the nuns in her order from learning how to alleviate symptoms of certain diseases. Mother Teresa would simply tell the poor and destitute: "You are suffering like Christ on the cross. So Jesus must be kissing you." She comforted most by reminding them that the ultimate reward for enduring such misery awaits in heaven.

In October 2003 she was canonized.

MOZI

PACIFIST; WARLORD

Most associate the concept "Love your neighbor as yourself" primarily with the teaching of Christ. It seems, though, that it was a trademark for another mystic and philosopher nearly five hundred years before Jesus made it his own. Mozi was born in China in 479 B.C. during a period of brutal warfare under changing regimes that made daily life precarious at best. Although Mozi came from a line of potters and weavers, he came to earn a respected government position. However, he ultimately abandoned it to become a wandering philosopher. His message was intended to bring peace and he recommended treating everyone as they wished to be treated. This stood in opposition to loyalty among only family lines, which was the reason for warring and societal disparity. He became a noted peacemaker and was commissioned as an emissary to courts of varying leaders who were preparing for war, whom he implored to apply the concept of "universal love." Mozi believed spirits watched humans closely and judged us at death by our actions. Mozi likewise admonished that none would be allowed passage into heaven if worldly gains were earned maliciously.

Later in life, unlike other mystics, he became such a noted pacifist that he formed communes that trained not only in his philosophical ideas but were hired out as specialists in countersiege techniques. He lived to the age of fifty-nine, dying in 438 B.C., supposedly turned to a puff of smoke during meditation. The Mohits, as his followers were called, were unable to fend off persecution without Mozi, and they were disbanded. Only a few textual fragments containing Mozi's original tenets were preserved, even if some of his ideas of how rulers should govern, for example, filtered into other schools of thought. Some believe that after his followers scattered into secret societies their descendants eventually made contact with Jesus, since a number of the ideas Jesus preached are quite similar to Mozi's. (*See also:* St. Issa)

A PACIFIST WITH AN ARMY?

In his heyday, Mozi gained enough followers to eventually control a vast army. Some reports said he could muster a million men at a moment's notice, in the hope of encouraging the stronger enemy to once again reconsider "universal love." Many times the mere sight of Mozi's million-man march sent the more powerful invaders into retreat. Although Mozi in a sense advocated turning the other cheek, he allowed violence under a number of circumstances: if provoked irrevocably into war, or persecuted by the tyranny of unjust rulers. His teaching deviated at this point from Jesus', since Mozi believed force was the only earthly recourse. His soldiers knew they would be judged well and allowed to enter heaven by dying while defending "good." If the warriors took a large number of the enemy with them on the way, it was even better, and so they fought without fear of death, with a wild-eyed bravery that thoroughly spooked even the most formidable armies. Some Triad gangs in Asia and the United States claim to trace their roots to the descendants of Mozi's "peaceful" warriors.

MUMMIES

ALL DRESSED UP AND NO PLACE TO GO

Imagine if the choice of suit or dress you wear in your coffin was the most important moment in your life. That's how it was for Egyptians—how they looked and were preserved after death was of the utmost significance and dictated whether they had a chance at rebirth in the afterlife. If the dead weren't prepared to look even better than they did in life, through embalming and other practices, which ended with their

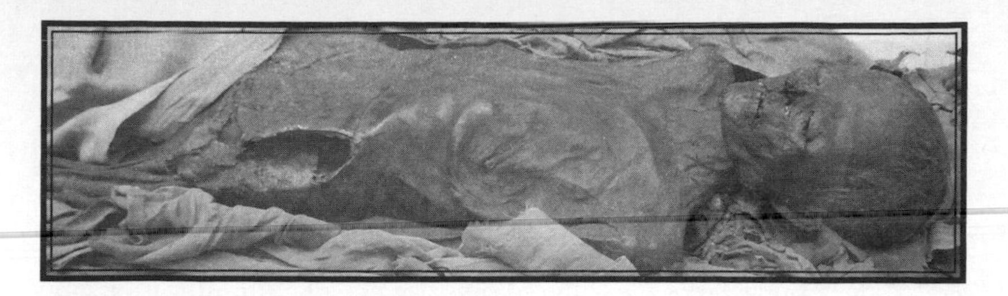

corpse bound with strip after strip of linen until mummified, then they would have zero chance of finding a good station on the other side. Linen, an integral aspect of the burial practices, and essential for mummification, was more valuable than gold. The money people put away for their burial was untouchable, even if they were faced with famine. To the ancient Egyptian it would be better to die of starvation and still have funds to be prepared properly at the moment of death. Nothing less than eternal paradise was at stake.

Egyptians learned of mummification by observing that people left to die in the desert remained remarkably preserved, a phenomenon caused primarily by low humidity and high concentrations of salt and other minerals in the sand. There were thriving businesses to handle the dead, with morgues that employed not only skillful embalmers but also priests who hovered about with incense and read from the sacred texts to help the departed be ready when quizzed in the afterlife. (*See also*: Book of the Dead) Mortuary technicians first used a hook inserted up the nostril to latch on to the brain and pull it out chunk by chunk. Then a four-inch incision was made with a stone knife near the abdomen. The hole was big enough for the practitioner to reach in and gather every organ, except the heart, which was left intact. Sometimes the organs were kept in jars to be placed inside the tomb. (The brain was considered worthless and discarded.) Then, to duplicate what the Sahara sands had done, the body was treated with natron, a natural salt and drying agent. The corpse's face was stuffed with everything from leaves to pottery fragments to help shape the deceased's countenance just so. If a bone protruded, a piece of calf flesh was sewn over it. Fingernails, lips, and cheeks were painted flush, with some mummies even found with hair extensions woven into their locks to add more beauty. Others got nicely painted gems for eyeballs. This process lasted—depending on the wealth of the dead—from a few weeks to more than two months. The mummy of a pharaoh used thousands of yards, more than a football-field-sized tarp, of gauze-thin linen, cut into two-inch strips soaked in a gluelike resin, scented with myrrh and cinnamon. Only 5 percent of the population could afford the deluxe treatment of these priest-embalmers. Budget morgues offered one-size-fits-all mummification, and wrapped the dead in rag fibers, or the equivalent of old newspapers.

MUMMY PILLS

To keep Egyptian tomb robbers away—since everyone knew the mummies were buried with abundant treasures—curses were written on the walls to dissuade looters looking for an easy mark. It was a capital offense to rob a tomb in Egyptian times, but centuries later they were regularly plundered. In the late 1800s, when Europeans became fascinated with all things Egyptian, locals gathered up mummies by the thousands and ground them into powder. They sold these as aphrodisiacs, or as an immortality supplement, though the concoctions were highly toxic to the liver and killed more than five thousand people who had been seeking the fast route to bliss.

THE GREAT REMEDY PROCLAIMED

TO ALL THE WORLD.

THE

ORIENTAL LIFE ELIXIR:

A GREAT

EGYPTIAN REMEDY

FOR ALL DISEASES.

RECIPE founded in the reign of Apophis, Shepherd King of Egypt, B. C. 2000 years!

RESTORED by the exhumation of a basaltic shaft in Lower Egypt, one thousand, eight hundred and twenty-eight years after the Christian Era, and explained by

Doctor Achille Ameilhon, of Paris.

And ye shall cause a shaft to be buried in the dust of Egypt, to HAMON, and his name ye shall write thereon, as the true God of Egypt, and his Elixir shall be written thereon. Thus decree I. APOPHIS, king.

The Mayor and Commonalty of the *City of N. York, of Boston, of Philadelphia, of Baltimore, of New Orleans*, are requested to test the virtues of the ORIENTAL LIFE ELIXIR each in their respective cities; and all public institutions including the hospitals of all the large cities, where infectious diseases, or epidemics, especially Asiatic Cholera, are likely at any time to exist.

The astonishing sensation which the recovery of this famous recipe has produced upon the *Medical Faculty*, both in this country and in Europe will be better under stood when it is stated that

£10.000

sterling was offered for it by a famous physician of London—one of the Medical advisers of the Queen.

The Oriental Life Elixir cleanses and purifies all the fluids of the body; it effects the most radical and healthy changes in the *system*; it expels all disease entirely, and *supports* and nourishes the whole animal frame.

No invalid should fail to use this wonderful Medicament. He who is crippled by

The above cut is a representation, in basalt, *of* the shaft thus "buried in the dust of Egypt" to the memory of HAMON, and which, covered with a hieroglyphic recipe of the Elixir, was discovered and deciphered by Doctor Ameilhon. *See history appended, over leaf.*

n

NATIVE AMERICAN CHURCH

THE PEYOTE RELIGION

Most indigenous Americans practiced religions particular to their tribe, and many still do. However, in 1918 a movement was formed to combine varying native beliefs with Christian dogma. This new religion urged the use of peyote (a cactus fruit that has psychoactive properties) along with a belief in Jesus. Apaches were thought to be among the first of the mainland American tribes to incorporate peyote in rituals. It was also considered sacred for more than ten thousand years among Mexican and Central American natives. Archeological evidence of its esteem was found in a South Texas cave dating to 3000 B.C. Peyote was believed to hold the voice of the creator spirit and to be a benevolent aid bestowed upon mankind to help in matters both spiritual and emotional.

In the 1880s, when the Great Plains nations—the Kiowa, Comanche, and Sioux—were relegated to reservations, a revival in peyote rituals spread. It was suppressed by the Bureau of Indian Affairs, most notably when Lakota mystics of the Sioux Nation were massacred at Wounded Knee for using it during their famous Ghost Dance ceremonies. Starting at first among Oklahoma reservations, peyote became a symbolic link to Native Americans' spiritual heritage. Christian missionaries and Indian Bureau agents tried desperately to outlaw peyote, but when a Winnebago Indian, Albert Hensley, testified before Congress in 1918, he finally got attention when he compared it to a Eucharist: "To us it is a portion of the body of Christ. Christians spoke of a Comforter who was to come. It never came to Indians until it was sent by God in the form of this Holy Medicine." He drafted a four-page document, citing that all the various peyote sects were part of the First Church of Christ, which came to be known as the Native American Church.

The movement still has no exact way to worship, and it is without centralized clergy or headquarters. Although ceremonial details and ritual practices vary among tribes, peyote is usually treated with respect, taken in tea, or chewed as "buttons," preferably in sweat lodges, teepees, or in group settings. The Bible is used as the church's moral authority, in conjunction with other traditional beliefs, though it can turn into a Pentecostal-type of ceremony, where Jesus' name is called out and implored—and when they are on peyote, this can go on for hours at a time. Those under peyote's influence claim to see Christ and hear his voice during its use. There are 250,000 members of the church, the largest being the Navajo Nation. One must have verifiable Native American lineage to join the church and to use peyote without arrest. Abstinence from alcohol is also recommended for membership, as well as advocating self-sufficiency, fidelity, and attending to family responsibility.

Peyote expert Omer C. Stewart noted, "[it] produces a warm and pleasant euphoria, an agreeable point of view, relaxation, colorful visual distortions, and a sense of timelessness that are conducive to all-night ceremony of the Native American Church." And Superintendent Frank A. Thackery, Bureau of Indian Affairs, remarked: "With peyote there is very rarely any violence shown from its use while quite the reverse is the case with alcohol."

NEBUCHADNEZZAR

THE KING WHO THOUGHT HE WAS A COW

For seven years, Nebuchadnezzar II, one of the most famous kings of ancient Babylon, thought he was a cow. Before this eccentric indulgence, he had earned respect and power by conquering a wide swath of rival cities, including Judea. But once in rule, and after Nebuchadnezzar had built an elaborate series of defensive walls around his city, he became dedicated to overseeing massive construction projects. Subsequently, Babylon became the envy of the civilized world, featuring great buildings, aqueducts, tunnels under rivers, and one of the Seven Wonders of the World, the Hanging Gardens of Babylon. Officially, he worshiped Marduk, and constructed a pyramid-temple called Etemenanki to honor the "bull calf of the Sun." He sought out magicians, sorcerers, and prophets from all regions and made a large entourage of these wise men in his court. When plagued by insomnia and a troubling dream, he called the learned men for analysis. However, he refused to tell them the dream's details, and reasoned that if they weren't clairvoyant enough to know what it was about, then their interpretations would be false. The wise men protested, saying no one had ever asked of such a thing, though Nebuchadnezzar was firm

and sentenced all to death if they couldn't guess his dream and divine its meaning. The wise men scoured the countryside to find someone to answer the king's demand, and hopefully for a scapegoat to take the brunt of Nebuchadnezzar's wrath.

The famous prophet Daniel, still a teenager, was then found among Jewish captives. Daniel claimed that if the king could give him a little time, God would tell him the dream. When brought before the court, Daniel's life and all the wise men's fates hung in the balance. Daniel, however, totally captivated Nebuchadnezzar with his eloquent storytelling and told of a dream the king liked, replete with a great statue festooned with a golden head. When Daniel interpreted its meaning, saying that Nebuchadnezzar was the shining headpiece and that his kingdom was "God's ideal government," the king came down from his throne and lay prostrate at Daniel's feet, ultimately making Daniel a prophet, and chief of all the wise men.

In the Hebrew scripture, Nebuchadnezzar is vilified for conquering Jerusalem and the destruction of its temple, but in the Middle East he's still regarded as a hero, even if it seems he was the first to have a mental illness known as boanthropy. During the height of his power Nebuchadnezzar would leave the palace to wander the fields. He was frequently observed walking naked on his hands and knees while grazing on grass and mooing. Boanthropy has since been identified as a mental disorder similar to lycanthropy, which is a condition that makes people believe they are wolfs or other canines. It's a neuroimaging malfunction, often caused by a brain tumor. Daniel later declared Nebuchadnezzar's cow period as God's punishment for the sin of pride. The king's notion that he was a cow apparently subsided, though he did die of a cerebral hemorrhage in 562 B.C., at the age of sixty-eight.

NEO-DRUIDISM

BACK TO NATURE

The Celts and their Druid priests spanned across Western Europe, Britain, and Ireland from eighth century B.C. until Roman times. Druid rituals and beliefs were strictly of an oral tradition. In the late 1700s romantic poet William Blake envisioned Druids as descendants of Adam and Noah, who he thought practiced a nature-worshiping religion. Ever since then, sects, fraternities, and clubs have formed to revitalize what some imagined to be the more harmonious of Celtic practices. (*See also*: Ritual Killings) Neo-Druidism movements generally focus on mankind's link to nature, with spirits inhabiting landscapes, plants, and animals; they honor gods of Celtic mythology. Most ceremonies are held outdoors, usually around a circle of stones, preferably under oak branches, considered a sacred symbol of the Mother Goddess. Participants dress in versions of Iron Age costumes: They wear hand-woven hooded robes, hold wooden staffs, wear flower garlands on their heads, and drink home-brewed mead and ale, especially at the celebration of the spring and autumn equinox. Pilgrimages to Stonehenge are also recommended. In the United States, approximately fifty thousand people practice some form of Druidry. (*See also*: Wicca)

NEW AEON, CHURCH OF THE

2029: YEAR OF THE GREAT CELESTIAL SHIFT

New Aeon groups believe that a major shift in mankind is going to occur relatively soon—in 2029. Many New Aeon followers uphold that master beings or "designers of the divine plan" have set this date for a major realignment, orchestrated for the purpose of accelerating human evolution. (*See also*: Akashic Record) Unfortunately, all persons unable to access a higher level of spiritual knowledge by 2029 will become extinct, since most will be unprepared to meet the drastic changes.

New Aeon leaders have calculated that the meteorite Aphosis will strike the planet on Friday, April 13, 2029. Their interpretation of the data suggests that the size of this object could wipe out all living things within a thousand miles of impact. The population of America, if that were the collision point, would be reduced by at least 75 percent within an hour. Many New Aeon enthusiasts believe an ice age will follow, with only pockets of humanity remaining. Some suggest that those who are spiritually prepared will be cloistered in subterranean communities.

NASA scientists, however, say this rock is only a thousand feet in diameter and merely capable of destroying a city the size of Los Angeles, for example, and is not of the magnitude that facilitated dinosaur extinction. It's also believed that the asteroid will come close, and fly under the satellites in 2029, but might swing back for a closer encounter in 2036. Distrustful of government calculations, New Aeon groups advocate clearing the mind of daily distractions and hurriedly making inner contact with angels who are noted for shielding earthlings during catastrophes. New Aeon leaders have devised a system of placing number values to letters in order to decipher the Bible and an assorted collection of "ancient texts," revealing clues on how to be prepared for the New Aeon, or the inevitable paradigm shift. In 2009, there were more than one million devotees of New Aeon theology. (*See also*: C.U.T.)

SHALL CHAOS TRIUMPH?

In ancient Greece, Plato's definition of *aeon* referred to the eternal rules of true knowledge behind the shadows, or superficial realism we mostly observe. He explained in his famous "Allegory of the Cave" that what we actually observe with our senses are shadow images that deceive us into thinking this is the extent of reality. In addition, Plato supposed the earth was periodically transformed by fire and flood every 10,800 years, with the next one scheduled, more or less, during the current time.

In the early twentieth century, occultist Aleister Crowley also predicted an impending New Aeon change. He reasoned that enlightened humans will survive by materializing needed supplies by the sheer force of their will, while the remaining hordes will devolve into scavengers and a lower form of human being.

AGE OF AQUARIUS

In astrology, time is divided into aeons lasting approximately 2,150 years and coinciding with the twelve signs of the zodiac. It's determined by the movement of star constellations as observed from earth, though few astrologers can agree when a new age begins. Some astrological experts believe we are entering the Age of Aquarius, which is noted for nonconformity, enlightenment, and fostering humanitarian ideals. This would replace the Age of Pisces, an era ruled by the sword, and when religion dominated mankind. However, the darker side of the Aquarian Age foretells of a civilization controlled by a handful of powerful elite, who will use science and technology instead of religion to control the masses.

ISAAC NEWTON

OCCULT PHYSICIST

Remembered as a brilliant physicist and mathematician, Isaac Newton devised "the Laws of Motion" and explained gravity, to name a few of his discoveries. He also had an alchemist's instinct and harbored highly unusual notions about the nature of the divine. In his old age, he admitted that despite his treatises on scientific matters, the daily study of the Bible was his greatest passion. Eventually the accolades he received made him believe he was selected by God to offer a new numerological interpretation of the Bible. Through mathematics he would reveal its hidden meanings. He liked the doomsday prophesies found in scripture, but according to his calculation he was certain the world would end no sooner than 2060. Newton intended to write a comprehensive study on prophets and mystics and their predictions, though he left it incomplete. He began his study of science as a search for the creator, and stated that "the sun, planets, and comets, could only proceed from the counsel and dominion of an intelligent and powerful Being." Math to him was an attempt to understand the mind of God. In public he was

vehemently anti-Catholic and a staunch Protestant, believing in the literal word of the Bible. However, the secret he kept until after his death was his keen interest in occult texts concerning magic and alchemy. He thought there was indeed a secret book of hidden knowledge passed down by angels, and he conducted experiments to find the Elixir of Life. The discovery of his extensive occult writings has made some reconsider whether Newton was the father of the Age of Reason, or actually the last of the great wizards.

NICAEA, COUNCIL OF

STANDARDIZING CHRISTIANITY

The Roman emperor Constantine ruled with a stern disciplinarian's hand and seemed resigned that Christianity had won out to become the most popular religion in the empire. He accepted it, rather than have Rome divided from within. Nevertheless, there were varying points of view concerning exactly who Jesus was, and other contradictory Christian customs. He invited 1,500 known leaders of various Christian sects to a council, all expenses paid. However, only three hundred bishops and church leaders attended the event, held at Nicaea, now Iznik, Turkey, in 325 A.D. The clergy gathered under a great tent to clear up inconsistencies and establish a universal creed, or belief system. These few men decided that God the Father, Jesus, and the Holy Ghost were one and the same, and although individuals, they existed together as a Trinity. Any books, gospels, and texts that diverged from this view were allegedly destroyed. The council also decided to make sure "Christian Passover," the celebration of the Resurrection, would be called Easter, and would never fall on the Jewish holiday. Instead Easter would occur on the first Sunday following the first full moon after the spring equinox. Other rules were also established, namely that no cleric should keep young women in their rectories; that Christians must stop the practice of usury, or lending money with high interest rates, which was a common way for clergy to make extra cash; that regular churchgoers should stand during services, and only sinners should kneel; and that priests should stop performing or encouraging self-castration. Incidentally, Constantine waited for his full conversion to the faith, and refused to be baptized until he was on his deathbed. He was never canonized a saint by the Roman Catholic Church.

The Second Council of Nicaea met in 787 A.D., with 350 attending, and participants tried to clarify issues further. The major accomplishment of these proceedings allowed for widespread veneration of saints. The council proclaimed that pictures of saints, the Virgin, angels, and crucifixes should be placed everywhere—on vestments, in homes, and on roadsides—"for reverent adoration," though not equal in worship afforded the Trinity.

WHY IS CHRISTMAS CELEBRATED ON DECEMBER 25?

Three hundred years after his death, no one remembered what day Jesus was born on. Prior to the Council of Nicaea, churches marked Christ's birthday randomly, on such dates as January 2, March 21, May 20, and November 17. Many cultures had a winter solstice celebration during the last week of December. The Greeks worshiped Dionysos on December 25 or thereabouts. The Romans also had a very festive day on December 25 called Sol Invictus, or Invincible Sun Day. In addition, many in Rome worshiped the Persian god Mirthas, the "god of light," on that day. Mirthas was born on December 25 from a "cosmic egg." Constantine decided that Christians should adopt that date in order to keep continuity with Roman customs. Thereafter, Jesus' birthday, Christmas, was celebrated on December 25.

ST. NICK

MODEL FOR SANTA CLAUS

The original Nicholas was not a toymaker, though he did bear gifts. Once, to save three young girls from being sold into prostitution by their desperately poor father, Nicholas dropped three bags of coins down their chimney to supply the girls a dowry and allow for a proper marriage. Nicholas was born in 270 A.D. in the Greek-ruled Mediterranean city of Patara, in Turkey, during a period when Christians were still regularly persecuted. While a young boy, both his parents died suddenly from an epidemic. Nicholas was taken in by an uncle, a Christian bishop, who taught Nicholas theology. From the beginning he was a pious student and noted for charitable acts. He gave away the considerable inheritance he received from his parents' fishing fleet business in dribs and drabs, most often anonymously.

By all accounts Nicholas truly did have a jolly disposition, and he eventually became a popular bishop. His reputation as lover of children stemmed from a miracle. During one famine, he had heard of a butcher who had lured three boys into his shop, murdered them, and pickled them in a barrel. Nicholas was so outraged that he stormed the butcher's storehouse, pulled out the dead remains, and resurrected the boys back to life. St. Nicholas is the patron saint of sailors, since he was noted for helping travelers visiting his port city, and is the patron saint for pawnbrokers. The three gold balls that hang from many pawnbroker stores symbolize the three bags of gold Nicholas had given the girls. He's also the official patron saint of prostitutes. Nicholas, unlike other saints from this era, who often died horribly, lived to the ripe age of seventy-three, bearing a long white beard and a paunch. He died on December 6, 343, and afterward, in a tradition to honor his lifetime of anonymous gift-giving, people began leaving presents to those in need, during the month of his feast day.

UNOFFICIAL PATRON SAINT OF PROSTITUTES

Over the centuries, the thirteenth-century martyr Santa Librada has been the one most prayed to by prostitutes, even if the saint's life was bizarre. Whether she was a medieval call girl is not mentioned, but in any case, Librada was brought before the authorities for getting pregnant while her husband was away. Afterward, she bore nine children at the same time, from what she claimed was a mysterious fertilization. She was promptly nailed to a cross at age twenty. Other reformed prostitutes pray to St. Mary of Egypt, a runaway at twelve who worked in the trade for seventeen years. When she went to Jerusalem to look for customers during the height of pilgrimage season, she had an awakening, particularly after an angel with a sword chased her away from a church. She became a hermit in the desert for the next three decades. If busted through entrapment, many prostitutes turn to neither saint, and instead implore St. Gerard, patron saint of those falsely accused of sexual crimes.

NOSTRADAMUS

PREDICTING THE FUTURE, VAGUELY

Michel de Notredame was born in 1503 to wealthy Jewish parents who had converted to Christianity to avoid persecution. He studied astronomy and medicine and practiced as a family physician for thirteen years in the south of France. When his wife and two children died in the bubonic plague of 1538, many came to doubt his skills as a doctor. Medicine then was a crude science, including sniffing stool samples and prescribing everything from wild pig penis to treat pleurisy, to a paste of pigeon dropping to soothe the eyes. In general, illness was viewed as a punishment from God. For Michel de Notredame, the death of his family was attributed to some moral lapse, such that his wife's relations requested the return of a substantial dowry. He was also ordered to the office of the Inquisition at Toulouse, concerning a seemingly sacrilegious comment Michel had made about a statue of the Virgin Mary. Not wishing to take chances during the feverish height of the church's hunt for heretics, he fled, and wandered from town to town for more than six years, remaining as anonymous as possible while practicing medicine wherever he could.

LES
PROPHETIES
DE M. MICHEL
NOSTRADAMVS.

***

Centuries VIII. IX. X.

Qui n'ont encores iamais esté imprimees.

During this period of grief and displacement, Notredame began to study astrology and cultivated the intuitive clairvoyance he believed he always had. He married again, to another affluent woman, and made a living by supplying herbal and chemical concoctions, which he sold as cosmetics to his wife's rich friends. However, once he settled down he was eyed suspiciously, especially as he began to focus more intently on occult, alchemy, and magic. Some report he returned to Jewish mysticism found in Kabbalah writings. Eventually he withdrew from further public contact, and considered much of society to be "barbarians." He began to write and publish almanacs and used astrological data to predict events for the coming year. He built a substantial reputation for his popular annual tomes.

In 1555, after five almanacs, fifty-one-year-old Michel de Notredame changed his name to the Latinized Nostradamus. His next work, *Centuries*, started a ten-book series containing predictions through the year 3797. He claimed the future was revealed to him by a "Divine Essence." He used scrying practices, that is, going into a trance while staring into a bowl of water. He also used "herbs" to achieve an hallucinogenic state. The archaic French-Latin language of his *Centuries* and its poetic style, filled with riddles and anagrams, was his way to avoid religious persecution. The populace associated futuristic predictions as no more than curses and devil worship, but his book became a rage with the upper class and was especially heralded by French queen Catherine de Medici and her court. He established fame during his lifetime when King Henry II died in a freak jousting accident; a lance pierced the king's visor, stabbing his eye. He died ten days later from brain injuries. What Nostradamus had written of the event three years earlier became proof of his astrological abilities: "The young lion will overcome the older one / On the field of combat in a single battle; He will pierce his eyes through a golden cage / Two wounds made one, then he dies a cruel death." Nostradamus died one year sooner than he predicted, in 1566, at age sixty-two, from gout. His verses were later interpreted to foretell of the French Revolution, Hitler, and the Kennedy brothers' assassinations. The following lines have been read as a prediction of the destruction of the World Trade Center in 2001: "In the year of the new century and nine months, from the sky will come a great king of terror." Though most of the images of the future are too vague to be of any practical use, he remains a champion of those seeking divine wisdom by deciphering astrological signs.

JOHN HUMPHREY NOYES

UTOPIA FOUNDER; RAPIST

In the 1960s, the counterculture generation thought they invented the idea of "free love." It was actually a term coined nearly 120 years before, by an upper-class New Englander named John Humphrey Noyes. He got in trouble by his mid-twenties; he was expelled from the Yale School of Divinity and had his preacher license yanked after he claimed that all religions were flawed. He believed that if God did indeed give mankind free will, then everything a person chose to do was, in his reasoning, divine. Noyes's path to God was achieved by striving toward "perfectionism," both in body and mind. Noyes wanted to emulate the example of first-century Christians, who in the face of persecution often survived in self-sustaining, tight-knit units in which everything was shared. Thus inspired, Noyes decided to form a similar utopian society. What really got Noyes in trouble was his broad definition of Christian love, which to him naturally included sex. After his arrest for adultery in 1847, at age thirty-six, Noyes and his key followers fled from their Vermont town to Oneida, New York, where they set up a commune for like-minded "Perfectionists," or Noyesians.

Financially, his experimental community was quite prosperous. Members worked diligently at either agriculture or manufacturing, including leather handbags, flatware, and animal traps. All worked without salary. He built a giant dormitory called "Mansion House," where more than fifty followers lived in "complex marriage" arrangements. Noyes considered sex with varying partners an antidote to selfishness. Sharing your wife or husband was likewise one of the noblest demonstrations of Christian love. Accordingly, he encouraged followers to engage in sex as often as possible, though he demanded men pull out upon ejaculation. (*See also*: Onan) There is no surprise that satellite branches of his commune were established from Ontario to Brooklyn. This withdrawal method of birth control was part of Noyes's bigger plan, an attempt at engineering a more perfect human being. He selected only those considered the most flawless (and good-looking) to produce children. Noyes subsequently fathered nine. The offspring from this eugenic experiment, which Noyes thought "an improvement of the race," were called "stirps" and were raised communally, with special attention paid to intensive early childhood education.

In 1879, a warrant was issued for Noyes's arrest for statutory rape. He fled the compound under the cloak of darkness to Canada, never to return to the States. He soon sent the various communes instructions to abandon "complex marriage" and accept traditional nuptial arrangements. He eventually gave ownership to the wide array of business enterprises to all commune members in the form of stocks. In the early 1900s, one of Noyes's sons turned the original communal enterprises into Oneida Unlimited, which became the largest manufacturer of high-quality forks, knives, and spoons for the rest of the century.

O

OBEAH

CARIBBEAN BLACK MAGIC

In many English-speaking Caribbean islands, including Jamaica and the Bahamas, Obeah is considered a potent folk religion that relies heavily on sorcery and magic. The world, according to Obeah followers, was created by a supreme god who planned in advance every facet of life, which will unfold in perfect equilibrium as designed. Even if some misfortune falls, it will later have a balancing effect, manifesting somewhere else, or in another time. The sect grew from an African religion that worshiped Orisha, the god of death, who is often depicted as a skeleton, similar to the Grim Reaper. Those who practiced worship tended to darker magic, and could call on Orisha to kill. The Obeah seers belonged to a secret shaman class. Selected for their special visionary powers, these men and women were then apprenticed to Obi-men, or Bone-men, to learn the spells and incantations needed to pierce the universal equilibrium of life and death, or alter events to their choosing. In modern times the feared power of an Obeah was used in the storyline in the *Pirates of the Caribbean* movie series, with the character Tia Dalma playing an Obeah witch doctor who has powers to bring dead buccaneers back to life. (*See also*: Witch Trials)

O-MING

SEXY BLISS

An offshoot of New Age thought is on the rise, in what's called the "slow-sex-movement." A San Francisco commune, the OneTaste Urban Retreat Center, has more than three dozen male and female followers practicing "O-Ming," a shortened buzzword for orgasmic meditation. Here, the seekers meet at 7 A.M. every morning to meditate on the meaning of the female orgasm. Women lie on yoga mats, nude from the waist down, while men wear white and remain fully clothed. The men are not permitted to look at the woman's face or touch any part of her body except the clitoris. Afterward, all sit in a circle on the floor and share their experience. One longtime resident stated, "We all knew it was a hard-core place, and we came to play hard." Another found personal transformation there after trying to live on a kibbutz in Israel to quell alcoholic tendencies. With O-Ming she found "access to the woman I am and the woman I want to be." (*See also*: Tantra)

ONAN

BIBLICAL MASTURBATOR

The first medical pamphlet about masturbation, published anonymously in London in 1715, was *Onania, or the heinous sin of self-pollution, and all its frightful consequences in both sexes considered.* Thereafter, masturbation became known as onanism, in honor of biblical Onan, noted "for spilling his seed" and subsequently slain by God for this capital offense. Onan was a Hebrew, ordered by his father to take his dead brother's wife to bed in order to produce an heir. He refused by following the withdrawal method of birth control. God explicitly killed Onan for this wasteful deed, one of the few he personally killed rather than dispatching assassin angels or natural disasters to do the job. The author of *Onania* cited biblical reasons why masturbation was evil, and thought it should be punished as sodomy was—with death. He also claimed it caused blindness, hairy palms, deformity in the hands and face, insanity, and ultimately premature fatality. The text also included end-page advertisements for an elixir and tincture the author sold on the side to help avoid self-fornication and gain strength to fight temptation to "spill the seed."

COITUS INTERRUPTUS

Many Catholics believe Onan's withdrawal method is the only birth control permitted by the church. They are mistaken, for the Vatican believes sex should be only for procreation and that masturbation is "an intrinsically and gravely disordered action." Islam also considers it *haram*, meaning forbidden. In Buddhism, masturbation is not permitted for monks, though it is allowed among followers, and monitoring the sudden urge is suggested as the path toward eliminating all desire. Wiccans, Neo-pagans, and earth-based religions have no problem with it, though they encourage doing it outdoors.

OOMOTO

APOCALYPTIC JAPANESE RELIGION

Deguchi Nao had, by any standards, a harsh and miserable life. Born in Fukuchiyama, Japan, in 1837, she had pawned nearly everything she

owned in order to feed her children and came to earn a living as a rag picker to care for her invalid husband. In 1892, during a New Year's celebration, when all seemed particularly hopeless, the fifty-six-year-old Nao believed she had become possessed. The spirit, or *kami*, that inhabited her body, identified as Konjin, was a deity who had been sleeping on an island for three thousand years. Emerging from a trance, Nao began to talk in a different voice and speak of things that were seemingly impossible for an uneducated woman to know. Within a few years, Nao had attracted a huge following, teaching that the earth and all of physical reality was actually a part of the god's body and healed by fostering peace. This, she advocated, was achieved immediately by acting ethically, including apologizing quickly, acting cordially, mutually helping each other, and frequent prayer. Nao wrote with an ink-dipped brush that moved spontaneously in her hand, and she eventually produced more than two hundred thousand pages. During these automatic writing sessions, the previously illiterate Nao created the founding documents of Oomoto, a religion whose name means "Great Source." Her teachings, however, and her sudden rise in popularity centered on Konjin's apocalyptic threat and urgency. She claimed that those who did not follow would be swept away in the rapidly approaching "Great Cleansing of the Three Thousand Worlds." Nao died wealthy in 1918 at age eighty-three, having spent the last part of her life treated as a living deity. It's estimated that about forty-five thousand still adhere to Oomoto practices, and although the religion is polytheistic, they concede that all the *kami* are part of the one "Great Source."

DO YOU FENG SHUI?

The Konjin spirit features prominently in the Japanese Shinto religion. The entity is in charge of many metal objects and guides the directional points of a compass. Until Nao gave him a new image, Konjin was previously considered a fickle evil spirit who was responsible for curses. Temples and houses were built where Konjin was thought to be unable to inflict too much damage. If a building collapsed during construction, or if a bridge fell—it was thought to be Konjin's doing.

Feng shui is of Chinese origin, and likewise advises on the proper position of structures and the arrangement of furniture in the home. A special compass is used to find the direction where the best positive energy flows. The Boxer Rebellion in China at the turn of the twentieth century was due in part to resentment of the fact that Westerners had crisscrossed the country with railroad tracks not adhering to feng shui. New Age feng shui decorators stress the importance of the smallest details to unblock this ever-present though unseen energy. For example, your bed should be arranged so your head faces south, and never let your feet point to a door. (*See also:* Qi)

ORENDA

IROQUOIS DREAM SPIRIT

The Iroquois were once the most powerful Native American tribe on the East Coast, centered originally near the Great Lakes. To the Iroquois, dreams held the true voice of the soul and were clues given directly from the Great Spirit, the creator, called Orenda. Dream interpretation had a tremendous impact on nearly all aspects of daily life. It was thought that the most reliable dreams occurred between midnight and 2 A.M. Trying to understand dreams was difficult, and they were often shared during a social activity known as "dream sharing." Ignoring a dream was foolish, thus communal dreaming ceremonies were encouraged and were an important ritual part of the religion. The most sacred was the Midwinter Dream Festival, which took place around New Year's Eve, when wooden masks were used to help attract dream spirits. Fasting, sometimes for more than a month, was a way to induce more divulging dreams. Most males entering puberty were often required to go off by themselves until experiencing a dream spoken to the "inner ear." Early missionaries dismissed their reverence of dream sharing and forbade mask dancing. In his masterwork *The Interpretation of Dreams*, Sigmund Freud, however, championed what the Iroquois had been doing for centuries, namely that the unconscious mind's repressed wishes are frequently revealed in dreams.

ORIGINAL SIN

BORN UNDER A BAD SIGN

When the first parents of humanity, Adam and Eve, disobeyed God, their "original sin" was then passed on to all their ancestors. However, few Christians believed this to be true until the third century, when Thomas Aquinas first introduced the theory. Thereafter, church scholars said the sin passed through lust and sexual intercourse, and for this reason, every person born is sinful by nature. By the twelfth century it was believed the only way to be absolved of it was through baptism. Scripture, namely the works of the apostle Paul, summed up original sin this way: "Sin entered the world through one man, and through sin—death, and thus death has spread through the whole human race because everyone has sinned." The forgiveness or nullification of this original sin is of great theological importance to many Christian religions, including Anglican, Roman Catholic, Lutheran, Methodist, Presbyterian, and Evangelical churches; all believe it must be absolved to enter heaven. Islam does not believe in original sin, and says the opposite, that man is born pure. Hebrews, and authors of the scriptures that described Adam's expulsion from paradise, do not believe in original sin, either. In Judaism, what Adam passed was not an original sin, but aging and death. (*See also*: Limbo)

OUIJA BOARD

DON'T PLAY WITH SPIRITS

A toy company trademarks the name Ouija for a "glow in the dark" board game recommended for ages ten or older. However, Chinese fortune-tellers invented the concept of the Ouija, or talking board, in the twelfth century B.C. Since antiquity, these flat boards inscribed with symbolically decorated markings, numbers, and letters have been considered a popular method of communicating with the dead. Fin-

gertips are placed on a pointer, or planchette, that glides on its own power to letters that spell messages from spirits. Spiritualists claim the dead inhabit their eyes and hands and guide the pointer to the words the departed wish to share. Scientists say the movement of the pointer could possibly be chosen by the subconscious, and that an "ideomotor effect" takes over, in the same way tears are "reflexed" into production by a sad thought. Others believe talking boards are a tool of evil spirits. Many religious officials and even respected demonologists assert that Ouija boards—in the wrong hands—are as perilous as street drugs. For others, the talking board is compared to a phone ringing in a public booth on the other side—such that any spirit floating by can pick it up, and converse with the living.

IF YOU DARE

Ouija toys carry a disclaimer that it should be used only for entertainment. But experienced users know there are unwritten warnings and rules. One is strongly advised never to use the board while alone. Do not let the spirits count backward, or go through the alphabet from Z to A, as spirits use this technique to escape. This may also happen if the pointer drops to the floor during a session. If you find the indicator quickly gliding toward the four corners, or tracing a figure eight, know you have an evil entity about to take possession of the board. Addiction to the board can also develop. For instance, a spirit might provide helpful answers to minor questions, making you come back for consultation, only to lead you astray by offering ruinous counsel. Handcrafted talking boards called "witchboards" are very dangerous; they are made from wood panels taken from a coffin, preferably used. They are most effective when consulted in a graveyard or where something bad has happened, and specifically are used to call forth demons. All talking boards should be discarded by breaking them into seven pieces, dousing them with holy water, and burying them. Burning is also a method of disposal. However, you must put your fingers in your ears as the flames destroy it, for if you hear spirits screaming you will be forever haunted, beyond the reach of exorcism.

In Ireland, as recently as 2004, a Ouija board figured in the strange doings of a ten-year-old boy, referred to in the press by a pseudonym, Gary Lyttle. He became possessed after finding a discarded talking board in a dump. As soon as he tugged the board loose from the heap and wiped off the debris, he was slammed to the ground on his butt, with the board settling firmly into position on his lap. The next moment he saw a pointer appear above and hover, before it settled into his trembling hand. Once it was in his locked grip, the ground trembled and split, until he found himself perched at the edge of a chasm, peering downward into hell. The boy was subsequently put under psychological examination, and during sessions he described how great-winged creatures, thick as a swarm of bats, rose from the foot of a throne. He said the devil he witnessed was ten times the size of a normal man. (*See also*: St. Teresa of Ávila) Strange outbursts, seizures, and trances followed. While possessed by a demonic entity identified as Tyrannus, Gary furiously filled composition books with odd symbols and archaic words that he inscribed with the calligraphic skill of an artist. When medical tests proved there were no physiological abnormalities, the boy's mother sought out the famous Protestant exorcist Canon William Lehan.

Reverend Lehan knew upon first glance that the boy had an "ancient and malevolent look to him," and that his eyes gazed with "oldness beyond his years." The priest successfully exorcised the spirit, though he warned it would return if the young man didn't go with another clergy to properly bless and burn the Ouija board. Feigning forgetfulness, the fifth-grader ultimately refused to reveal its hiding spot. As of 2009, Lyttle, now a teenager, was still reportedly contacting "evil spirits" via the board he retrieved from a dump. His grandmother, her house, and other places where this boy visited have since required exorcism rites as well.

In literature, poets Sylvia Plath, Ted Hughes (Plath's husband), and James Merrill claimed to have been inspired during Ouija board sessions. The writer John G. Fuller cited it as a research source in his book *The Ghost Flight of 401*. Through the Ouija board Fuller contacted the deceased flight engineer on the plane that crashed into the Miami Everglades in 1972—with 101 fatalities—and discovered facts that trained investigators said were impossible to know. In cinema, *The Exorcist* portrayed the trials of a demon-possessed child, and was based on the incident of an evil force occupying the body of a thirteen-year-old Maryland boy. He had been distraught when a favorite aunt died and decided to consult the Ouija to speak to her once again. Soon after, all hell, literally, broke loose. The Professional Board of American Psychics currently lists hundreds of members specializing in flat-board readings, and it's estimated that in the United States one hundred thousand people still use Ouija boards for one reason or another.

THE QUEEN OF OUIJA

Pearl Curran, a high school dropout, admitted she was never fond of books. In 1912, the thirty-year-old housewife from St. Louis decided to use a neighbor's Ouija board for fun. Then it was typical to place your fingers on an upside-down shot glass as the moving pointer. The woman caused a sensation after communicating with a spirit identified via letters spelled rapidly on the board as "Patience Worth." This spirit was a good one, and by chance, had considerable literary talent. Via automatic writing, Mrs. Curran produced over a million legible words, seemingly impossible for someone with her academic background to compose. Many of her books became bestsellers. Curran called her Ouija connection "one of the most beautiful relationships that can be the privilege of a human being to experience." Until 1937 she continued having the ear of Patience Worth, a spirit identified as last living in the Victorian era, and quite fond of literature. At one final session Patience said, "You'll have to carry on the best you can." From that day the Ouija board went silent in Curran's hands.

She died a month later from a sudden case of pneumonia at age fifty-four. It's uncertain if she asked the only question forbidden of board users—the date of your own death.

THE OXFORD PLAN

HOW TO ACHIEVE VICTORY OVER SIN (AND COMMUNISTS)

Frank Buchman, a onetime Lutheran minister, came to the notion that for confessions to really work they had to be done communally. In 1909, while sitting in a small chapel, he had a religious experience that he described as "a vibrant feeling, as though a strong current of life had suddenly been poured into me." It motivated Buchman to start a new spiritual movement for living a prosperous and moral life. However, it took more than ten years before he perfected an original recruitment technique, which he billed as "House Parties."

> *The secret is God control. The only sane people in an insane world are those controlled by God. God-controlled personalities make God-controlled nationalities. This is the aim of the Oxford Group.*
>
> —Frank Buchman

Buchman presented an aura of intellectual acumen and an understated zeal; he was described as acting more like a CEO than a preacher. He initially attracted the upper classes, especially after he actively sought celebrities to join his cause. He held the "House Parties" at prestigious locations, such as Princeton University, where he advocated "sharing" as the key to "victory over sin" and living by the five C's: "confidence, confession, conviction, conversion, and continuance." At some events participants formed a circle by holding hands while a new member publicly confessed his sins. The new recruit was monitored by a handful of assigned "sponsors" until he "surrendered his life over to the care of God." After promptly making restitution for wrongs done to others, the initiate was sent out to convert more to the Oxford Plan. Each day one was advised to spend "quiet time" to ask God what the future had in store, and then check with the group to see if the message received was accurate. Buchman said his was a Christian organization and not a religion, with no hierarchy, places of worship, or salaries for its workers, and that the group supported itself from donations of its members. By the mid-1930s Buchman was considered an internationally recognized moral authority, with more than fifteen thousand followers attending each of the group's events. At the height of the Great Depression, he had a skillful knack for thoughtful phrases, such as "Spiritual recovery must precede economic recovery," and, "Interesting sinners make compelling saints." He built lavish headquarters and traveled exclusively, first class. When criticized for his extravagant and un-Christ-like example, Buchman retorted, "Isn't God a millionaire?"

However, his fame took a nosedive when he claimed he could prevent the outbreak of World War II if only he could get the Nazis to follow his Christian ideals. He was quoted as saying, "Thank Heaven for a man like Adolf Hitler, who built a front line of defense against the anti-Christ of Communism." In addition, he received backlash for condemning sex. (He eventually renamed his group the Moral Re-Armament.) Although he suffered a stroke in 1942, Buchman continued urging what he called a "Christian revolution for remaking the world," until his death in 1961, at age eighty-three. During his later years, especially during the 1950s, Buchman focused on fighting communism and labor unions, on the grounds that they were not in line with Christian ideals. In recent times the group went on a campaign to quell homosexuality with its "It Can Be Cured" platform. In 2001 the group changed its name again and became known as Initiatives for Change. In 2008 it attracted only three hundred people to its annual event.

ALCOHOLICS ANONYMOUS

Ebby Thacher, a friend and former drinking buddy of AA co-founder Bill Wilson, had "found religion" by attending Oxford Group meetings. Attempt-

ing to spread the message, Thacher introduced Wilson to the "conversion cure." Wilson went to a number of Oxford meetings, but he had his spiritual awakening in a detox ward instead. While under the care of alcoholism researcher Dr. Charles B. Towns, Wilson saw "the room lit up with a great white light," and ultimately left the hospital never to drink again. Wilson had been given Dr. Towns's "Belladonna Cure," which contained ingredients specifically designed to cause hallucinations. The practice was later condemned, and Towns was accused of medical quackery. When Wilson wrote the Big Book, his outline for the "Twelve-Step" recovery program, he acknowledged that "ideas of self-examination, acknowledgment of character defects, restitution for harm done, and working with others [came] straight from the Oxford Group and from nowhere else." Wilson never gained the lavish lifestyle Buchman achieved, though he became an icon for recovery. The AA program Wilson developed required one to find a Higher Power and turn one's will and life over to its care. Meetings usually end with members forming a circle and reciting the Lord's Prayer, or the Serenity Prayer. Ebby Thacher, incidentally, suffered repeated bouts of binge drinking after introducing Wilson to the Oxford cure, and died of emphysema in 1965. Wilson succumbed to the same disease in 1971. He requested to have one final drink on his deathbed, but the AA members in attendance did not allow Wilson to indulge in his last wish. (*See also:* Church of Synanon).

P

ST. PAUL

THE APOSTLE WHO HAD PERSECUTED CHRISTIANS

Paul, then known as Saul of Tarsus, spent the first few years after Jesus of Nazareth's crucifixion sniffing out new Christians for persecution and beheading. While on horseback to Damascus in 35 A.D., his stallion reared at an unusually brilliant ball of light in his path. The pulsating beam revealed an image of Jesus, whom Paul had never met. Paul instantly converted, transforming from persecutor to ardent promoter of the Christian cause. He eventually spread the word abroad and was a hands-on church builder throughout Asia Minor. When returning for a quiet and nostalgic visit to Jerusalem nearly thirty years later, Paul was arrested, and about to be stoned. But since he had papers proving his Roman citizenship, he demanded an audience with the emperor. He was extradited to Rome, shipwrecked once along the way, but Paul was arrested again in Malta and sent to Rome. He lingered in prison for two years until finally beheaded by Nero in 67 A.D.

Paul's friend St. Peter was also nabbed by the notorious Roman emperor Nero, and died in 64 A.D. He was crucified, though requested to have his head pointed downward, which was obliged and thought very bizarre, and entertaining to the Romans. Peter supposedly didn't believe his death should be of the same status as Jesus' right-side-up crucifixion.

WILLIAM PELLEY

"STAR-SPANGLED FASCIST"

Son of a New England Methodist minister, William D. Pelley earned a reputation as a foreign correspondent while covering the Bolshevik Revolution. He eventually found success buying and selling newspapers, writing fiction and Hollywood screenplays. During his stint in Hollywood, Pelley owned restaurants, ad agencies, and real estate. However, by 1927 he admitted to finding no "zest" in life, and he claimed to be

"spiritually tired" of what he described as the "unceasing struggle for money or acclaim." Pelley had attended a few séances, though he admitted he had found spiritualism out of sync with his practical-minded

New England and Christian upbringing. Nevertheless, while writing a book about racial history at his secluded bungalow in the foothills of the Sierra Madre mountains in Pasadena, California, Pelley had an out-of-body experience. He explained in the booklet *My Seven Minutes in Eternity* how he thought he was having a heart attack and screamed out "I am dying!" Instead he fell down a big blue hole and found himself lying naked on a marble slab. Pelley discovered he wasn't dead, but rather taken to the spirit realm, where he was reunited with deceased people he had known. They explained that he had once lived in this mystical dominion, but that he had elected to be reborn to become his current self on earth.

Most interestingly, the spirits explained the issue he had found confusing while writing his latest book: why there are people of different skin colors and races. His spiritual guides told him "each race is an earthly classroom to which people go to get certain lessons in specific things." Subsequently, he was instructed to spread what he had learned. In the early 1930s he began an exhaustive lecture tour and formed a Christian, right-wing sect, which he dubbed the Silver Legion of America. His celestial handlers had informed him that communists and Jews were bad, and that blacks were inferior to whites. Pelley claimed these were the "truths" he was compelled to share. By 1934 he had fifteen thousand supporters; and following the example of Hitler, who Pelley thought was also inspired by spirits, he dressed his group in a version of a Nazi uniform that became known as the Silver Shirts. Even though ridiculed as a "Star-Spangled Fascist," Pelley commanded tremendous attention during the 1930s, especially after he condemned President Roosevelt and his New Deal policies. He actually ran for president as a Christian Party nominee in 1936.

WHAT WAS HITLER'S RELIGION?

Hitler was raised, or at least indoctrinated as a Catholic, but when he came to power he banned crucifixes in schools and Catholic youth leagues. Instead, he favored a neo-pagan type of wedding ceremony for SS officers, though he likewise condemned occult groups and secret societies, other than those loyal to the Nazi Party. The swastika, he believed, was the symbol for immortality, and he knew it was origi-

nally a holy Hindu icon thought to be linked to the first Aryan race. When speaking before a Christian group on Christmas Day in 1926, he stated: "Christ was the greatest early fighter in the battle against the world enemy, the Jews. The work that Christ started but could not finish, I—Adolf Hitler—will conclude." In his manifesto, *Mein Kampf*, Hitler elaborated on how he believed himself divinely called: "I believe today that my conduct is in accordance with the will of the Almighty Creator."

In 1939, the FBI put out a wanted poster for Pelley, citing "un-American activities." "Dark gray eyes, very penetrating; good talker; interested in psychic research," read the FBI profile. By 1941, when the United States entered World War II, his group immediately disbanded, and he was sent to prison the following year for treason and sedition. He was paroled in 1950 under the condition that he would never again use his psychic insights in the political arena. Pelley moved to Asheville, North Carolina, and continued to write on metaphysics until his death in 1962. Pelley often compared himself to Jesus, as he felt he had been likewise persecuted. He predicted that the religious right wing would eventually gain control of America, since it was, he said, borrowing Hitler's line, "the will of the Almighty Creator."

PENTECOSTALISM

SPEAKING IN TONGUES

In 1906, in a run-down warehouse district of Los Angeles, there were rumors of strange religious happenings. African-American preacher William J. Seymour was leading a ragtag congregation meeting in a former livery stable and Methodist mission at 312 Azusa Street. An unassuming, quiet man, he became even more incongruous when transformed overnight into an electrifying speaker who encouraged everyone to become filled with the power of the Holy Spirit. In this makeshift church, the one-eyed Seymour sat on a shoe box, and put his head inside another box while he prayed; suddenly, he would leap up inspired, energized, and seemingly possessed. He touched people and they fell away, while others trembled and began calling out in religious-sounding gibberish, some remaining in a trancelike state, babbling for hours.

The day after the *Los Angeles Times* ran an article about Seymour, "Weird Tongues of Babel," an earthquake virtually destroyed San Francisco. Shortly after, Seymour found 1,500 or more people of all races cramped into his makeshift mission church, looking for spiritual revival. Some say that when transfixed by a heightened state, the collective unconscious remembers foreign dialects, though most who speak in tongues during such incidents do so in languages no one can understand. Seymour's movement flourished and more branches of Pentecostals, Evangelist, and charismatic Christian groups proliferated during the twentieth century. Today, religions that focus on revival ceremonies, stressing a personal infusion of a "holy spirit," have more than 100 million parishioners worldwide.

PHALLIC CULTS

ORGAN WORSHIP

Many consider the phallus to have been the first of all religious symbols. Even the most primitive societies realized that a man's organ seemed to have a mind of its own—and that it clearly played a major part in the mystery of reproduction. From phallic objects found in Amazon rain forests, to a dildo discovered in a German cave (known as the Hohle phallus, apparently a sacred sex aid to Ice Age humans twenty-eight thousand years ago), phallic worship of some type was universal in all of the earliest religions. Hindus used worship of the snake as a phallic symbol; it was capable of bringing life as well as destruction. The Egyptians considered Osiris, the god primarily associated with regeneration and rebirth, in phallic terms; thus the moisture in the ground, which helped crops grow, was thought to be his semen. The goat also became a symbol of the male creator among Egyptians, and it was considered quite normal during ceremonies for women to copulate publicly with animals to increase their own fertility. The Greeks created the half goat, half man, called Pan, depicted with an oversize male organ, as a symbol of reproductive powers. The temple honoring the Greek goddess Aphrodite had a 180-foot phallic statue at its gates and was climbed annually by a priest who remained at its pinnacle for seven days in prayer. In Rome, Venus took the place of Aphrodite, and her celebrations during spring included a huge phallic sculpture paraded in a cart that was touched reverently to gain luck and fertility. The Romans also wore penis-shaped jewelry to ward off evil eyes that might wish them impotent or infertile. The most intense phallic worship among Roman cults was the devotees of Cybele and Attis. Cybele was a mother goddess, and Attis her chariot driver, who was rendered mad by his unrequited love for her; Attis subsequently cut off his penis. During the Cybele and Attis sect's main holiday, the Day of Blood, frenzied crowds and wild celebrations took place; initiates into the religion's priesthood were required to castrate themselves during the festival. When Christianity rose to power, it destroyed most ancient phallic statues or iconic symbols. It erected crosses and church steeples instead.

CLEAN MONDAY

Every year on the first Monday before the beginning of Lent, the Greek town of Tyrnavos still openly celebrates the phallus. In its version of the Mardi Gras, more than twenty thousand tourists gather to celebrate the male organ, eating phallic-shaped pastries and drinking through penis-like straws from souvenir phallic mugs. Many carry phallic amulets that are urged by all to kiss for good luck.

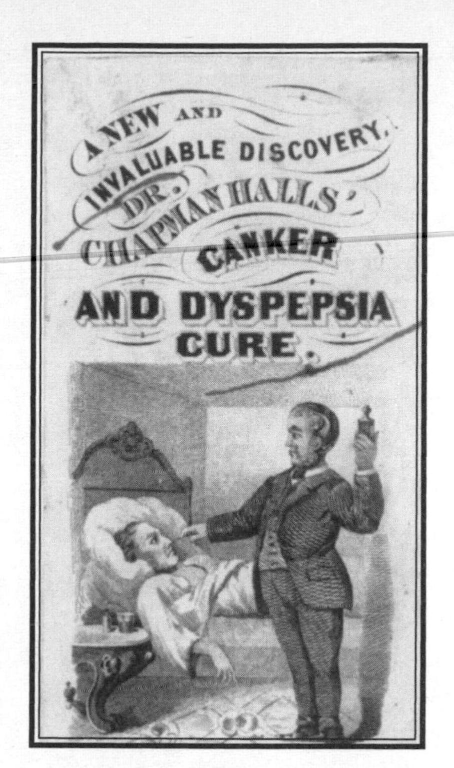

PHARMACOLOGISM

FAITH IN DRUGS

Certain beliefs permeate a society and are gradually not only accepted, but also considered a conduit to salvation. Since the early 1900s, a steady reliance on pharmaceuticals has become the answer to many issues that once only belonged to the realm of shamans and medicine men. Prescription drugs are a nearly unquestioned path to find healing. In addition to health concerns, faith is placed on medicines to address a wide range of needs, everything from curing visions and hallucinations to fertility and sexual virility. In Western cultures, drug formulas and brand names are guarded with more intent than any alchemists' potion of the past. When a new drug is offered, it is quickly accepted and taken with a deep-seated belief that it will deliver on the promises of its effects. Today, pharmacology, if measured by religious-like devotion to its powers, is accepted by a vastly larger population than any spiritual dogma relating to health and well-being throughout history. Even though more than one hundred thousand people die each year from taking prescribed drugs exactly as directed, there are more than 400 billion pills swallowed religiously every day.

GOD VERSUS DOCTOR

A Harvard Medical School survey in 2007 concluded through clinical evidence that prayer did not improve health. Fifty-seven percent said they did it anyway and believed that miracles or divine intervention was still possible even if physicians had declared all possible treatments for an illness to be futile. Dr. Joanne Lynne, director of the Center to Improve Care of the Dying at George Washington University Medical Center, studied two thousand coma deaths and found that most doctors gave inaccurate predictions on how long a person might live: 20 percent, after being told that they had two months to live, died the very next day. Many Pentecostal religions forbid taking medicines, and Jehovah's Witnesses refuse blood transfusions, no matter how dire the health situation. (*See also:* Mary Baker Eddy)

PIETER PLOCKHOY

SEVENTEENTH-CENTURY UTOPIA DISASTER

The ill-fated dreamer Pieter Plockhoy arrived from the Netherlands to the Delaware Bay area in 1663 with forty-one followers. He hoped to form a model utopia based on economic equality and division of church and state. An adherent of many Mennonite beliefs, Plockhoy wished to expand the religion's advocacy of tolerance. He recruited primarily from the urban poor. After years of writing pamphlets, he achieved some fame when he had English prime minister Oliver Cromwell interested in his idea of establishing a universal Christian church that would be tolerant of all sects. However, when Cromwell was out of power, Plockhoy became disenchanted and decided to shoot for smaller goals: "Let us reduce our friendship and society to a few in number, that we might truly be distinguished from the Barbarous and Savage people."

His American colony, Zwaanendael (Dutch for "Valley of the Swans"), moved rapidly toward self-sufficiency. Members shared everything and donated six hours of labor each day to the good of the community. Plockhoy's dream got stomped by the turn of history, only thirteen months later, when the English seized Dutch-controlled settlements from New Netherland (New York) to Delaware. The English military policy usually allowed Dutch citizens to swear allegiance to the Crown to avoid bloodshed, but when the English came across Plockhoy's group they decided not to give them a chance. Pacifists and offering no resistance, the utopians of Zwaanendael stood helpless as the "Barbarous and Savage" reduced the compound to ashes; many were taken to be sold as slaves in Virginia. Plockhoy apparently survived, though his exact fate remains cloudy. Some reports say he went blind, but in any case it has been confirmed he never attempted another utopia.

MENNONITES

Originally an ordained Catholic priest from the Netherlands, Menno Simons resigned from the church and became a popular leader in the Protestant Reformation. He was born and lived through war-ravaged years, witnessing cruelty and killings. Menno advised Christians to physically separate themselves from the secular world, become steadfast pacifists, and do good deeds in quiet. His followers became known as Mennonites. They also banned infant baptism, the taking of oaths to governments, owning guns, filing lawsuits, and holding civil office. Today, the well-known Amish Mennonites of Pennsylvania are among the more than 1.5 million worldwide members of this religion. Menno died "peacefully" in 1561, at age sixty-five.

ST. PIUS X SOCIETY

FRATERNITY OF RELIGIOUS CONSERVATIVES

In 1970, Archbishop Marcel Lefebvre founded the St. Pius X Society in opposition to the liberal changes made at the Second Vatican Council (1962–65). An association for priests, the society was organized for the purposes of "praying for the intentions of the Holy Father and the welfare of the local Ordinary at every Mass." But primarily, the archbishop was among those who were angry to learn that the liturgy of the mass was to be said in native tongues, rather than Latin, as had been done for centuries. The Vatican Council also urged priests to downplay the previous importance of "Death, Judgment, Heaven, and Hell," and foster an open-minded stance toward other religions. Lefebvre and his society of priests believed that Catholics should be taught as they always were, namely that theirs is the one true universal "catholic" religion, ordained by God. The intellectual arm of the society publishes papers and insights regarding many questionable ideas, such as its piece "Defense of the Inquisition." The group garnered headlines when one member, Bishop Richard Williamson, declared on television that "historical evidence is hugely against 6 million Jews having been deliberately gassed in gas chambers as a deliberate policy of Adolf Hitler," which led to the priest's excommunication in 1988. In 2009, German-born Pope Benedict XVI controversially reversed the excommunication verdict.

OPUS DEI

Another Catholic society, Opus Dei, received bad press when portrayed in the bestselling book *The Da Vinci Code* as a scheming, self-flagellating class of clergy. It was formed in 1928 by since-canonized Spanish priest Josemaría Escrivá. However, from its inception, the secret society was accused of foul play and of involvement in clandestine plots to influence both political and religious issues. It remains the most powerful of Catholic societies, with nearly eighty thousand laymen, under the guidance of select priests and bishops. It claims to operate with complete transparency and is noted for a substantial track record of supporting charitable deeds, or simply fulfilling its mission to do "The Work of God." Self-denial and corporal mortification are still recommended, and have been explained by society spokesperson Father Michael Geisler: "It conquers the insidious demons of softness, pessimism and lukewarm faith that dominate the lives of so many today." Methods include "the discipline," or self-lashing while reciting the Lord's Prayer, or by wearing the cilice, a barbed chain wrapped around the leg for a minimum of two hours. For beginners, sleeping on the floor or not drinking when thirsty also qualifies as self-denial.

PYTHAGORAS

FATHER OF MYSTICAL NUMBERS

The soul of man is divided into three parts, intelligence, reason and passion. Intelligence and passion are possessed by other animals, but reason by man alone. . . . Reason is immortal, all else mortal.

—Pythagoras

Pythagoras was a sixth-century B.C. Greek mathematician who so believed in the powers of numbers that he thought nothing else could better unravel the mysteries of the universe. He founded a belief system, Pythagoreanism, based on the idea that numbers and mathematics are mystical, and that everything in reality is related to their harmony, pattern, or dissonance. To him math was the means to understand the true boundlessness of nature and the Supreme Being. He also placed great value in the number 1 and thought it generated all other numbers, represented reason, and allowed the material world to appear. He

added significance to prime numbers, and odd and even numbers, and considered the simple equation 1+2+3+4=10, to be holy, the secret algebraic formula, of sorts, that created the universe. His followers lived in communes, working on math problems; they were said to be pacifists and were encouraged to act brotherly toward one another. They were a lively lot, though, and very much into music, which Pythagoras said exemplified math. Pythagoras was considered a radical in his own time, since he allowed women, at the time only thought worthy of housekeeping, to join his organization, and he encouraged them to study math as well. He believed in transmigration of the soul and hoped that when he died, as he did at age ninety, in 500 B.C., he would join the solar system, which to him was actually moving in place as musical notes, following some greater universal symphony.

Q

PHINEAS PARKHURST QUIMBY

FATHER OF THE "NEW THOUGHT" MOVEMENT

Phineas Parkhurst Quimby was born in New Hampshire in 1802. As a teen, he was apprenticed to a watchmaker, and had little formal education. His life changed when he fell ill with tuberculosis and began to seek ways to cure the disease. In the 1880s, one in four died from this form of "consumption"; treatments ranged from strenuous outdoor activity, or getting "fresh air" in the cold, damp climate of the Adirondacks, to wearing a beard. Park, as Quimby was known, became interested in Mesmerism to cure himself, and when he seemed to facilitate his own miraculous recovery, he set up an office as a magnetizer in Portland, Maine. He came to see that a suggestion for healing worked better than any medicine available and he put his theories into a book, *The Science of Man*. He treated more than twelve thousand people and influenced Mary Baker Eddy, founder of Christian Science. He ultimately has come to be considered the father of "New Thought" religions. There are varying interpretations of his beliefs, but by 1915 the International New Thought Alliance was formed to "teach the Infinitude of the Supreme One, the Divinity of Man and the Infinite possibilities through the creative power of constructive thinking." Quimby died in 1866, at age sixty-four, of "exhaustion," then considered a noble ending.

If your mind has been deceived by some invisible enemy into a belief, you have put it into the form of a disease, with or without your knowledge.

—P. P. Quimby

QUR'AN

MUSLIM HOLY BOOK

One year after the death of Muslim founder, Muhammad, the first texts of his orally transmitted beliefs was written and became known as the Qur'an, or Koran. Muhammad was inspired by the angel Gabriel (*See also:* Incubation, Cave of), and Muslims believe the Qur'an is a divine guide to "God's final revelation." As with many portions of the Qur'an, there are numerous interpretations of the meaning of the title itself. Most believe it was a word Muhammad used to instruct followers to recite and repeat, or *qur'an* what he said. That is why there are general repetitions throughout its 114 chapters, which do not read in a linear fashion. It contains ideas that simply appear like a stream of consciousness, which might have come to Muhammad throughout his days. The Qur'an states the Torah and the Gospels were also divinely inspired, but corrupted, while the Muslim holy book "has no flaws, contradictions, or inconsistencies." The completed version was inspired by God, revealed by Muhammad, and written by his companions. Any Qur'an is considered false if not translated from the seventh-century edition.

Since Muhammad had spread his religion primarily by conquest, the Qur'an also contains more than one hundred of his verses that call for violence against nonbelievers.

R

RAËLISM

EXTRATERRESTRIAL RELIGION

Former race car test-driver Claude Vorilhon, known as Raël, founded the Raëlian Church in 1974, having been compelled to do so after a six-day extraterrestrial encounter. According to his testimony, Raël met with aliens for an hour each day on the Puy de Lassolas volcanic crater near Clermond-Ferrand, France. They conversed openly with Vorilhon in perfect French. He was informed that historic sightings of angels were actually visits from members of this otherworldly species. They admitted to periodically checking in on human civilization, since they were responsible for creating man twenty-five thousand years ago through DNA experimentation. In 1975, Vorilhon was taken back to their planet and met Jesus, Buddha, and Muhammad, and he was further instructed in wisdom he was obliged to share.

Raëlians believe free love and nudity is the answer to many of the world's problems, reasoning there would be no wars, strife, or suicidal terrorists, for example, if everyone were naked. At Raëlian conferences, attendees choose a colored ribbon that indicates the gender of sex one prefers. His movement now has more than eighty thousand members in ninety countries. Afterlife is not possible to Raëlians just yet, but when cloning is perfected through an "accelerated-growth process," the mind will be transferred to one's duplicated self, or into whatever body you'd like to have.

PASCHAL B. RANDOLPH

"SEX MAGICIAN"

Paschal B. Randolph was a nineteenth-century physician specializing in magnetic healing and was a devotee of occult practices. Born in 1825 to a wealthy Virginian merchant and an African-American mother, he was orphaned by age five and reduced to begging on the streets. Except for one semester in grammar school, Randolph had no formal training, though he was later heralded as one of the most learned men of his times. Randolph eventually traveled throughout the world, from Central and South America to Europe, Greece, Egypt, and Arabia on a mystical quest, gravitating toward the occult.

When he returned to the States he worked as a barber while studying medicine and soon made a living as a "trance medium" and healer. Randolph was also an outspoken abolitionist. Initially well regarded for his views on social transformation, he even became a close friend of Abraham Lincoln, who considered Randolph a patriot for encouraging blacks to join the Union army. In addition to his political activities, Randolph believed he communicated directly with various Zoroaster spirits and a number of deceased guides. From 1858 until his death in 1875, he was the Supreme Master of the first U.S. Rosicrucian mystical order, Fraternitas Rosae Crucis. (Lincoln was also a member.)

True sex power is God power.
—P. B. Randolph

Randolph's trouble began in the early 1870s after he developed an elaborate system of rituals that included sex and the use of hashish to achieve higher levels of insight. Randolph was brought to trial and briefly imprisoned for advocating "free love," though he was freed when he clarified that he recommended sex only among married couples. He was one of the first to introduce the idea of finding a "soul mate" (*see also*: Emanuel Swedenborg), and believed it possible to identify this special partner only by mixing sexual fluids. He became a pioneer on treating sexual dysfunctions; his cures were eventually deemed "sex magic" or "affectional alchemy." He came to believe that through certain prayers and incantations, partners could work to achieve simultaneous orgasms, or as he called it, "the nuptive moment." All knowledge of the divine was attainable at that brief but powerful co-orgasmic experience. He also credited the nuptial moment with promoting wealth, health, social harmony, and even immortality. His theory stemmed from the notion that lack of sex, or poorly performed lovemaking, sapped and maligned the magnetic field of the individual, which was instantly restored with proper intercourse.

In the end, Randolph fell heavily into drinking and shot himself in the head at age forty-nine. He had been suspicious that his younger wife was unfaithful and it became too much. Others thought the death too untimely, and some followers suspected murder, part of a conspiracy to remove Randolph, with his sex scandals, as leader of his Rosicrucian lodge.

ROSICRUCIANS—SECRET ORDER OF THE ROSE

Thousands of various Rosicrucian societies of mystics formed throughout history under different names. Many groups trace their origin to a secretive Egyptian priestly class. Most sects combine elements of the occult but frequently incorporate aspects of Christianity by using the symbol of a rose inside a cross. At present, more than 250,000 belong to Rosicrucian orders. The more traditional still seek to strengthen one's psychic abilities, as they believe in mental projection, teleportation, and metaphysical healing. However, palm readers, astrologers, or hypnotists cannot join. The basic notion is to master life and reunite with a Supreme Being. In addition to Abraham Lincoln and Paschal B. Randolph, other famous members of the Fraternitas Rosae Crucis included George Washington, Benjamin Franklin, Marquis de Lafayette, and Thomas Paine.

RASPUTIN

INDESTRUCTABLE RUSSIAN MYSTIC

Born a peasant in Siberia in 1869, Grigory Rasputin as a child had failed to save two siblings from drowning in a river. At age eighteen he was sent for three months as punishment and penance for stealing to a mon-

astery, where he did his time as a custodian. He liked the rituals he observed while there and became impressed at the amount of time the monks spent in prayer. Rasputin reportedly had demonstrated clairvoyance as a child, especially after the death of his brother and sister, and he became interested in understanding his powers. He spent time with a well-known hermit and holy man upon leaving the monastery, and soon after, while walking back to his hometown, he claimed a "spiritual overwhelming" in the form of a vision, which Rasputin believed to be the Virgin Mary. Nevertheless, he married two years later, had three children, and he fathered a fourth with another woman, then decided to become a pilgrim, a wanderer, and a full-time mystical healer. During his home years he became interested in a Pentecostal Christian sect, particularly the Khlysts, who used flagellation and sought salvation through sex rituals known as "group sin." (*See also*: Khlysts) Rasputin, however, opposed physical pain as a means to seek divine revelations, although he approved of the sex part. Eventually he set off for Jerusalem and Greece, then returned to Russia, all the while making a name for himself as one with special powers. He claimed he received his mystical gifts through his influential use of prayer and by channeling.

Despite his ragtag look of a holy man, women in general were enamored of him, which some say was due to his knowledge learned while with the Khlyst sect. Others began to wonder if he was in cahoots with the devil and dismissed him as the "Mad Monk." His reputation was such, however, that while at home in Siberia he received a call from Czar Nicholas II's wife, Alexandra. She asked him to save her only son, a hemophiliac, who was seemingly dying from internal bleeding after a slight fall. Rasputin said he would cure the boy by prayer from where he was, miles from the royal residence in St. Petersburg. When the prince roused and recovered, both the czar and his wife considered Rasputin a true holy man and prophet. Within no time, he became a fixture at the court, and eventually was so trusted that he became the gatekeeper, deciding who should see the royal family, and when. His central messianic message was that salvation could only be achieved through a cycle of repeated sin and repetitive repentance. By this thinking, he justified his own public drunkenness and rampant sex among his disciples. If, for example, he exposed himself at a society party, to the embarrassment of all, he repented and pointed to it as an example of how best to reduce the sin of vanity. Nevertheless, his words on everything from financial to military matters were considered prophesies, and were often followed.

In the end, he was murdered at age forty-six in 1916 by marginal members of the royal family who had grown tired of his interference. He was poisoned, shot four times, clubbed, and when he refused to die, was finally drowned in a river, meeting the same fate as his siblings. During the cremation two weeks later, his body sprung up to a sitting position and blinked once before turning to ash. He predicted the date of his own death fairly accurately—to within days. His last prophesy, that the inner royals would be slain within two years, was likewise fulfilled, as all were executed during the Bolshevik Revolution. Empress Alexandra, Czar Nicholas Romanov, and their children were later canonized as saints by the Russian Orthodox Church. Rasputin has not been venerated yet.

RASPUTIN'S LUCKY CHARM

Rasputin was castrated before being dumped into the river, though one of his young female admirers allegedly retrieved his penis. Although many searched for its whereabouts, Rasputin's organ eventually was sold to a group of wealthy Russian expatriates living in Paris in 1920. A cult was formed to worship Rasputin's thirteen-inch member, which featured a birthmark that resembled a silhouette of his distinctive portrait. The organ was said to radiate a spiritual aura so strong it caused women to achieve a state of ecstasy simply by touching it. Others claimed it cured impotence in men. Rasputin's daughter, once she heard of its existence, laid legal claim and kept it in a hanging wire fruit basket until her death in 1977. However, that proved a fake, as were other versions. It is believed by some that this "bait and switch" of the mystic's penis was part of a successful conspiracy to divert attention from its true whereabouts, perpetrated by the secret international cult that is still in possession of the original.

RASTAFARIANS

SMOKERS OF SACRED GANJA

Rastafarianism rose from the ideas of varying black nationalist movements in the United States during the 1930s, though it was especially

influenced by the African Communities League, an organization run by activist Marcus Garvey. Three years after Garvey predicted "Look to Africa, where a black king shall be crowned, for the day of deliverance is here," Haile Selassie I was made Ethiopian emperor, and it was seen as a fulfillment of Garvey's prophesy. Ruling Ethiopia from 1930 to 1974 (with the exception of 1936–41), Selassie was heralded as the messiah and eventually came to be revered as a god in his own lifetime. He became the spiritual leader for Rastafarians, and was thought to usher in a golden age of African prosperity. According to Rastas, the emperor and Jesus are part of the Holy Trinity now, including the Father, the Son(s), and the Holy Spirit, called Jah—short for Jehovah. Selassie believed Ethiopians to be the true descendants of the original twelve tribes of Israel, with his own lineage traced back to the offspring of King Solomon and the queen of Sheba. Rastas believe in many Christian teachings but pick and choose primarily from Hebrew texts, particularly the book of Revelation. Rastas generally consider the King James Version of the Bible corrupted and influenced by "Babylon," a catchall word referring to the negative influence of Western ideas and culture. Others believe Selassie never died and that he remains their god and king who will lead them to Zion, or the Promised Land. This concept adheres to Rastafarian belief that death can be overcome for a select few. Differing from Christian and Jewish dogma, Rastas see the phenomenon as "Everliving," where both the physical self and the soul remain conjoined for all eternity. For example, when reggae musician Bob Marley knew he was certain to die from cancer, stemming from not allowing an amputation (as it was against Rastafarian views), he refused to write a will, reasoning it would affect his chances of becoming an "Everliving." By not doing so, however, he left his vast royalties and estates in question.

Emancipate yourselves from mental slavery, none but ourselves can free our minds!

—Bob Marley

Rastas generally do not cut their hair, and favor dreadlocks (*See also*: Divine Hair). This aversion to the use of combs, scissors, or razors stems from their interpretation of passages in the Talmud and coincides with attempts to live in a more natural state rather than following trends devised by "Babylon." It is not a requirement to have dreadlocks to be a Rastafarian, but it is regarded as a mark of spiritual growth. Growing long dreads is seen as a lesson of patience. Those who do cut their hair are considered "baldheads." There are few formal services other than gatherings for smoking of cannabis, called ganja (from a Sanskrit word for hemp), which they never refer to as "weed." At these religious gatherings, called "Reasonings," a joint is lit by one informally chosen to lead, and it is passed around clockwise during peacetime, or counterclockwise during periods of strife. While smoking ganja, Bible passages might be read, or stories about Rasta prophets told. Another Rasta holiday is called "Grounation," marked by feasts, music, dance, and more ganja smoking, lasting a few days.

Rationale for cannabis's sacred use was found in Bible sources. Rastas believe the herb was the first plant that grew on Solomon's grave and that the original "Tree of Life" was actually the marijuana plant. Many point to a line in Psalms that states, "He causeth the grass to grow for the cattle, and *herb* for the service of man." For Rastafarians, ganja (also referred to as "lamb's bread") is similar to a Christian Eucharist. However, it helps one plan life's journey, with the purpose of ultimately becoming closer to Jah. Alcohol, in contrast, was promoted by Babylon to further enslave, while ganja was God's gift to offer glimpses of paradise. Illegalization of marijuana is seen by the more than one million practicing Rastas as religious persecution.

MARIJUANA AND RELIGION

Cannabis first grew naturally throughout the Asian central steppes, from as far south as Iran to the northern tundra and Siberia. Archeological evidence suggests that it was used for religious purposes as early as 6,000 B.C. The nomadic Scythians used it regularly, and digs at early Bronze Age sites have found ceremonial pipes and cannabis seeds. The Assyrians referred to the plant as a "fumigant," and they smoked it to relieve the soul's sorrow and grief, believing it sacred incense. Among Germanic people, cannabis was used during celebrations to Freya, the goddesses of love; harvesting of the plant was marked

by feasts and erotically wild times. Egyptians also used it as incense and believed the prayers offered among the fumes became more divine. Biblical sources indicate cannabis was an ingredient in holy anointing oils used by Jews. In China, Taoists sought its hallucinogenic properties, and in Chinese folklore it was thought to allow glimpses into the future when mixed with ginseng. Sufis and whirling dervishes particularly found it a useful aid in achieving mystical understanding and they thought it was helpful in keeping balanced while dancing. When nineteenth-century English explorer David Livingstone went deep into the African continent, he came across the Makololo people and witnessed how they used special bamboo pipes and rituals centered around smoking cannabis, which they called *matokwane*. Philosopher and spiritual leader Ram Dass, after falling ill, said this of the plant: "I use marijuana because the stroke captures my consciousness—and I use it to free my consciousness from the stroke. I use it to free my words." (*See also:* Sadhus)

RAVENS

RELIGIOUS BIRDS

The raven, a black-feathered bird with a pointy, blood-veined orange beak, has remained a religious and mystically symbolic creature for centuries. In the Middle Ages, it was common knowledge that a wizard could turn a person into a raven with the simplest of incantations, and alchemists thought it the bird best capable of shape-shifting. In biblical times, the raven was considered an olfactory genius, able to smell rotting flesh, and its cawing over a house was thought to be a premonition of death. In the Noah story, some now believe it was ravens and not doves that Noah released from the ark to spy for dry land. Instead of the genteel image of a twig in a dove's claws, it was likely a raven returning with a piece of car-

rion, such as an eyeball dangling by a gruesome tendril, that told Noah where to anchor his ark. The Prophet Elijah, while in hiding for years in a cave, showed his divine connection by having ravens bring him food. To people of those times this was a miracle, since the raven, in flocks of a hundred strong, were commonly known to peck and devour—in piranha style—a lone or fallen traveler in record speed. The Dane and Viking gods Badb and Odin had ravenlike qualities and attributes. For Druids, a person who could listen to a raven's caw and interpret its language and messages was held in high esteem. The Druids saw the raven as an embodiment of the goddess Morrigna, noted for her trinity of powers overseeing war, death, and romantic love. During the Inquisition, the accused were often found to have some sort of raven connection, under the assumption that Satan and his minions transformed into ravens more than they did any other creature. One could be deemed guilty of heresy or witchcraft for merely liking the blackbird. It was considered serious bad luck to be seen feeding ravens, and various forms of scarecrows were fashioned to protect the soul, more than their current use of guarding a newly seeded field or a garden.

Other birds that have religious symbolism include the stork, which some believe was the bird that Angel Gabriel transformed into when he informed Mary of her pregnancy, and is seen in paintings representing the Annunciation. The stork also likes to eat snakes, which are frequently considered demonic. Peacocks, with an eyelike design on their feathers, have symbolized God's watchfulness when one becomes too vain. The eagle, during medieval times, was Christ, watching earth from above with an "eagle eye." (That's one reason why Ben Franklin objected to the bald eagle as America's mascot, and preferred the wild turkey, an important and more practical food source to early settlers.) The sparrow represented the insignificant everyman but was always provided for with crumbs. The dove was chosen to portray the Holy Spirit, the peacemaker of the Trinity.

TEMPTATION OF THE RAVEN

St. Benedict lived in the fifth century and was noted for writing a rulebook for monks that is still in use in many monasteries and convents. In his time, a priest who was envious of Benedict's notoriety cast a spell on a raven. In a plot to kill the pious monk, a raven was sent to bring a piece of poison bread in its beak to offer St. Benedict. However, this seemingly friendly bird somehow reminded St. Benedict of a beautiful village girl he saw frequently, and roused his desires. Instead of taking the tainted morsel, the saint commanded the bird to leave, and to further quell temptation Benedict then stripped off all his clothes and threw himself naked into a thornbush. Although nude entanglement in a briar patch worked for St. Benedict, this method is not in the current handbooks on preserving the dedicated from lustful temptation.

RELICS

SACRED SOUVENIRS

Every Catholic altar has to have some relic from a saint in order for it to be sanctified. The cross used to crucify Jesus was supposedly found in in the early fourth century and its splinters were then spread around the world. However, if all the churches claiming to have part of the true cross were combined, it would make the cross's original weight close to three tons, or as heavy as an SUV. Saints' bones, pieces of fabrics they wore, part of the blanket used to swaddle baby Jesus, and even a vial of Virgin Mary's breast milk are entombed in altars throughout the world. In the early years, around 70 A.D., Christians were regularly killed and turned into instant martyrs. Even back then there was a prosperous trade, with resourceful entrepreneurs eager to dissemble the bones and sell them to new churches, especially after it was noted that other Christians risked their own lives to dab a cloth with the blood of a fallen comrade and passed it among the faithful. Up until 900 A.D. any human remains found around Rome were easily marketed and sold as holy. St.

Augustine tried to squash fraudulent relic sales in the fourth century, but it persisted, with objects that merely touched holy relics themselves gaining venerable status. Now relics must have some miraculous happening attributed to them to be deemed as such. With demand exceeding supply, the Vatican's secret vaults where some official relics remain are still under heavy guard.

RITUAL KILLINGS

MURDER IN THE NAME OF GOD

In the Bible's book of Judges, a warrior and official named Jepthah is noted to have offered his daughter as a human sacrifice. He made a bargain with God that if he won a battle he would offer the first person he saw upon his triumphant return as a gift. Unluckily, his daughter and only child heard of his homecoming and raced to meet her dad. Jepthah tore at his own clothes and cried out when he saw her: "Alas, my daughter! You have brought me very low!" His daughter was stunned when Jepthah explained she must be sacrificed, since he had made a promise to God and couldn't renege. When informed he was bound to kill her, she asked, curiously, for two months to "bewail her virginity." Whether that meant she intended to have sex before her death remains unclear, but she is in fact the only human sacrifice to Yahweh, as noted in the Old Testament.

God asked Abraham to sacrifice his only son Isaac, and although absolutely upset by the request, Abraham planned to do it. When Abraham raised the knife to kill his beloved son, he was stopped at the last second by an angel. Abraham was relieved to learn it was only a fire drill of sorts that God had devised to test Abraham's obedience.

In many early religions, human sacrifice was considered the highest offering and significantly more effective than killing goats or lambs. The Phoenicians and many cultures throughout the Middle East worshiped Moloch (noted by varying names), the king of gods who extracted the most demanding payment. The ancient city of Byblos had a three-story-tall Moloch statue, which featured a slide that went from the bull's head to a pit of hot coals below. Babies were rolled down it and burst into

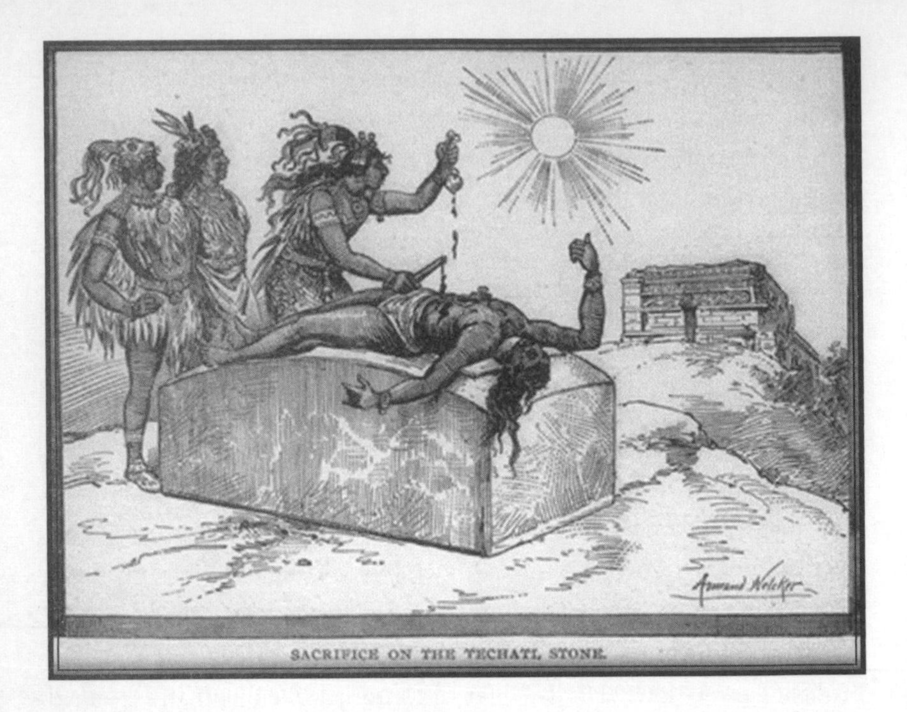
SACRIFICE ON THE TECHATL STONE.

flames. At the same time, Druidic religions farther north in Europe and the British Isles also used human life as offerings. Many of these bodies have been discovered relatively preserved in peat bogs, where they were placed ceremoniously. Various Druid gods preferred particular methods to make the sacrifice acceptable. For example, Toutatis, a Celtic god, deemed that his sacrifices must be drowned, while Esus, god of war, favored the human gift to be bound to a tree and whipped to death. Taranis, the thunder god, liked the human offerings to be burnt. One account of Taranis worship describes how a huge figure of the god was sculpted from saplings. Chosen sacrifices were placed inside this crisscrossed, cagelike monument: Once filled with people, the wooden figurine was then set on fire.

In Hawaii, the ancient religion made human offerings at *luakinis*, or plateau-like temples spread among the islands. Prisoners of war, losing politicians, or lawbreakers supplied the blood for the gods, Kane, the main creator, Ku, the war god, and Lono, the weather god. Hawaiian priests ended sacrificial ceremonies by holding aloft one eyeball of the slain, and in the other hand an eye from a bonito fish. When he put them in his mouth and was able to swallow both without regurgitating, it meant the gods accepted those last human offerings, to the great relief of all. Among Native Americans, the Pawnees sacrificed a young girl at the annual Morning Star Festival, and Iroquois were likewise known to fatally surrender the occasional maiden to please the Great Spirit.

ROMANY

GYPSY RELIGION

When a nomadic people originally from India migrated throughout Europe, beginning in the eleventh century, many thought they were from Egypt, and so they were first called E-gypsies. They had formulated a distinct religion, which was assimilated from contact with numerous cultures; at the core was the belief that one must avoid the evil entities of the universe, called *marime*. For a gypsy, all actions must adhere to codes of purity, or *wuzho*. Dealings with Gaje, or nongypsies, is impure, as is the lower half of the body, bathing, and owning pets. They have an elaborate system of curses, spells, and rituals to avoid *marime*, such as keeping a piece of bread in a pocket to ward off bad luck. Women are generally the healers, and they consider spit as a curative, which they claim works better than antibiotics. Ghost vomit, called *johai*, does the job even better, though it's a hard-to-recognize substance often found in Dumpsters, a place, for reasons unknown, contemporary ghosts favor. Most gypsies do not believe in dieting, and consider obesity a sign of wealth. Marriages are usually arranged, with most girls betrothed by sixteen. Birth into a Romany family is the only way to convert, though one can be banished through a community ceremony called *kris*. One example of banishment recently occurred when a Romany man was convicted of filing two insurance claims for the same stolen car. He wasn't banished for committing fraud, but rather because he was caught. Worship is centered on female gods and ghosts, and they believe in reincarnation. Many hope to become a *mulo*, a living dead who can linger on earth to seek revenge on all who had done them wrong during life. Today there are more than ten million Romany, as gypsies are now called. Many still offer psychic readings, and frequently post a sign advertising this service in front of their homes. Throughout history Romany people have been repeatedly subjected to discrimination and persecution. During World War II, Hitler killed 1.5 million Romany in the gas chambers, even though their bloodlines were Aryan; he had executed them for "cultural impurity."

RUTH AND NAOMI

THE FIRST SAME-SEX UNION

In about 900 B.C., Naomi, an Israelite woman from Bethlehem, lost her husband and two sons during a famine. Her son's wives, Oprah and Ruth, were encouraged to go back to their homelands and remarry. Oprah departed, but apparently Ruth and Naomi had a special bond. Ruth offered a pledge to her mother-in-law: "Where you go I will go, and where you stay I will stay. Your people will be my people and your God my God. Where you die I will die, and there I will be buried. May the Lord deal with me, be it ever so severely, if anything but death separates you and me." Today these lines are frequently used in same-sex marriage services. Whether their relation was romantic is open to dispute, even if the original words chosen to describe the union are the same used in other scriptural references that relate to sex. Nevertheless, Naomi seemed to be sexually astute and instructed Ruth on how to seduce her next husband, Boaz, an elderly rich man, with her naked body while the man was taking a nap during a wheat harvest. Ruth is also noted as the great-grandmother of King David. When Ruth left to remarry, Naomi (meaning "delightful") then changed her name: "Do not call me Naomi, call me Mara," a word translated as meaning "bitter."

KING DAVID AND HIS MALE LOVER

Those seeking a biblical example of accepted homosexual associations turn to young David and his relationship with King Saul's son Jonathan. Regardless of how various translations have tried to soften the details, it has been noted that "the soul of Jonathan was knit with the soul of David," and during a time before people wore undergarments, "Jonathan took off the robe he was wearing and gave it to David, along with his tunic, and even his sword, his bow and his belt." Jonathan was the one who warned David of his father's plans to kill the young warrior, and when he tried to plead with King Saul to have David returned to good standings, Saul seemed fully aware of Jonathan's leanings: "Do I not know that you have chosen the son of Jesse to your own shame, and to the shame of your mother's nakedness?" The Bible also described David and Jonathan kissing, and when David heard that Jonathan died in battle, David was devastated, saying, "Jonathan's love to me was wonderful, passing the love of women."

The most direct biblical reference to homosexuality is noted in Leviticus, the third book of the Torah: "If a man lie with mankind, as with womankind, both of them have committed abomination; they shall surely be put to death; their blood shall be upon them." While the passage has been used by antigay groups, American Orthodox rabbi Shmuley Boteach offered a contemporary interpretation: "Homosexuality and sodomy are not ethical sins. No one is being hurt, no one is being cheated, nobody's rights are being infringed upon. Homosexuality is a religious sin, analogous to other Biblical prohibitions, like not eating the carcass of a dead animal, or not sleeping with a woman during her menstrual cycle."

S

MARIA SABINA

THE WOMAN WITH THE MAGIC MUSHROOMS

The celebrated medicine woman—or *curanda*—Maria Sabina was born in 1888 in the Sierra Mazateca mountains of southern Mexico. As a child she had a fondness for eating mushrooms, particularly a certain type referred to by her Native American tribe as *teo-nanacatl*, which was later discovered to contain the hallucinogen psilocybin. When she was as young as eight years old the spirits she envisioned while entranced directed her to find different herbs, boil them, and restore to health an ailing relative whom other healers, known as *curanderas*, had failed to cure. From then on her spiritual insights were widely sought. She called the psychedelic fungi "sacred children," which she ingested with reverence and respect for the mushrooms' dangerous potency.

Her fame spread when American researcher R. Gordon Wasson hiked to her small remote village and befriended the demure, kind-hearted Maria. He had heard that Maria was considered a true shaman and seer, and Wasson only found her after months of arduous searching. Maria had silky black hair that reached to her feet and lived in a small mud hut with a thatched roof. In 1955, Wasson was allowed to partake in a ceremony that had previously been kept secret since the Spanish Conquistadors' arrival five hundred years earlier. His article on the experience was published in *Life* magazine (May 13, 1957), and it marked a turning point in drug history, spurring modern interest in mushrooms as a means to insight. Within years, Hoffman Laboratories (creators of heroin and Bayer aspirins) had isolated the hallucinogenic chemicals of the mushrooms and used it in their drugs, including one that is currently prescribed to treat cluster headaches.

Through the 1960s, Maria grew disappointed as droves of "young people with long hair came in search of God" and invaded her small village, which soon resembled a scene from a Hollywood frat movie more than a spiritual journey. Before, the mushrooms were medicine and the means to open a portal where spirits would offer the true diagnosis of the ill and indicate remedies, like a kind of spiritual MRI. In her elder years, Maria became bitter, sorry she had unleashed the secret upon "the unready." She also saw imitators proliferate and offer rituals to tourists looking for the mushrooms that grew wild everywhere. Many shaman-impostors earned a considerable sum selling séances and "trips" to spiritual-minded stoners for a stiff sum. Carlos Castaneda, John Lennon, Bob Dylan, and Timothy Leary were among the many celebrities rumored to have sought Maria's mushrooms and insights.

Later in life, Wasson also claimed to have nightmares for having revealed the secret of the mushrooms on the world, and for the resulting commercialization it brought to the villages. Maria Sabina lamented the lost power of the "sacred children," although her inability to experience more revealing hallucinations may have a medical explanation. Although psilocybin is not addictive, the body builds tolerance if it is ingested frequently, as Maria did. She eventually tried to accommodate the onslaught of visitors, and like her imitators, she charged whatever pesos she could earn for her services. Nevertheless, when she died at age ninety-one, in 1985, she was still revered, with restaurants, hotels, and even taxicabs displaying a logo with her image.

SHROOMING TOWARD GOD

In 2006, a research project conducted by the Johns Hopkins School of Medicine recruited three dozen volunteers who had never tripped before to eat psilocybin mushrooms. Sixty-one percent of the group reported a "complete mystical experience." The alkaloid chemical compound found in certain mushrooms takes ten to forty minutes to digest and ultimately affect the brain by mimicking serotonin, a neurotransmitter with the keys to mood, aggression, happiness, and other essential metabolic functions. What's termed "closed-eye visuals" can produce an altered reality. Some reported seeing concealed geometric shapes hidden in ordinary things. One experimenter discovered the origin of the spiderweb, for example, but could not put it into words. However, most described the effect of eating mushrooms as both cerebral and spiritually euphoric, usually bestowing a feeling of benevolence toward humanity. Psilocybin, whether from living mushrooms or in powder form, is illegal, classified along with cocaine and heroin as Schedule I drugs. Deaths have occurred after eating the wrong kind of wild mushrooms, which contain a different toxin, one that causes organ failure. As for its medical use, new studies have administered mushroom-size doses of the hallucinogen to patients with terminal cancer. Many of the terminally ill expressed a greater understanding of their disease in an entirely profound and different way, thereby ultimately easing their anxiety about death. According to the National Council on Drug Abuse, 3 percent of the adult population use some form of psychedelics each year.

SADHUS

CANNABIS-SMOKING HINDU MYSTICS

There are an estimated five million sadhus in India who are confirmed ascetics, dedicated to nothing else but meditation and contemplation of God. Hindus believe that for most people three other phases must be accomplished before one becomes a meditative sadhus, or achieves the *mosksha* stage, defined as complete liberation. Sadhus opt for the fast track and bypass the *kama* phase, defined as experiencing life's enjoyments, as well as the *artha* level, the period of life when goals are set and achieved. In addition, the sadhus likewise ignore the *dharma* requirement that Hinduism regards as character development and showing a sense of duty to something. Instead, sadhus find a location not yet claimed by sadhus in a town or city and basically set up what Westerners would call a homeless person's camp. The sadhus, however, are considered extremely holy and wear an ochre-colored gown to mark their calling. Sometimes, if they feel like it, they walk around naked and often paint their faces with ritual markings to signify the gods they worship. They also smoke cannabis and burn it as incense to appease the god Shiva; others do it for Vishnu. Each neighborhood or village comes to revere their local sadhu, though they secretly fear them, since many sadhus are known to hand out blessings or curses with equal abandon.

Sometimes the sadhus are called upon to make *bhang*, a drink made from cannabis buds and flowers that is consumed heartily by many during Holi, a Hindu feast marking the beginning of spring. In this atypical Hindu celebration, all seriousness is discarded and people party in the streets in a style similar to a New Orleans Mardi Gras. Ten percent of sadhus are female. (*See also*: Sun Gazing)

SALVATION ARMY

"GENERAL" BOOTH'S CHRISTIAN CAMPAIGN

Though the Salvation Army Santas who seek donations during Christmastime are often paid minimum wage for their daylong shift, the organization has one of the best nonprofit track records, according to *Forbes*: Nearly sixty cents of every donated dollar is actually used to help unfortunates. The group's founder, William Booth, was born into a rich English family in 1829, though at age thirteen he found himself yanked from a privileged boarding school and instead apprenticed to a pawnbroker. With eyes accustomed to a different world, the life of England's poor during the Industrial Revolution thoroughly shocked Booth and molded his religious and political views. While still pawnbrokering, he began to make street-corner preaching his evening pastime, and he promoted the notion that salvation can only come from the willful

and public repentance of the sinner. Along with his wife, Catherine, he formed the Christian Revival Society in 1865. They changed the organization's name to the Salvation Army a dozen years later. Booth was named "the General" of this new army of volunteers.

Booth invented his own military-style uniforms, flag, and marching songs, and published a newspaper, the *War Cry*, which he used to express his thoughts on class restructuring and communal societies. The Salvation Army brought deliverance to the poor through the "three-S program"—first soup, then soap, then salvation. Resistance to his efforts was great, especially after he targeted alcohol as one of the main problems among the poor. Subsequently a tavern owner started a retaliatory organization, "The Skeleton Army," which advocated the three B's instead: more beer, beef, and bacca (wine). More than four thousand joined that movement to torment Booth's uniformed converts wherever they went, and the harassment even led to the death of a few Salvation Army members.

Nevertheless, Booth's organization had gained a quarter million followers by 1881, and it is still going strong. Continuing to wear nineteenth-century-style military uniforms, the Salvation Army members are responsible for annually serving fifteen million meals to the poor around the world. The idea behind the uniform was intended by Booth to signal that his "officers" and "soldiers" were ready to serve. New Salvationists must purchase their own military costumes, though ministers who are required to wear it throughout the day are given a small petty cash fund for dry cleaning. Top commanders are also provided free housing. Santas ringing the bells are often hired from homeless shelters, but must promise not to drink or smell of alcohol while on duty. Ultimate salvation for a Salvationist depends on faith and obedience to Jesus.

William Booth

SAMSON

BIBLICAL STRONGMAN

Samson didn't get his extraordinary strength from biblical steroids; he thought it had to do with his long hair. He could wrestle a lion, or wipe out an entire army in a berserk rampage. The Israelites were under Philistine control when Samson's mother, unable to bear children, had a visitation from an angel. She was told to refrain from drinking wine in order to conceive. The angel also told her never to cut the newborn's hair and that her child would eventually free the Israelites from Philistine domination. However, when Samson was in his twenties he fell in love with a Philistine woman. At the wedding he told the mostly Philistine guests that before they could eat they must solve a riddle about one of his adventures—an incident in which he killed a lion and ate honey from bees that nested in the carcass. He posed this: "What is sweeter than honey? What is stronger than a lion?" Samson also promised to give the thirty groomsmen in attendance fine linens if they cracked this obscure brainteaser. Of course, no one came close, and before long insults were exchanged, which sent Samson into a rage. His crying bride-to-be implored Samson to tell her the answer, which she then whispered to the guests. When Samson was forced to keep his promise of giving away linen and garments, he acquired the "gifts" by murdering and stripping thirty passersby. In the commotion, the

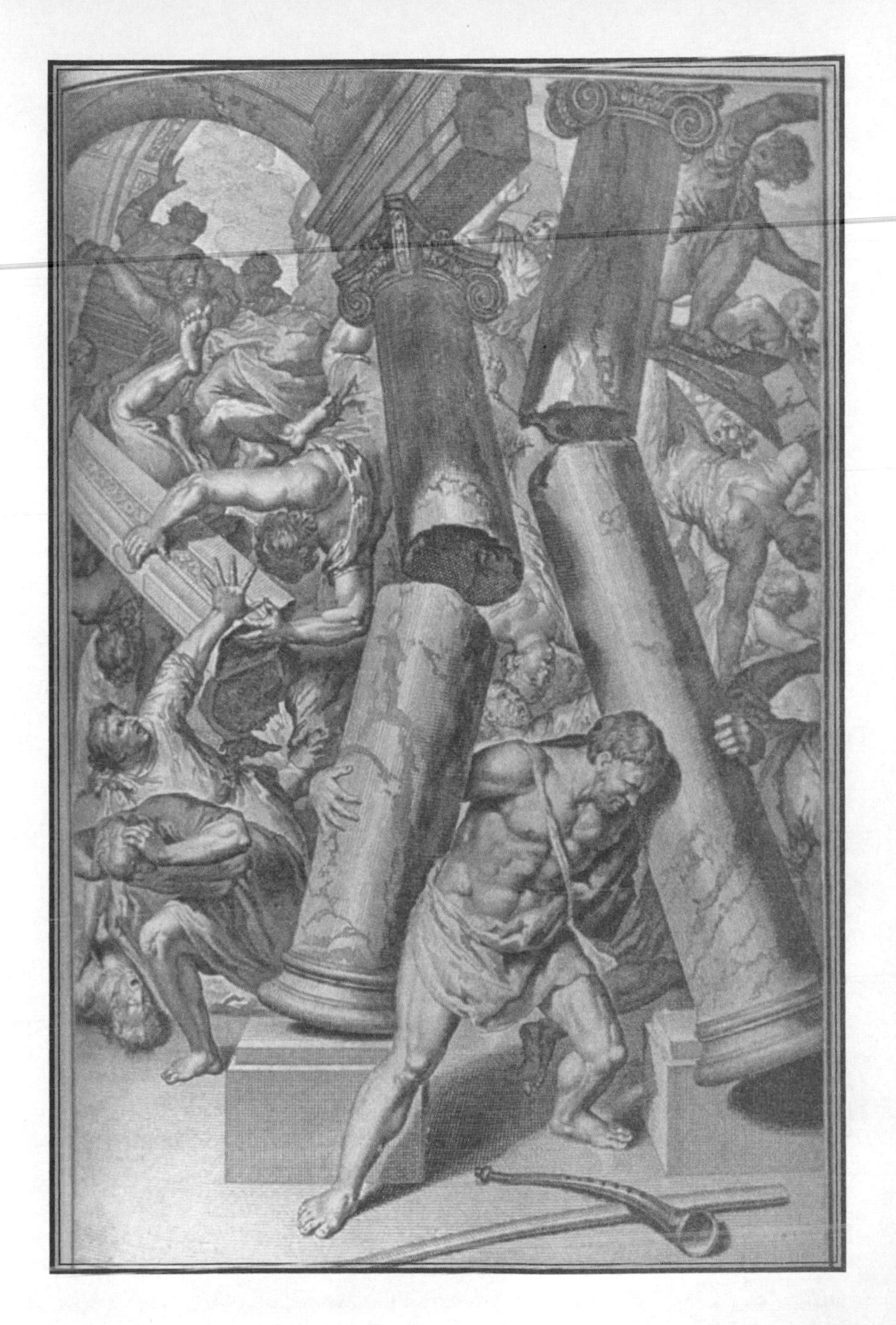

bride was instead given to the best man. Samson then captured three hundred foxes, and in one of the most bizarre cases of biblical animal abuse, Samson tied flaming torches to the foxes' tails and let them loose through the Philistine fields, burning everything. When Samson was discovered to be behind the scheme, he hid in a cave, but to spare his countrymen from revenge, he allowed himself to be handed over bound in ropes. However, when in the Philistines' midst he busted free and killed a thousand soldiers with the jawbone of a donkey.

The Philistines then bribed Samson's new girlfriend, Delilah (who was also a Philistine), to find out the secret of his strength. At first he teased that if he were tied in bowstrings or woven human hair, then he might be unable to break free; whereafter, on two separate mornings, he found himself bound in each. Delilah's beauty and means of seduction must have been incredible, since Samson eventually divulged his real secret. When he woke the next morning, Samson found his head shaved; the divine hair that had given him strength was gone. Delilah signaled to the Philistine soldiers in waiting to come for Samson, and she reportedly collected a treasure in silver. In short, Samson had both eyes singed out with hot irons, and he was then chained to a manually powered wheat-grinding wheel. Plans were made for a feast to thank the Philistine god for delivering the strongman, and in a few months thousands gathered on a temple rooftop in the city of Gaza for a party to watch the blinded Samson delivered for a spectacle of humiliation. However, by then Samson was for some reason allowed to grow his hair long again, possibly assuming he could do no damage without sight. They were wrong, for after requesting God to give him one more burst of strength to commit suicide and destroy his enemies, Samson felt his power return. He feigned weakness and asked his handlers for a brief rest against a pillar. He then bear-hugged the temple's main support column and knocked it over, killing more than three thousand and himself in the destruction. The fate of Delilah remains unknown.

THE NAZARITE SECT

The angel that Samson's mother met specifically told her to follow the rules of an ascetic sect called Nazarites, who required the avoidance of alcohol, vinegar, grapes, and raisins. Cutting hair or using a comb was also forbidden. Some Nazarites believed it was impure to look upon a corpse, and even shielded their eyes when passing a graveyard. Samson, Joseph, and John the Baptist were among biblical figures who at various points had taken the Nazarite vows.

SANTERÍA

AFRO-CUBAN RELIGION

When the Yorubas, an ethnic group originally from West Africa, were displaced as slaves to work on Cuban and Caribbean plantations, they secretly continued to practice their religion, called Lukumi. After being indoctrinated with Catholic dogma, they melded veneration of saints into their native spiritual ideas, and they eventually used drums, dances, and animal sacrifices to communicate with both Christian saints and Yoruba gods. *Santería* means "way of the saints," and the name allowed them to retain many of the ancient practices under Catholicism's watchful eye. Accordingly, every one is born to a particular guardian saint who serves as the intermediary to other deities. The main god is Obatala, creator and father of all the other gods and guardians. Prayers and magic rituals are used to gain favor with the saints and employ certain herbs, roots, and iconic charms.

Today, pictures and statues of saints, and many other items used in the religion, are sold in stores called Botanicas. Animal sacrifices, primarily beheading chickens, are still practiced, and carcasses are commonly found along railroad tracks or in secluded parks throughout South Florida, where it's estimated that nearly one in four Hispanics practices or "dabbles" in the religion. Even for those who do not believe entirely in Santería, many are respectful of its power to produce both beneficial and sometimes malevolent results. For example, in 2006, a forty-two-year-old Cuban man was on trial for drug dealing. While out on bail and awaiting sentencing, he visited a *santera* who he hoped would cleanse his spirit. After performing a number of rituals, she threw dice, which were made from coconut shells. The readings she got indicated things would not go well for the man. He was found dead not far from the *santera*'s house the next morning, resulting from hyperventilation and subsequently cardiac arrest. In another incident that same year, a twenty-five-year-old man was found shot to death. He was covered in glitter, wore two different-colored shoes, and had on eyeglasses with one lens clear and the other darkened, all signs that he was practicing Santería. When the killer was found, he confessed to the shooting and said he did so because the victim was powerful enough, he believed, to carry out a threat to cut off his head and offer it to the saints. The exact number of *santeros* and *santeras* are unknown, since the religion recommends strict secrecy among adherents. However, in 2008, in greater Miami alone, there were more than sixty Botanicas in operation.

JUDGES RULE

In 1993, the U.S. Supreme Court decision protected Santería priests' right to perform rituals: "Although the practice of animal sacrifice may seem abhorrent to some, religious beliefs need not be acceptable, logical, consistent, or comprehensible to others in order to merit First Amendment protection."

SATAN

MASTER OF HELL

Satan is sort of like the self-appointed prosecuting attorney of mankind, always devising tests and various entrapment schemes to prove God was wrong in giving humans free will. According to popular tradition, he has horns, red skin, a small goatee, sprouts batlike wings, and has a tail with an arrow tip. However, he's supposedly been known to transform into both male and female human form and wears disguises ranging from a skirt or business suit to even overalls. Other times he becomes a snake, a raven, a dragon, or simply any object that becomes a temptation. Many religious traditions also call him Lucifer, considered originally a cherub, a high-ranking angel who rebelled and thought he could do a better job

than God. Other versions say he was so prideful with his own self that he refused to worship God or anyone. Lucifer and his angels were beaten by Michael the Archangel and cast from heaven, although retaining all their angelic powers. According to Dante's *Inferno* he resides in the ninth circle and deepest part of hell, rejoicing every time he catches another human in his web to suffer eternity in his dark realm.

CHURCH OF SATAN

Howard Levey, a musician and onetime circus hand who was previously interested in occult practices and philosophies that advocated materialism and individualism, formed the Church of Satan on June 6, 1966. He called that date day one in the new Age of Satan. (*See also:* Six-Six-Six) A few years later he wrote the *Satanic Bible*, which explained his dogma and beliefs, and gained him wide media attention—he even once sat in the guest chair of *The Tonight*

Show, interviewed by Johnny Carson. One tenet of the church specifically forbids harming children, though it states: "When walking in open territory, bother no one. If someone bothers you, ask him to stop. If he does not stop, destroy him." Levey, who changed his name to Anton LaVey, died at age sixty-seven of respiratory ailments, in 1997. He was taken by chance to St. Mary's Hospital in San Francisco, and to some, understandably died there. His ashes were divided among family and are believed to be potent in performing satanic rituals. According to the website of the Modern Church of Satan, an offshoot of LaVey's original movement, "You are God. This is your world and everyone and everything in it has been placed here for your enjoyment."

SCIENCE OF THE MIND

TAPPING THE POWER OF POSITIVE THINKING

Ernest Holmes was a grocery clerk and supposedly asked three questions of everyone he met: "What is God? Who am I? Why am I here?" After studying Mary Baker Eddy's Christian Science, and the writings of Ralph Waldo Emerson, he formulated his version of metaphysics, centering on pantheism, the idea that God is in all things. In 1919, he wrote his first book on the subject, *The Creative Mind*. Decades before *The Secret* (2006) or Norman Vincent Peale's *Power of Positive Thinking* (1952), he believed the power of positive thinking not only healed diseases but could also actually change physical reality. He abandoned the notions of heaven and hell and believed both were a state of mind; sins are correctable mistakes. He didn't start out to make a religion, but by 1958 he noted, "We have launched a Movement which, in the next 100 years, will be the great new religious impulsion of modern times, far exceeding, in its capacity to envelop the world, anything that has happened since Mohammedanism started." Repetition of prosperity mantras are recommended, as is attendance at "Spiritual Mind Treatments," and calling a twenty-four-hour prayer hotline. Holmes died in 1960 at age seventy-three and left a movement that has seventy thousand followers who still practice his "Laws of the Mind." Members are asked to donate 2 to 10 percent of their income, try to remake themselves and the world a better place through focused thought, and strive to attain unity with the "Infinite Intelligence," or "Universal Mind."

HOLY SEPULCHER

THE CHURCH OF LONG-LASTING DISAGREEMENTS

In 325 A.D. a hill and the area near the gates of the original Jerusalem were selected as the historic site of Jesus' crucifixion, burial, and resurrection. Emperor Constantine's mother, Queen Helena, believed it to be so, and she urged for a church to be erected there, after first demolishing a temple to Venus that had been built on the site. The first Church of the Holy Sepulcher was funded by Constantine to appease his mother, though it was knocked down three hundred years later by invading Muslims. Another was rebuilt, only to be destroyed itself in 1009 by Egyptian rulers. In 1099 the Crusaders constructed another version, which still stands, though numerous parts have been added through the centuries.

Many different religious orders lay claim to the church, divided architecturally into a patchwork of parts, rarely agreeing, such that a ladder placed outside a window during a routine repair more than a hundred years ago has yet to be removed, lacking consensus if it should be taken down, and by whom. In 1901, monks in disagreement over who should sweep a certain flight of stairs caused bloodshed when clerical combatants attacked each other with broken broom handles and stones. In 2009, two monks from different orders got in a fistfight and were arrested after squabbling about a procession to honor the discovery of the true cross. Israeli riot police were eventually needed to calm the agitated clergy.

This most revered religious pilgrim site holds claim to many Christian wonders, including the supposed Tomb of Joseph of Arimathea, where some claim Jesus was buried, even if archeologists identified a garden tomb site farther away as the more likely setting. Another altar marks the hill where Jesus was crucified; now the roof of the church covers the entire hill. It was here that some think the true cross was found by legendary relic collector St. Helena. Accordingly, when three identical crosses were unearthed, she thought to devise a test by dragging a woman from her deathbed to the excavation area. When one of the crosses touched this poor lady and caused spontaneous healing—the startled woman was said to have gotten up and fled—it was determined which was the true one. Helena also claimed to have found the very nails used for Jesus' execution. She gave her emperor son a few and had them welded to his battle helmet.

THE TIMBER OF THE CROSS

When Adam was dying he called out to his faithful son Seth and asked that a seed from the Tree of Life—still seen growing in the cordoned-off Garden of Eden—be picked and dropped down his throat. Michael the Archangel was asked to retrieve this special nut, reportedly the size of a walnut, which Seth then placed inside Adam's mouth at burial. In no time, a giant tree with the strongest grain and texture sprouted from Adam's corpse; it eventually formed a forest. The timber from the Tree of Life woodlands was so good that Solomon built a mighty bridge out of its lumber. Even the queen of Sheba, while en route to Judea, stopped her entourage to marvel at the quality of the bridge's wood, and she was said to have knelt down and kissed it. When the queen of Sheba foretold that a piece of the wood from that bridge would somehow figure in Solomon's downfall, he burnt the bridge and buried all the cut lumber, as well as clear-cutting to stumps the forest where it grew. Some of the wood was later found to be in the Temple of Jerusalem, which during Jesus' time was undergoing minor remodeling. The temple's discarded lumber was used by Roman subcontractors, namely the thriving crucifix makers, thus tying Jesus' death to

Adam's tree-sprouting corpse. Analysis of all splinters said to have been part of the true cross proved them to be primarily pinewood, and not an extinct wonder lumber.

SERPENT HANDLERS

BITTEN BY FAITH

In 1910, George W. Hensley was a Tennessee bootlegger trying to turn over a new leaf, and he thought prayer might be a way to get out of the moonshine racket. He achieved success, for a while, by attending revival meetings at the Church of God in Cleveland. He was soon given a shot at the pulpit as a way to testify about how he changed into a better man by getting filled with faith, and by believing in the literal interpretation of the Bible.

One day, a bunch of Hensley's old cronies heard that he had gotten religious, and perhaps fearful he would start naming names, the moonshiners dumped a basket of rattlesnakes at the foot of the pulpit where Hensley spoke. Naturally, as the snakes slithered throughout the congregation, panic ensued, but Hensley called for order and went about rounding up each snake "like a boy picking up firewood." He was not bitten once. Hensley found the appropriate Bible passage to explain the phenomena: "They shall take up serpents; and if they drink any deadly thing, it shall not hurt them; they shall lay hands on the sick, and they shall recover," as found in the Gospel of Mark, and it proved to the congregation that everything in the scriptures was indeed literally true.

By the early 1920s, Pentecostal serpent-handling churches had sprung up throughout the South and into Florida. Hensley would be known to hold five wiggling rattlers and two water moccasins in one hand and shove the bunch at a member in the congregation, commanding the person to grab them or be "doomed to hell." Newspapers began to bring attention to deaths and thousands of injuries as a result of the practice. An Alabama man narrowly escaped a lynching for allowing his toddler to be bitten and die at a ceremony. After the bad publicity, mainstream churches distanced themselves from Hensley and his brand of preaching, even if it had once yielded great turnouts.

Hensley returned to his native Appalachia, where serpent-handling churches remained vibrant, beyond the view of scoffers and those wishing to ban the practice. "Little Fiery George" Hensley died as expected, from a snakebite, in 1955.

Contrary to popular belief, serpent handlers do not assume they are protected by the Holy Ghost or are immune to venom. They do it to prove that they have absolute faith in the Bible and are thoroughly obedient to God; they point out that unlike others who agree in words, they alone prove it by handling deadly snakes as instructed to do so by scripture. All preacher-handlers get bitten, most multiple times, as do those in the congregations, with many bearing atrophied hands, missing fingers, and amputations caused by snake bites. Recently, not many have converted to serpent-handling churches, and the more than three thousand current members of the thirty or so registered congregations in the United States usually comprise families and those who through marriage have been involved with the "calling" for four generations or more. A preacher in a serpent church located outside Atlanta commented about a death caused by a copperhead used during a religious ceremony: "Some people were bit, and I believe God was ready for them and their time had come." The same pastor, Junior McCormick, was quoted in the *National Geographic:* "I was bit 14 times, by rattlesnakes, copperheads, water moccasins, and I never used anti venom—all I had was just Jesus. I've been bitten badly, but I'll go back the next week and take them [serpents] out again."

SNAKE CHARMERS

Hypnotizing a cobra with a flute had been practiced in ancient Egypt, with a branch of theology dedicated to studying snakes and reptiles associated with various gods. Even earlier origins of snake charming, dated at 5000 B.C., occurred in India and stemmed from Hindu beliefs: The god Shiva is often depicted wearing cobra earrings and a lair of living cobras as a belt. In ancient times, poisonous snakes were a tremendous cause of fatalities, and so those who learned of their habits and could control them were considered mystically enlightened. (*See also:* Phallic Cults) In Indian folklore snake charming is often traced to a snake-handling guru, Gogol Vir, who became high priest of a cult that worshiped snake deities known as Nagas. He was killed when a snake hiding above a threshold caught him off guard and sunk its fangs into the back of his head. As he lay dying, he instructed his son to cook the snake and eat it, but instead seven hungry thieves rushed in and feasted on the meat. They obtained the magic and power over snakes, and their descendants became the wandering snake charmers of India ever since. It was a tradition usually handed down from father to son, but snake charmers of recent times have been reduced to a class of street performers. Since a 1972 law in India prohibited harboring poisonous reptiles, many of the cobras used in public have had their fangs and venomous glands removed, or their mouths sewn shut.

ST. SETON

FOUNDER OF PAROCHIAL SCHOOLS

Catholic schools in the United States can trace their origins to Elizabeth Ann Seton, who in the early 1800s was dedicated to feeding poor girls and giving them an education. This saint bore five children before abandoning her life to sisterhood soon after her husband died, leaving the family bankrupt. Previously, she had prayed to God to spare her ill father, and offered the life of her infant daughter, Catherine, in exchange. Her father died anyway, and Catherine grew to become Mother Mary Catherine of the Sisters of Mercy, dying at age ninety, in 1891. (Three of Seton's children died before the age of twenty-five.) Seton was a Protestant who had converted to Catholicism before she founded the first Catholic parochial school in the United States, in Maryland. Soon after she formed her own order, called the Sisters of St. Joseph. She established systems for administration and curriculum that are still used in modified form in many parochial schools. It took her 150 years to be canonized. By 2008, according to the National Catholic Educational Association, enrollment in Catholic schools nationwide had dropped to 2.3 million from its high mark in 1960 of 5.2 million. Some blame this on the lack of prayers said to St. Seton.

SEVENTH-DAY ADVENTISTS

CORN FLAKES AND DOOMSDAY

When the world didn't end as was predicted by the doomsday prophet William Miller in 1844, his disappointed remaining followers banded together to form a new movement known today as the Seventh-Day Adventists. The seven-day part of their name means they rested on Saturday instead of Sunday, unlike most Christian groups. Adventists still believed the world faced an imminent end, but after several bogus updates passed uneventfully, they wisely never again gave an exact date.

In the 1840s, one teenager in Maine, Ellen Haron, clung to Miller's original belief so reverently that she began to announce visions and prophesies concerning the matter. She claimed angels had taken her on a very arduous journey to heaven, though she was sent back to start a new religion. She eventually married James White, a fellow apocalyptic advocate, and together they gathered many new followers through preaching, claiming of new visions, and extensive writings. They made the center of their church in Battle Creek, Michigan, and established many creeds regarding worship and health issues. Ellen had been a sickly child, some attributing it to working with toxic chemicals in her parents' hat business; at that time hatmakers used ample amounts of mercury to form supple brims, though the chemical was later discovered to cause varying degrees of dementia. In some ways Ellen laid the foundation for what is now termed holistic health care. In one vision she was told to avoid doctors and prescription drugs—commonly opium elixirs—and instead said, "God's great medicine is water." Getting hosed down and other hydropathy remedies were advocated. Her recommendation for eating simple foods could be said to have started the breakfast cereal industry. John Kellogg was a doctor in charge of an Adventists' health-food division and invented corn flakes. Sylvester Graham, interested in Ellen's vision regarding flour, promoted whole wheat products and invented the Graham cracker. By the turn of the century there were sixty-seven thousand members in her church, and by the year 2000, the congregation had hit the five-million mark.

TUNGUS SHAMANS

SIBERIAN MYSTICS

Russian Cossacks called the holy men living among northern tribes *šaman*, a word meaning "to know." These shamans were held in high esteem by the Evenki tribe and other native Siberian peoples. To the Evenki, reality was divided into three parts: The lower world was for ancestors and evil spirits, the middle section contained animals and humans, and gods and divine spirits resided in the upper realm. Tungus shamans were called to heal, cast hexes, and communicate with spirits. Colorful costumes loaded with shiny objects, or feathers, fur patches, and draped animal hides as headwear were needed to fly into these other realms. They danced in this ritual outfit while banging a tambourine-size drum, eventually going white-eyed and falling into a trance while they transported to the other worlds. The shaman was also in charge of taking the dead's soul to the proper resting place, though usually not until the withered corpse was two to three years old. As harsh as their environment was, these Siberian tribes believed a suicide had no rest and wandered the earth for eternity.

SHAKTI

HINDU MOTHER GOD

In Hindu traditions, Shakti is the supreme mother goddess. As Hindu scholar Swami Abhedananda noted, "India is in fact the only place in the world where God is worshiped as Mother." Originally, Shakti was easy to contact, the way a mother attends to the call of a crying baby. But traditions grew to perform rituals to receive her blessings, especially the nine-day ceremony called Navaratri. Shakti is an all-encompassing word meaning power. Hindus see her in everything, from a lightning bolt to the seedpod that pushes its head from the soil. Women who worship her recite certain sacred syllables while touching various body parts, in the belief that the ritual will leave them purified. Hymns, sand paintings, and art drawn on skins, as well as the tossing of saris, or gowns, into raging bonfires, are part of the nine-day feast. The most popular destination for mother-worshiping pilgrims is the Kamakhya temple in Assam, India. Here the fertility aspect of Shakti is more prevalent, since it lays claim to a giant boulder, shaped like a vulva. A spring flows near it, which always keeps the vulva stone moist, though it's recommended to touch it when the water miraculously turns red. The colored droplets are sopped up and considered to contain powerful energy.

MENSTRUAL RELIGIOUS CUSTOMS

Many religions have addressed a woman's menses and thought the time unclean. In some Jewish sects, detailed laws concerning the matter were established, forbidding all physical contact with a woman in *niddah.* Merely accepting an object that touched her hand was sinful. In addition, men were not permitted to sleep in the same bed as her, and the women had to be wrapped in special sheets. There are ritual cleansing baths, and other customs, including slapping a girl in the face when she experiences her first menses. In Islam, avoidance is also recommended, and although a woman can be touched, no sex is permitted. A Muslim woman is not allowed to hold the Qur'an during her period. Early Christians forbade a woman during her monthly time from entering a church; menstruation was one of the arguments used to ban women from becoming priests or handling the Eucharist, since it was thought a drop of her blood might contaminate the sacrament. Numerous Christian theologians warned men that having sex during a wife's period "will father children that are leprous." The Romans felt the same way, and according to historian Pliny, "Contact with the monthly flux of women turns new wine sour, makes crops wither and dims the bright surface of mirrors." Many Native American tribes also practiced some form of menstrual seclusion, with only a few religions seeing it as positive. The Khoisan women of the Kalahari region of Africa, although sent to a special hut, were thought powerful enough to kill a disrespectful male with a snap of the fingers during their time. And in India, the Bauls, a mystical sect, believed drinking some menstrual blood a holy act. In satanic circles, a "Red Mass," or sex with a menstruating woman, was believed the most effective way to exploit demonic forces. In Celtic-styled religions, however, although not quite celebrated, the period was nonetheless seen as a bond to the moon cycles and fostered greater union with the Mother Goddess.

MOTHER SHIPTON

THE FEMALE NOSTRADAMUS

A woman known as Mother Shipton lived in England during the sixteenth century and is credited for predicting the destruction of the Spanish Armada, the Great Fire of London, and that people would someday travel by carriages without horses. Her diaries and prophesies were not published until eighty years after her death, which made many

Mother Shipton

believe her very existence was a literary invention and a myth. There was one Ursula Southeil, who is considered the model for Mother Shipton, and by many reports she was a noted fortune-teller. Ursula was born in a cave, one that had a mystical wellspring and smelled of sulfur. Her unwed mother left the child at the doorstep of neighbors; at age twenty-four Ursula married a carpenter, Toby Shipton, of York, in 1512. Mother Shipton grew to be a woman of exceptionally large stature. She was also extremely ugly; however, she was kind and told fortunes for townsfolk that proved to be true. Some believe she was describing World War II when she wrote: "In nineteen hundred and twenty six build houses light of straw and sticks, for then shall mighty wars be planned, and fire and sword shall sweep the land. Boats like fishes will swim beneath the sea, and men like birds shall scour the sky. Then half the world, deep drenched in blood shall die." She also predicted the Internet five hundred years before it happened: "Around the world thoughts shall fly in the twinkling of an eye." She knew the date and year she was to die and did so as predicted in 1561. Her grave site was marked by a stone, though its whereabouts are unknown, likely below a row of houses outside York.

SIX-SIX-SIX

BEASTLY NUMBER

This number became infamous due to its mention in the book of Revelation: "And that no man might buy or sell, save he that had the mark, or the name of the beast, or the number of his name. Here is wisdom. Let him that hath understanding count the number of the beast: for it is the number of a man; and his number is Six hundred threescore and six." Ever since, the number was considered demonic, with some believing that when the Antichrist or devil showed up at doomsday he would be known by 666 names. It was actually a measurement of taxes placed on land during King Solomon's reign, and was originally referenced in the book of Revelation as a criticism of Solomon's interest in acquiring wealth. Others deciphering biblical numbers suggest it was a number code placed on Emperor Nero. Recent Bible translations, however, cite the real number of the mark of the beast to be 616, throwing centuries of numerology into confusion.

The belief that numbers are codes has always fascinated man. (*See also*: Pythagoras) St. Augustine thought numbers were a universal language given by God, and many church scholars, including the Venerable Bede, went to great lengths to decipher the deep meaning of scripture through number values. Numerology is still considered an important part of certain occult practices. Many architects continue to omit the number 13 from floor plans, and it's rare for any building to have its address listed as 666. (*See also*: Tarot)

SIMEON OF STYLITES

PATRON SAINT OF FLAGPOLE SITTERS

In 390, when Simeon's Syrian mother, Martha, was told of her arranged marriage, she planned to run away. Instead the ghostly severed head of St. John the Baptist appeared to her, and told her that the pending nuptial would produce a saintly son. It turned out to be true, for by the age of thirteen, her son Simeon was into inflicting pain on himself, enduring long fasts, and going around the Arabian countryside reciting the Beatitudes. When he entered a monastery his drastic measures got him the boot, especially after he was found unconscious with a palm leaf girdle around his waist bound so tight that it made his waist only ten inches wide. Afterward, Simeon shut himself in a hut for three years, and learned to stand in one position without a flinch, until dropping. When he came back to town, people watched this frozen-standing routine in amazement, and began to consider him a saint. Wherever he went, people crowded to hear his words and imitate his murmured prayers. All the attention was too much for Simeon, so he climbed atop a pillar that was about sixty feet in the air. He wouldn't come down, and stayed there for thirty-seven years. He allowed no women to come near his pillar, not even his mother, who then retreated to a more sanctified life and became a saint as well. Simeon's fame spread throughout the world, drawing more people than he had ever hoped, until his death in 459 A.D., at age sixty-nine; even on his high-altitude deathbed he refused to come down to be attended by physicians. From then on, and for about a century later, thousands of ascetics took to sitting on pillars and became known as Stylites.

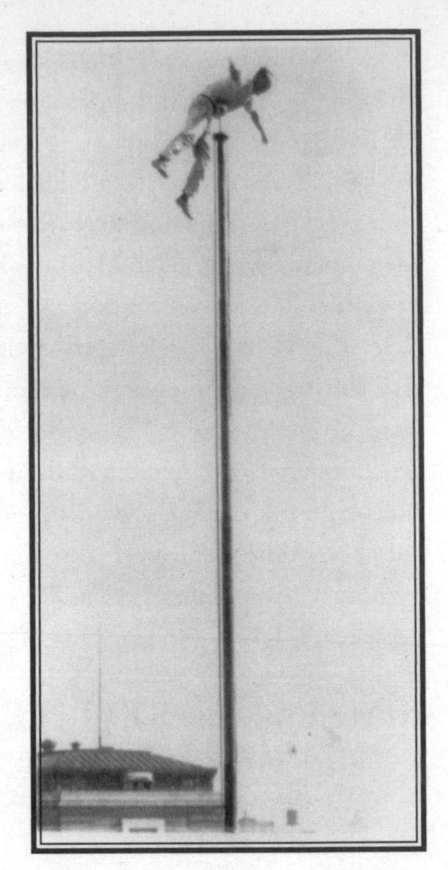

During various flagpole-sitting crazes, particularly during the 1920s, many of the bystanders prayed to Simeon that the high-perched showman wouldn't fall.

SKOPTSY

THE CASTRATION SECT

The first castration sect is attributed to Origen, a second-century Christian scholar from Egypt who was taken in and protected by a wealthy widow after his father was martyred. To limit the old woman's advances and focus solely on scripture study, he castrated himself, believing Jesus' words in the book of Matthew: "Eunuchs . . . have renounced marriage because of the kingdom of heaven. The one who can accept this should accept it." (*See also*: Jesus on Sex)

Origen opened a school, and from what can be discerned from woodcuts, he apparently had castrated male students lay nude on carpets while he sat at a small desk reading his interpretations of scriptures. He also lectured at convents and all-girl schools, considered acceptable employment because of his eunuchness. In the end, Origen and many other Christians were blamed for a plague that hit Rome in 251 A.D., and he subsequently got swept up in a round of tortures. Origen was chained to a post and beaten, and though saddened that he didn't die a martyr, he succumbed to complications from torture, dying a few years later. Church officials played down the castration part of his bio, saying it was fabricated by destractors, though various castration sects popped up throughout Christian history, often citing Origen as an inspiration.

The most widespread castration sect began in Russia during the late 1700s. (Castration had been sporadically popular ever since Adrian, a castrated monk, preached there in 1000 A.D.) In 1771, a Russian peasant, Kondratii Selivanov, was convicted of persuading thirteen other men to castrate themselves and was sent to Siberia for the crime. When he returned to St. Petersburg he declared himself the son of God and the incarnate soul of the recently assassinated Russian emperor Peter III, who was suspected of being murdered by a conspiracy instigated by his wife. To Selivanov, Peter's downfall, and that of all mankind, was due to lust, sex, and any expression of libido. After claiming he was "God of Gods and King of Kings," he unveiled his simple belief system, stating that the fastest route to salvation was through castration. Russian rulers were baffled when the movement attracted tens of thousands of followers, who became known as the Skoptsy sect, a Russian word for "a castrated one." Selivanov was eventually locked in an insane asylum, though in old age he was released to a monastery, where he died at the age of one hundred, in 1832. By then the movement had spread from being predominantly peasants to include leading members of royalty and other prominent citizens.

Before admittance into the sect, some were allowed to father one child. Castration was broken into degrees among the faithful, with a Lesser Skoptic only severing the testicles, while a Greater Skoptic removed the penis as well. Women Skoptics had both breasts removed, in addition to the labia and clitoris. Ceremonies took place from dusk to dawn in secret locations, and they consisted of hymn singing and swirling around for hours, until achieving a state of ecstasy. Everyone wore white and they called themselves "White Doves." The men, if not castrated, had special girdles that displayed their genitals, which by the ceremony's end would be cut off. By 1877 nearly six thousand Skoptics were identified and made to register with the state as being castrated; they were given a stamp on their ID papers that classified them as eunuchs. Further attempts to stem this bizarre practice included public humiliation, such that the men were dressed in female garb, made to wear dunce caps, and paraded through the streets. However, the movement went unabated until more than one hundred thousand people in Russia and Romania were classified as such. Often they were employed as taxi drivers in Romania, and others became known for operating moneylending businesses in St. Petersburg after many attested that they went from poverty to riches once they castrated themselves. In 1910 Russian authorities made a strident effort to eliminate the sect for good, rounding up 141 eunuchs, ranging in age from fourteen to eighty-five, and held a well-publicized trial and conviction. By 1929 only a few remained alive, and today the sect is nearly unheard of.

JOSEPH SMITH, JR.

FOUNDER OF MORMONISM

Joseph Smith's father could never make a go of it as a farmer in Vermont, primarily because he spent as much time plowing as he did treasure hunting. Going on unfounded hunches, the elder Smith dug holes and tunnels at the unluckiest of spots, though he did make money by hiring out his services to find lost articles and valuables with divining rods and magic stones. His son, Joseph Jr., likewise took up the pursuit, even if he ultimately found a different type of gold.

As a child, while Smith's father favored magic, his mother went with equal enthusiasm from one religious fad to the next, everything from spiritualism to revivalism to doomsday Evangelical notions. Naturally, Smith was uncertain which religion to follow, and so with little formal education he began to study the Bible for clues. One day while he was praying in the forest a suffocating power suddenly seized and engulfed the then-fifteen-year-old Smith in a "vortex of darkness." When Smith called out for help, "a pillar of light," described as a divine spotlight, appeared. Two ghostly apparitions hovered above Smith and were later said to be angels, Moroni and Nephi. Smith then asked the question that haunted him: "Which of the sects was right, and which one should

The Youthful Prophet, Joseph Smith, Jr., and Oliver Cowdery, Receiving the Aaronic Priesthood under the hands of John the Baptist, May 15, 1829.

I join?" Smith said the "Personage" told him to join none, and it would show him how to start a better one. Three years later Moroni, identified both as an angel and a resurrected "Native American historian," told Smith that sacred texts and further instructions were written on gold plates buried nearby. No one believed the boy except his father and both went straightaway to the location. But just as father and son began to dig, Moroni showed up again and told Smith to wait four more years.

Eventually Smith Jr. was convicted of being a "glass-looker, disorderly person and an imposter" for bilking neighbors with the promise of finding lost gold mines on their properties by using a peep stone. In 1827, when all seemed bleak, Smith claimed he had finally gotten his hands on the angelic tablets. To help him read the illegible text, the angels gave Smith a pair of ancient eyeglasses called Urim and Thummim. Over a few years, before returning the tablets to the angels, Smith, along with a schoolteacher, Oliver Crowdery, transcribed the tablets, and they wrote *The Book of Mormon: An Account Written by the Hand of Mormon upon Plates Taken from the Plates of Nephi*. Both Crowley and Smith met the angels again and were ordained in the Priesthood of Aaron, giving them the power to "lay hands" and transfer the Holy Spirit to others. Moving westward, Smith and his followers looked for the new Zion, and although they were persecuted wherever they went, Smith landed in Commerce, Illinois, and renamed the town Nauvoo. By the early 1840s he became the mayor of twenty thousand new Mormon citizens. In 1844, Smith thought to run for president of the United States and rule the nation through Mormon theocracy, but when the Nauvoo newspaper criticized his ambitions, he had the publisher's presses dismantled with sledgehammers. Arrested for violating freedom of speech and accused of treason, a mob stormed the jail and murdered Smith, which to his followers raised him to martyr status. When the Mormons were ordered to leave Illinois, Brigham Young, who had an uncanny ability to mimic Smith's voice, succeeded in laying claim to Smith's divine mantle. In 1847 Young led a wagon train of 150 people to settle near Salt Lake City, Utah. Within a decade, seventy thousand more followed to eventually make the Utah Territory the Mormons' Promised Land.

Smith's Book of Mormon was considered the literal word of God. It encouraged faith in Jesus, baptism by submersion, speaking in tongues, and virtuous living; it also contained two ideas, namely "Blood Atonement" and "Celestial Marriage," which from the start caused problems. Blood atonement is required if one commits a grievous sin. As a favor of sorts, the sinners need to have their "blood spilt on the ground" for their souls to be saved. The sentence of capital punishment was levied for the following sins: adultery, murder, using God's name in vain, lying, marrying a black person, and "resisting the Gospel."

THE WIVES OF BRIGHAM YOUNG

Marriage, according to Young, was to be a patriarchal polygamy, and the institution was seen as God's means of building a strong family on earth. The more wives a man gathered, the better his status would be in heaven. Young said: "I prophesy to you that the principle of polygamy will make its way, and will triumph over the prejudices and all the priestcraft of the day; it will be embraced by the most intelligent parts of the world as one of the best doctrines ever proclaimed to any people." Young had fifty-six wives, but he fought off government interference from the beginning. In 1860, the U.S. Army sent an expedition to the Utah Territory to rout polygamists. In an attempt to get statehood, Mormons eventually conceded on the concept and in 1890 passed the "Woodruff Manifesto," banning plural marriages. Nevertheless, they secretly persisted, with more than one hundred thousand people in 2008 committed to Smith's original polygamy revelation. Even some of the monogamously Mormon faithful look forward to enjoying polygamy in heaven.

CHURCH OF JESUS CHRIST OF LATTER-DAY SAINTS

Not all Mormons left for Utah, and power struggles ensued, ultimately leading to new schisms, including a church under the guidance of one of Joseph Smith's thirty-three widows, and son Joseph III. The followers never called themselves Mormons, and it was Brigham Young who incorporated the religion as the Church of Jesus Christ of Latter-Day Saints (LDS). All rely on the Book of Mormon for guidance; there are more than twelve million followers throughout the world. This relatively large number proves to the Mormon faithful that the prophetic words of Joseph Smith, Jr., and other Mormon prophets are still coming true. The alphabet Smith found on the plates that relate most of LDS's dearly held prophesies was said to be from a language called "reformed Egyptian." Linguists consider it pure invention. This argument is dismissed by believers; it is known that the spirit Moroni used that dialect on purpose, since it took up less space than Hebrew or Latin would on the golden tablets.

TOO MUCH, TOO LITTLE CHURCH

In 2006, a Massachusetts man walked into a police station and confessed to killing his thirty-seven-year-old wife and eleven-year-old stepson with a hammer. He offered a plea, blaming the LDS church for making them spend too much time at services and for nagging him about his drinking coffee, which the church bans. He got two life sentences.

SOKA GAKKAI

GREEDY BUDDHISTS

A Buddhist sect called Soka Gakkai was formed in the 1930s in Japan and was dedicated primarily to chanting. It was believed that a person could achieve enlightened status by repeating "Nam-myoho-renge-kyo." Not only can a person become transformed, but if more people chant this simultaneously, the entire world can change. The religion is based on the teaching of a thirteenth-century Japanese monk, Nichiren, and his belief in the power of the Lotus Sutras. This document was said to be dictated by the original Buddha at the end of his life, but it was thought so advanced that none would be able to understand it for at least five hundred years. Buddha had it hidden in a cave, guarded by dragons. Soka Gakkai members believe the chant will help them understand the secrets of the Lotus Sutras. Since destiny is changed by the chant, it's thought to be more effective if many are recruited to do likewise. This notion has led to criticism of the group's aggressive proselytizing and its use of questionable recruitment tactics, such as a method known as "break and subdue." Their goal of creating world peace is seen by others as an ultimate plan for world domination, where every morning and evening the entirety of humanity will repeat—anywhere from five minutes to three hours—the "Nam-kyo" chant. Followers believe gaining material success is fine, and a product of good chanting. There are currently eighteen million Soka Gakkai members, with a sizable percentage most likely chanting at this very moment.

SOUL'S WEIGHT

21 GRAMS?

A soul is generally defined as the spiritual essence of a human, the part that many believe survives after physical death. For believers in reincarnation, the soul returns in another physical form. For most others, including Christians, a soul may go to heaven or hell, while a smaller group thinks that if there is a soul, it simply ends with death. Scientists throughout history have tried to find a tangible way to prove the soul's existence and learn more of its attributes by conducting experiments. At the turn of the twentieth century, one physician, Duncan MacDougall, set off to prove that the soul was a tangible, measurable entity. MacDougall's hypothesis: The soul must have weight. MacDougall reportedly worked on the project for more than five years, constructing huge, counterbalancing, bedlike scales that were sensitive enough to measure weight fluctuations to one-tenth of an ounce. The result of his research made the *New York Times* (March 10, 1907, and the front page on July 24, 1911). MacDougall described results of six dying people, whom he somehow managed to convince to spend their final moments in his laboratory. He described one man, dying of tuberculosis, as "an ordinary type . . . not marked with a phlegmatic disposition," meaning perhaps that he was not evil, or not without a soul. The doctor did the same thing with five other subjects and noted that as soon as the heart stopped beating an instantaneous weight change occurred. Each lost approximately twenty-one grams of mass upon death. As a control to verify his research, he weighed dozens of dying dogs, and upon expiration, found no weight change, or no soul in the canine species. When naysayers said his measurements were flawed (and that he did not consider calculations for the naturally occurring weight loss that happens to a freshly dead body due to evaporation), MacDougall resorted to the then-new X-ray machine. Measuring even more dying subjects, he observed the soul in the X-ray as light, similar to "interstellar ether."

In 1998, Donald Gilbert Carpenter went in the other direction and tried to pinpoint the moment when the soul entered a fetus. According to his book *Physically Weighing the Soul*, this event occurred at forty-three days after conception, or when electroactivity begins in the brain of the fetus. Carpenter allowed that the soul might get bigger as a person grows and gains life experiences.

A number of contemporary scientists, with infinitely better measuring devices than MacDougall's, hope to search for the weight of the soul under the guidance of the First Law of Thermodynamics, which states that no energy can be created or destroyed. The soul, to these objective researchers, is treated as a measurement of energy, which should change at the moment of death. When calibrated electromagnetically, death produces a detectable loss of less than a billionth of a kilogram.

ANNA AND HORATIO SPAFFORD

EUNICHS FOR THE KINGDOM OF HEAVEN

Born in 1842 in Norway, Anna Oglende was an infant when she immigrated to Chicago with her parents. By age seven she was an orphan; her mother and father died within a year of cholera. In her teen years Anna caught the eye of Horatio Spafford, a lawyer fifteen years her senior. Horatio was a pious Evangelist and met Anna while she at-

Dec. 17, 1904. **AMERICAN COLONY, JERUSALEM**

tended a Sunday Bible class he taught. He paid for Anna's education and waited until she was eighteen before marrying her. They had a comfortable life in Chicago, and survived the Great Fire of 1871, which turned a four-mile stretch of the city to ash. In 1873, for a vacation and reprieve, Anna took her four daughters on a trip abroad, while Horatio remained at work, trying to recoup financially since he had invested in much of the real estate that was destroyed. While Anna and the girls headed to Europe on the paddle-steamer, SS *Ville du Havre*, the ship was rammed and sank in less than twelve minutes. All four daughters and 222 other passengers drowned. Anna was plucked unconscious from the sea, found with her arms wrapped around a floating timber. Soon after her return she met more hardship when Horatio was accused of fraud—he was suspected of dipping into trust funds in his charge. However, Horatio explained that it was not sinful because it enabled him to do less attorney work and spend more time at Bible studies, his true calling. When the scandal got deeper, he quit law and became an apocalyptic preacher, declaring the end of the world was at hand. In his view, every sinner, even trust fund pillagers, would be saved. His brand of followers would become known as "Overcomers," in reference to their attitudes toward sin. They referred to themselves as "the Saints."

While hosting small Bible study groups in a shack-size church they built next to their home, Anna soon began to get heavenly "sniffles" that she claimed were signs from God. As the law was closing in on Horatio, he declared that the best place to be when the world ended was Jerusalem. Along with thirteen adults and three children, the Spaffords set up a commune they called the American Colony in a ruin of a building near the walls of the ancient city. Horatio knew from scriptures that the Jews would have to return to the Holy Land before apocalypse could happen, and he intended to convert them to Christianity when they arrived. The concept was to start a Christian utopia: living simply as early Christians once did, sharing everything.

Their marriage produced seven children, though only two survived, yet Horatio declared affections and lust to be the most grievous sin. He proclaimed: "Henceforward I live a eunuch for the Kingdom of Heaven's sake. I rely exclusively on the power and grace of God."

From the beginning, the practices within the compound raised concerns. Converts and followers worked diligently in a variety of businesses Anna set up and controlled. Every penny was turned over to the Spaffords. Marriage was abolished and couples were broken up in hopes of keeping members chaste. Children were also separated from parents and raised communally. When Horatio died of malaria in 1888, six years after their arrival in Jerusalem, Anna took tighter control, demanding everyone live a more austere ascetic life, except her. She traveled back to the United States to entice more converts, and by 1896 the colony had more than 150 members. When Anna died in 1923, her children kept the commune going, though some said it was merely to keep a horde of workers that received no pay. The utopian Christian community disbanded in the 1950s, though the luxurious American Colony Hotel, once the site of their commune, still operates, and is owned by Anna and Horatio's descendants, who now pay all their employees a salary.

SUN GAZING

BLINDED BY THE LIGHT

Ophthalmologists caution that staring into the sun for more than two minutes can cause retina damage. Nevertheless, since ancient times, staring purposefully into the sun has been practiced as religious devotion, commonly called sun gazing. Mayans and Aztecs worshiped the sun, as did Egyptians, Sumerians, and Hindus. For example, the Aztec sun god, Huitzilopochtli, demanded human sacrifice to keep it rising and setting. During the fourteenth and fifteenth centuries, high priests stared into the sun to determine the number of human offerings that were needed on a daily basis, and it's conservatively estimated that more than ten thousand people were killed annually to keep the sun rising and setting. Mesoamerican priests, in addition to looking at the sun for answers, often did so while they drained some of their own blood, or inflicted other self-tortures, such as sticking barbed cords into their ears.

Among Hindus, sun gazing has been popular for centuries, and it has now been revitalized by a movement called Surya (solar) yoga. Devotees of this practice claim it helps one commune with nature, gives spiritual enlightenment, and allows the more dedicated to live without food, getting nutrients instead in a photosynthetic manner, just as a plant does. Some gaze at the sun through the slit made by two fingers, while others prefer looking through an aperture in Swiss cheese, or even the hole in a doughnut. The more serious recommend the use of no props, but to avoid eye damage, they caution the novice to stare only during sunrise or sunset, slowly working up to full gaze at high noon. For the religious sun gazer, claims of healing and even immortality have been attributed to a dedicated practice of staring into the sun. In the 1990s, one retired Indian businessman, Hira Manek, set out to prove the value of sun gazing. He claimed it allowed him to live primarily on water for more than seven months, and that by the end of a decade he had gone more than ten years without solid food, getting nutrients only from the sun. He stated that he turned his body into a "solar chip." In 2007, another solar-gazing guru, Sunyogi Umasankar, likewise became an *inedia*, per his Web site, existing on little else but sunlight. When pressed to demonstrate this ability, he announced he was moving into the Buddhist stage of silence and would not speak for many years to come. Sun gazers strive to achieve the Hindu concept of *prana*, or life's vital life force, where food or water is no longer needed and one can survive through "Breatharianism," defined as living on air and sunlight alone.

In 569 B.C., a thirty-year-old Indian royal, Lord Mahavir, gave up all his possessions, clothes included, and spent the next twelve years staring at the sun. Once enlightened, he walked barefoot throughout India, preaching about respect for all living things, and reportedly rarely ate, so as not to harm anything. He became an important figure in Jainist religion and it is believed Mahavir attained complete liberation, or nirvana, when he died. Many current sun gazers look for spiritual guidance to the Jainist religion, which began to practice sun gazing 2,600 years ago.

EMANUEL SWEDENBORG

THE SCIENTIST WHO MET JESUS AT A PUB

For most of his life Emanuel Swedenborg was considered a brilliant engineer, inventor, and scientist, noted for offering the first published drawing of a working flying machine in 1716. At age fifty-six he switched careers after becoming plagued by strange dreams. While he was stuffing himself at a London tavern one evening in 1745, an odd man approached and told Swedenborg to put down the fork and go home. That night in a dream the same man appeared and revealed that he was Jesus, and that he needed Swedenborg to rewrite the Bible the correct way. God allowed the scientist to take day trips to both heaven and hell and converse with angels. For the next twenty-eight years Swedenborg wrote nearly twenty hefty volumes detailing his theological notions. He believed there was no Trinity and that the divine trio had moved into one supreme deity when Jesus of Nazareth was born. He also said that faith was not a requirement for salvation, since only works of charity mattered. Although he never married, he offered extensive advice on the subject, claiming that "righteous" couples stayed betrothed for all eternity, for better or worse. If a person dies unmarried, then a spouse will be found for him or her in heaven. He had hoped that a religion might form under the guidance of his insights, but he never went about proselytizing or seeking recruits. It wasn't until nearly fifteen years after he died, in 1772 at age eighty-four from complications of a stroke, that a new form of Christianity called Swedenborgianism arose. It was based on the belief that Swedenborg actually had conversed with God. His writings were deemed the "Third Testament." In 2009, Swedenborgians numbered about twenty-five thousand.

Man was so created by the Lord as to be able while living in the body to speak with spirits and angels, as in fact was done in the most ancient times; for, being a spirit clothed with a body, he is one with them.

—Emanuel Swedenborg

THE MYSTICAL JOHNNY APPLESEED

Massachusetts-born John Chapman was eighteen years old when he went west and began to plant and give away apple seeds: The seeds were provided to him for free by cider mills. He was never interested in running a huge orchard and only nursed the plants until they were large enough to be used for barter. By 1800 he was commonly known as Johnny Appleseed. He became a member of the Swedenborg Church of the New Jerusalem and committed himself to living an austere life, adhering to Swedenborg's principle that the more one endures in this life, the easier it will be in the afterlife. He never actually wore a pot on his head as folk legend has it, but he did prefer a ragtag outfit and was known to walk barefoot even in winter. His motivation for giving away seeds and many of the items he received from bartering was based on Swedenborg tenets of living charitably. With every apple seed he gave away, however, one usually also received a sermon about Swedenborg's teachings, Chapman stating that he had "news fresh from heaven." When he died in 1845 at the age of sixty, he had remained a wanderer and a "primitive Christian," although he had acquired title to land worth millions. It is conservatively estimated that Johnny Appleseed personally planted at least one million trees, including the 1,200 acres with its nearly 330,000 apple trees that he left to his sister. As for vices, Johnny seemed to have few, except for an addiction to snuff. Like Swedenborg, Chapman never married, and he believed if he remained single in this world he would be rewarded with two female spirits in heaven.

CHURCH OF SYNANON

DRUG REHABILITATION RELIGION

Charles Dederich, Sr., was known as Chuck D. to members of his Santa Monica, California, Alcoholics Anonymous group. Chuck told how he started drinking at age twelve because of the early death of his father, his rage over his mother's remarriage, and his subsequent failure at school, jobs, and marriage. At age forty-three he arrived in Southern California from Ohio with the intention to "become a beach bum"; instead he joined AA. Before long Chuck became a popular guest lecturer on the recovery circuit. However, he came to view the third step of the AA program—"turn your will and your life over to the care of a Higher Power"—as too religious.

In 1958, he began to hold alternative meetings in his apartment, and he then opened a storefront in Venice, California, welcoming not only alcoholics but also drug addicts, who were usually shunned at regular

AA meetings. After volunteering to partake in an LSD trial conducted at the Neuropsychiatric Clinic at UCLA, during which he read Ralph Waldo Emerson's essay "Self-Reliance," Dederich had an epiphany about how to finally cure alcoholism and drug addiction. He came to create a locked-down rehab called "Tender Loving Care," which eventually became a paramilitary cult he defiantly named the Church of Synanon.

By the mid-1960s, Chuck D. had become a nationally recognized rehab guru, known for his unorthodox, though seemingly successful means of removing the rapidly growing numbers of drug-addicted youths off the streets. Criminal courts began to send offenders to his institution in lieu of jail, and city officials even overrode code variances, the need for medical licensing, and other obstacles to keep Dederich's rehab in business. Chuck believed an addict's ego had to be deflated by extreme means of humiliation and control, geared to making members "turn their life over to the care of a Higher Power"—namely himself.

WITCHING HOUR

One technique Dederich devised for recovery was to bring two "witches," women dressed in occultist robes, into the rehab to conduct Ouija board sessions. The message always spelled out passages from Emerson's "Self Reliance," reinforcing the notion that even spirits put their trust in Synanon. (See also: *Ouija*)

He built a $10 million a year empire staffed by his shaved-headed rehab residents who worked for no pay at a multitude of tax-exempt businesses, everything from gas stations to renovating and selling valuable real estate. In 1967, Dederich grew dissatisfied with the results of his program; he saw that addicts, no matter how long they spent at his facility, often returned to drug use once back on the streets. Hence he determined that no one should ever leave the "Synanon family," and that "containment," or severing contact with family and the outside world, was the only long-term solution for "lifetime rehabilitation." When it was learned that mothers had their babies taken from them to be raised communally in a "hatchery," and that Chuck had ordered abortions and vasectomies, mandated divorces, and reassigned marital partners, critics tried to shut him down and remove tax-exempt status. He then went on a "Holy War," renaming his rehab a religion and declaring that he hoped to see "the ear [of an enemy] in a glass of alcohol on his desk." He formed an enforcement division, named "Imperial Marines," which went about beating and intimidating "splittees," and his troops once even put a rattlesnake inside the mailbox of a prosecutor. In 1978, after his own arrest for drunkenness, he avoided jail for a mounting list of allegations by simply stepping down as chairman of Synanon. Dederich left a legacy (in addition to acquiring $50 million in assets), however, and ushered in the era of "behavior modification facilities," after which the famous Day Top Village was originally modeled, and which continues in various forms at a number of "residential" drug and alcohol treatment programs. Dederich died in 1997, at the age of eighty-three.

I say this with as much humility as I am capable, which isn't very much, but when I sit down and start to talk, people start gathering.

—Charles Dederich, Sr.

T

TANTRA

SEXUAL ENLIGHTENMENT

Tantra applies to Hindu writings, originally in Sanskrit, dating to at least 1500 B.C., that offered guides on how to act and worship, and frequently employed elements of magic. Early tantric religious rites explained how certain body fluids were significant to unite clans, and to establish a holy connection to gods or goddesses. One ritual described how the orgasmic fluid of a female was mixed with the semen of a guru, and placed enema-style into young male initiates, who thereby became intimately part of the group and blessed by the clan's "nectar." This was a common practice for centuries. Sects stressing *maithuna*, described as various methods for sexual union, became popular as alternative paths toward enlightenment. Unlike the ascetic and meditative approach, this required intense involvement in the physical plane, including enjoyment of music and dance, though achieved best by passionate sexual encounters. In this theory, sex has two functions other than procreation, namely gratification and spiritual emancipation. Tantra monks had sex with many female partners, whom they dressed up in costumes to make it easier to visualize as goddesses, claiming it was not for kinky pleasure but as a way to understand the divine mysteries of the universe. Western cultures focused on the sexual aspect of tantra translations, especially as it advocated not only the mechanics of sex but deep intimacy as the only way to achieve another level of bliss or ecstasy. It's now used less for attaining divine knowledge than as a way to bond couples through prolonged passion. This is achieved by a locked embrace and by touching chakra points, or the erotic energy zones on the partner's body. The use of certain religious symbolism, mantras, or prayers is intended to give the sex a spiritual context.

In 2008, a prostitute advertised in an Internet classified ad that her rate for regular sex was two hundred dollars per hour, while tantra sex required a "donation" three times that amount. (See also: *Kama Sutra*)

TAROT

WHAT'S IN THE CARDS?

The earliest complete sets of playing cards date to sixth-century China. They featured carved or painted figures and symbols made on thin ivory tiles that were about the size of a domino. They were used primarily for gambling, and records banning them for such use began appearing in the late 1300s throughout Europe. Tarot cards used for divining and fortune-telling coincide with the migratory history of the nomadic Romany people, who became known as gypsies. Through antiquity, they were reputed to be the best at fortune-telling. While traveling over the centuries, the gypsies gathered the cards with varying iconic arts, and considered them pages of an unbound book that contained wisdoms synthesized from the many belief systems they encountered. The first recognizable deck of tarot cards decorated with medieval kings, queens, and such were produced in the Taro River valley in northern Italy in the fifteenth century. At first they were only a plaything for royalty, but eventually "readers" were employed to interpret what came to be considered a universal language of ancient knowledge told through allegory. Life for Tarot disciples is considered a "holographic experience," for which the cards provide answers. Each card can be drawn randomly, though it's believed to be chosen not by chance but by the unconscious, mirroring what lies hidden in dreams. For example, cards containing images of cups or chalices can have multiple meanings, depending on how they are drawn from the deck. Each card can yield unlimited interpretations—everything from urging the card drawer to seek inner reflection, or to avoid a friend who is concealing a dangerous flaw. Tarot experts claim the cards can stimulate philosophical inquiry and clairvoyance and help navigate one through the impending chaos that waits. Philip Young, Ph.D., a professional tarot reader, stated: "Tarot helped me work through to my authentic self and continue to help me maintain my authentic self through each evolution of my life." Many new versions of tarot cards appear regularly, but most adhere to twenty-two cards per set.

TECHNOPAGANISM

GOD IN CYBERSPACE

For this new religion, formed in the 1980s, ancient things and high-tech gadgetry are not incongruent, but connected by a spiritual energy intrinsic in all things. The digital realm is of an identical reality as that of a forest, and devotees believe the ancient elements of earth, fire, wind, and water are now accessed more proficiently in cyberspace. Technopagans usually put a gemstone on the computer, and some even have a USB cable draped ceremoniously into a pot of soil next to the monitor. The goal is to seek harmony between technology and nature. An ancient spell book or a rerun of a *SpongeBob* episode is examined with equal intent for hidden meanings. Gods vary, but Eris, the Goddess of Discord, is often worshiped to avoid viruses or program crashes. There are annual events for the more than one hundred thousand followers to meet—where no "mundanes" are allowed—though most group rituals are done online. The organizations broadly classified under technopaganism are without a coherent leadership, though it is generally believed that the collective information comprising the Internet is actually developing into a spiritual entity.

NOOSPHERE—THE COLLECTIVE BRAIN

Pierre Teilhard de Chardin (left) was a twentieth-century Jesuit theologian, scientist, and philosopher. In his famous book, *The Phenomenon of Man,* he argued for the existence of a "noosphere," or a cosmic intelligence that broke into parts at the beginning of time. These "fragments" formed each human mind. Redemption, and even the ultimate salvation of humanity, would require a coming together of all these fragmented individual consciousnesses into one source—into one way of like thinking. Many contemporary observers have begun to note that cyberspace, although still in its infant stages, may eventually fulfill Teilhard de Chardin's prediction. Incidentally, shortly before his death, at age seventy-three, Teilhard reportedly asked for a sign from the divine consciousness: "If in my life I haven't been wrong, I beg God to allow me to die on Easter Sunday." He expired on Easter Sunday in 1955.

In Dan Brown's Lost Symbol, *Teilhard's philosophy and Noetics Science play a key role in the thriller's mystery. At the Consciousness Research Laboratory of the Institute of Noetic Sciences, founded in 1973 by* Apollo 14 *astronaut, Dr. Edgar* Mitchell, *members have been trying to prove how human thought can change reality. Test subjects stare at computer screens that randomly generate odd and even numbers and they then try to break the pattern by thinking it to do so. Results are mixed. The mission of Noetics is dedicated to the "collective transformation" of humanity.*

TELEKINESIS

MIND OVER MATTER

Many who seek astral wisdom point to tapping into subliminal mind power as the key to defying the laws of physics. Put a toothpick in a bowl of water and the mind can move it, or cause a spoon to bend, or levitate a chair. Although most parapsychological demonstrations of such activities have proved to be mostly magician tricks and fail most scientifically controlled tests, proponents claim it has to do with quark theory, or yet-undiscovered mysterious atomic particles that hover aura-like around objects. Some say laboratory testing to study mystical happenings are useless and that such things can only be observed "in the wild," or by chance. Neuroscientists think altering gravity or electromagnetism through brain waves is an anatomical anomaly.

Nevertheless, Uri Geller achieved fame during the 1970s as a master of telekinesis, claiming that his gifts manifested at age four in 1950 after a "mysterious encounter with a sphere of light." However, in 1973, a demonstration on a national TV show proved disastrous, especially when producers switched the objects to which his powers were to be applied; Geller admitted he wasn't "feeling strong" enough to make it happen.

Certain miracles fall in the realm of psychokinesis and are defined as the ability to gather and manipulate free-floating energy through the power of the mind. Examples of its power have been noted throughout history. In the early twentieth century, Eusapia Palladino, an Italian medium, was famous for levitating tables. Indian Swami Rama, was observed during the 1970s and '80s moving knitting needles, demonstrating how he mastered the laws of friction and gravity. In 2001, sixty students at the University of Arizona attended a "spoon-bending party." Psychology professor Gary Schwartz instructed the group to visualize a force moving down their arms and then direct it to the hand holding a spoon or fork. They had to say "bend" three times. What would normally require a thousand degrees of heat was achieved, accordingly, by the mind. One chemical engineering student, Tiffany Nakajima, said, "I came into it pretty skeptical. But then, watching people's spoons bend, it was pretty cool."

TELEVANGELISTS

GOD ON THE TUBE

One of the main tenets of Evangelical thought is to convert as many others as possible to one's beliefs. Since radio began, many have turned to the airwaves to reach candidates, and with no FCC regulations to oversee religious content programs, anyone with a microphone and access to a studio could become a preacher. The phenomenon likewise flourished during the early days of TV. The first national Evangelical shows were made popular by Billy Graham, Rex Humbard, and Oral Roberts. Since the preacher's message can be individual, with no need to answer to a governing hierarchy, media-Evangelist ministers can and will say anything, though inevitably they ask for donations to keep the costly airwave "gospel conversions" going. In 1977, Roberts wanted to construct a hospital complex that employed prayer and medicine as a method of healing, after he claimed a visitation from a nine-hundred-foot-tall Jesus. Although no one else saw this gargantuan miracle, viewers of his TV program entirely funded the construction of the City of Faith Medical and Research Center in Tulsa, Oklahoma, the first hospital that used faith healing as part of a patient's treatment. The facility went out of business after eight years. In 1983, Roberts went on TV and claimed that Jesus told him to find a cure for cancer; he thus received another windfall of donations. Breaking records for most donations received in the shortest time, he told listeners in 1987 that he had to raise $8 million in a few months or "God would call him home." The televangelist got $1 million more than he requested.

The most successful, in terms of number of viewers and dollars earned, were and are televangelists who specialize in what's called "prosperity gospel," which provides an upbeat and a hopeful message. A prosperity preacher states, for example, that he knows that it's God's will for every viewer to be healthy and wealthy. Some will reach into the audience and heal an apparently infirm person balancing on crutches, and with a touch and an impromptu prayer send the healed running up the aisle and out of the theater church at full speed. Many use the passage from Corinthians "Remember this: Whoever sows sparingly will also reap sparingly, and whoever sows generously will also reap generously" as the basis for asking for money and donations, implying that every dollar the viewers send will be multiplied "in God's time." Since most are registered as religious organizations, TV ministries are tax-exempt. In addition, the churches need not file various tax forms, as nonprofit organizations do. A number of televangelists have favored luxury cars and lavish lifestyles, and were subjected to national disgrace when implicated in scandals, including being caught at the very "sins" they had advocated avoiding.

In 2008, the top televangelist preacher of the "prosperity gospel" category, Joel Osteen, had known revenue of more than $75 million. Although he received no formal theological training, Osteen has since been called upon to offer his new interpretation of the Bible to more than seven million viewers every week. (*See also*: Sister Aimee)

POINTING RIGHTEOUS FINGERS

Jimmy Swaggart, a pioneer of the televangelist movement, exposed the sexual infidelities of two competitors he believed were taking his audiences and ratings. In 1987, Swaggart effectively squashed the career of televangelists Jim and Tammy Bakker, after revealing Jim's extramarital affair. He publicly called the former friend "a cancer in the body of Christ." Retribution came when Swaggart was caught with a prostitute two years later, and eventually lost credibility with his viewers when found with yet another hooker in 1989. At that point he went before his congregation and brushed away the indiscretion, saying, "The Lord told me it [the scandal] is flat none of your business."

God knows where the money is, and he knows how to get the money to you.
— Gloria Copeland, Prosperity Preacher

ST. TERESA OF ÁVILA

PATRON SAINT OF SPAIN; ECSTATIC VIRGIN

During the early 1500s, Teresa's Jewish grandfather from Toledo, Spain, had converted to Christianity, but the Inquisition still killed him for secretly studying the mystical Kabbalah. Thereafter, her mother and father made extra strides to appear unquestionably Christian. Teresa's mother diligently drilled stories about the lives of saints into the child's head from the moment she was born. No wonder by age seven Teresa had run away with an older brother, both traipsing toward the Moorish-held territories with the hope of being beheaded and turned into martyrs. Her father retrieved his children, and all seemed well until Teresa's mother died. Soon after, the widowed father watched as his thirteen-year-old daughter made a scene by asking various statues of the Virgin Mary throughout town to be her mother; her father placed Teresa into a convent.

Said to be extremely beautiful, Teresa worried about the sin of her own vanity and sought to purge the desires and temptations of vainglorious thoughts by inflicting pain upon herself. She became obsessed with studying Christian doctrines on ways to achieve ecstasy, and she learned of all the popular meditative techniques and assorted methods of practicing corporal punishment. Anorexia, bulimia, enemas, or other odd means of purging were employed as the path to achieve the fastest "union with God." She began to claim divine visions, often involving angels and devils, until she ultimately announced that Jesus had inhabited her body. Teresa claimed that the savior's spirit occupied her body for hours on end. Typically, Teresa was observed smiling as she chatted with her Jesus within, though ultimately she became white-eyed, trembling in rapture. Fellow nuns attested to witnessing Teresa levitate. She eventually came to be known as Teresa de Jesus and wrote long and passionate books on mystical ways to reach God, primarily through the use of pain and extreme asceticism. On finding God she wrote: "The pain was so severe that it made me utter several moans. The sweetness caused by this intense pain is so extreme that one can not possibly wish it to cease, nor is one's soul content with anything but God. This is not a physical but a spiritual pain, though the body has some share in it—even a considerable share." Teresa died in 1582, at age sixty-seven. She was canonized sixty years after her death and is one of only two women recognized by the Vatican as Doctors of the Church. Her body was dug up numerous times; it remained nearly intact and was said to give off a sweet aroma. Over the centuries her corpse was cut up and divided. Her hands, her right foot, her right arm, her left eye, and her heart are currently kept in separate churches for display.

THUGGEE CULT

MUGGERS FOR GOD

The English word *thug*, as in brutish thief, derives from a Hindu cult devoted to the goddess Kali, a deity associated with death, destruction, as well as time and change. In the 1300s a secret society of men took to worshiping this dark figure, who is often portrayed as wearing a necklace of skulls and brandishing a sword in one hand and a severed head in the other. A violent goddess, savage in battle, Kali was noted as drinking the blood of her conquests and growing stronger in the frenzy. The cult of Kali, whose members the English simply referred to as "Thuggees," had worshiped the goddess by killing an estimated two million during their seven-hundred-year reign. Their method was to befriend a traveler, then accompany them for days, seemingly as friendly guides, until the victim was led to a certain sacred spot and offered to Kali by strangulation. Of course, they only picked victims with cash, and if a male child accompanied the dead, those would be raised by the secret cult to learn the ways of Thuggees. They were nearly impossible to bring to justice, since the only weapon they used was a ceremonial yellow scarf, and they brutally punished those who testified against them. When Thuggees were apprehended, they remained religiously silent, and if capture seemed imminent, they often cut off their own tongues to protect their secrecy.

By the 1800s, the group's religious connections seemed to wane as many joined the cult primarily to rob rather than worship. Eventually a leader was tricked into being captured in 1830 and identified by the ruling British authorities as the notorious Thug Behram. With promises of leniency, Thug caved and gave details of the cult's means of operation. He later became known by some accounts as the world's most infamous serial killer, when he admitted to more than 930 murders, even if he referred to it as worship to his goddess. He later offered the following correction: He had personally strangled only two hundred or so but had been present when his cult members, always working in twos and threes, had killed the rest. He was promised immunity for testifying against his cult, but Thug was hanged nevertheless in 1840. Kali cults still exist, though they have tried to distance themselves from their questionable past, worshiping their goddess in ways other than by strangling strangers.

TRANSMIGRATION OF SOULS

DIVINE POPULATION CONTROL

The ancient Greeks believed that after death the eternal soul drank from the Lethe River in Hades, which contained an ingredient to cause complete amnesia. When reborn, the Greek had zero memory of past lives. Hindus believed in transmigration, called reincarnation, though these rebirthed souls retained some memories of previous lives, with the purposes of working out defects to eventually achieve a state of enlightenment. Without this, one would have to endure an endless cycle of death and rebirth. Proponents of this theory point to population statistics: The world population is estimated to be approximately six billion, while the total number of humans who have died since records were kept in the 1600s is also six billion, or close to it. This has led some to surmise that nearly all relatively recent and once-living souls are back again to live in this time, albeit in new bodies. Variations on this belief include ancestral incarnation, in which the rebirths of the dead follow genetic lines. In theory, you travel in the endless dying and rebirthing with the same group of individuals. With this program, you may have been the father in one past life, for example, though now you are the daughter, or instead, your current mother may have once been your brother during one of your shared repeating lives. Every person one meets and feels a "bond" with was of this endless spiritual recycling group and you will continue to meet time after time, until conflicts are resolved and enlightenment is attained.

THE WHEEL OF 84

Paul Twitchell founded a spiritual movement in 1965 called Eckankar (ECK), which teaches methods on how to achieve higher levels of consciousness through "soul travel." Mantras, breathing exercises, and keeping dream journals are said to allow the soul to leave the body and travel freely to other planes, where divine wisdom is accessible and less conflicted by daily distractions. The Eckankars' bible, *Shariyat-Ki-Sugmad*, translated as "Way of the Eternal," emphatically states it is the only path to God. Karma and reincarnation are part of ECK's belief system, with the Wheel of 84 referring to the endless cycle of rebirths. Twitchell believed the soul came to earth first as a mineral, himself being some sort of stone eight million years ago, who had to go through the life of a fish, then a reptile, then animal, until reaching humanhood. He said he got off the Wheel of 84 during his present life and expected no more reincarnations when he died, which he did in 1971 from a heart attack, at the age of 159. Twitchell had previously claimed he was born into his last body in 1812, though conflicting birth records were discovered that indicated he was either born in 1908 or 1912, meaning he took his last spin on the wheel instead at age sixty-three or fifty-nine.

TRANSUBSTANTIATION

SPIRITUAL ALCHEMY

Central to many Christian religions, transubstantiation refers to the process by which an ordinary wheat wafer, or Eucharist, along with sac-

ramental wine is changed into the literal body and blood of Jesus Christ. The wafer does not, however, transform into a miniature Jesus; instead it retains its size, shape, color, and taste. After the wafers have been concentrated, all unused ones must be kept in a vault, since transubstantiation is not reversible. The unused wine-turned-blood is usually ingested by the attending priest. The transmutation principle, turning a substance such as iron into gold, was a dubious though popular alchemist quest for eons. Nevertheless, early church leaders used this notion to explain how bread and wine could be transformed. The Eucharist's change, however, did not rely on an alchemist's magic, but rather on a divine hand. Transubstantiation is solely credited to a miracle that God willingly does every single time one wafer is offered for consecration. Nothing had been formally written about this subject for ten centuries, even if the transubstantiation "sacrifice" was done among Christians nearly continuously since the first century. The Council of Trent, convened in the 1550s, was held to get straight all Catholic Church ideas on the subject and put it down on paper, primarily to squash the uptick of the Protestant Reformation. Council members first used the word *transubstantiation* to mean that Jesus is "really, truly, substantially present" in the concentrated bread and wine. Church dogma remains relatively unchanged from that time. (*See also*: Original Sin)

The scholars at Trent ultimately relied on scriptures and Jesus' words, as they were recorded by the Gospel of John, as final proof. Jesus was noted to say, "I tell you the truth, unless you eat the flesh of the Son of

Man and drink His blood; you have no life in you. Whoever eats my flesh and drinks my blood has eternal life. For my flesh is real food and my blood is real drink." Jesus reiterated the same message with a bread and wine sermon at his last meal, and said, "Do this in memory of me." Problems arose where some scholars even then argued that this sounded as if Jesus had advocated cannibalism. Others simply wondered if the changed wafer was the body of Jesus while he lived on earth, or if it was transformed into his divine form as he was in heaven. For those who believe in the process, many attest to feeling warmth after ingesting a consecrated Eucharist, even when medical monitoring of such activity shows no change in body temperature.

RECOMMENDED DOSAGE

The Catholic Church does not have a regulation concerning the number of times one is allowed to receive the Eucharist. Priests can consume it in unlimited quantities, and they are required to do so during as many masses they may happen to serve in one day. For laypersons, receiving it more than once a day is considered abuse of the sacrament. However, a loophole does exist. The church still adheres to a calendar that counts a new day beginning at 4 P.M. Thus one is allowed to receive communion at a morning Mass, and again at services conducted in the evening—without being accused of Eucharist overconsumption.

JOHANNES TRITHEMIUS

THE MONK WHO INVENTED CRYPTOGRAPHY

During a blizzard in 1482, a young man sought refuge in an abbey. It was a poorly run Benedictine German monastery, but twenty-one-year-old Johannes Trithemius decided to remain and was subsequently elevated to the status of an abbot within a year. At once he was fascinated by the abbey's meager collection of a few dozen books, which he increased to more than two thousand volumes during his tenure. Although a priest, he was a thorn to church officials, since he seemed most interested in magic. He wrote tremendously lengthy volumes, over fifty books, the most famous being *Steganographia*, filled with secret codes and messages detailing how to communicate with angels. He devised a system that

identified spirits who governed various aspects of reality and he showed readers how to reach them through coded cover letters, presumably read by intermediate spirits, who in turn delivered them to an angel higher up the chain.

Despite his tendency toward nonstandard theology, Trithemius was valued and gave advice to European monarchs, and became legendary as one of the largest ancient book collectors of his time. He's regarded as the father of cryptography and spurred interest in mysticism and magic during his time and to this day. Trithemius admitted many of his works were about magic, though meant for the astute, saying he encrypted it purposely with many layers, buried within subtexts, predicting that for the "thick-skinned turnip-eaters it might for all time remain a hidden secret, and be to their dull intellects a sealed book forever." In the late 1880s interest resurged and a group of intellectuals and socialites formed the Hermetic Order of the Golden Dawn using Trithemius's writing as a basis, as well as other encrypted texts. However, when founding members MacGregor Mathers and Aleister Crowley used Trithemius's spells to conjure up vampires and a regiment of demons to attack each other, splintering occurred, until the group disbanded in 1903. (*See also:* Aleister Crowley) Trithemius died in 1516, at the age of fifty-six, while stationed in the drafty Abbey of St. Jakob in Würzburg, most likely of pneumonia.

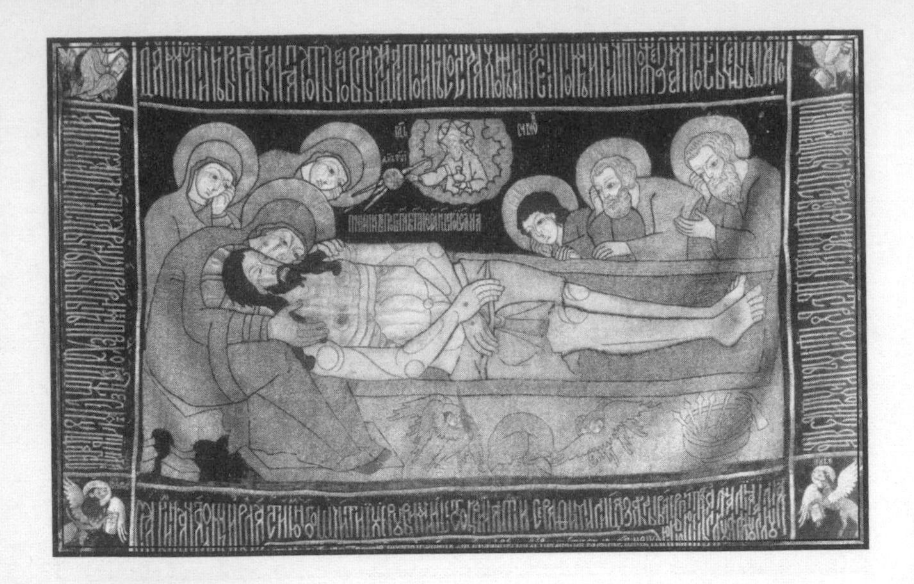

SHROUD OF TURIN

SACRED LAUNDRY

One linen cloth, measuring 14.3 by 3.7 feet, bears the death mask of a man with wounds matching punctures made by a crown of thorns, as well as showing a body outline with stains near the wrists, insteps, and chest area, typical of a body crucified. It is said to be the burial garment Jesus had been wrapped in after he was taken from the cross and laid in the tomb. It depicts a bearded man with shoulder-length hair, parted in the middle, and approximately five feet, nine inches in height. It was said to be discovered by Mary Magdalene when she found the burial vault empty, and it has passed among the faithful as a sacred relic ever since.

References to the existence of such an iconic artifact were noted as early as 40 A.D., and all subsequent art renderings depicting Jesus as he often is were taken from the shroud's image. Some believe Jesus left this impression of his tortured body to prove to doubters that he was once really here on earth. By the 1300s the original linen was lost, or hidden so well that more than forty-three people claimed to hold the true shroud. But the one that is housed in Turin, Italy, remains to this day the shroud considered most authentic. It was taken from Jerusalem by a crusader in 1204 and hidden by the Knights Templars. After the group was disbanded as heretics, the shroud was made available to the public in the mid-fourteenth century, though from the start it was questioned as a fraud and a fabrication. Some radiocarbon dating has corroborated that it was probably created in that era, while other tests refute those findings and suggest the carbon found was caused by a fire that damaged the cloth during the Middle Ages. In 1958, Pope Pius XII made its celebration a feast day, which greatly legitimized the relic's authenticity. Forensic studies have indicated the red stains were actually formed in part by blood, specifically the type AB. New researchers are requesting to try a cloning experiment with whatever DNA can be salvaged from the cloth, but the Catholic Church has refused such requests. Vatican officials remain silent on the theological implications of creating a cloned or scientifically enhanced Second Coming of Jesus of Nazareth.

FRIDAY THE 13TH

In 1119, a shabby bunch of nine Frenchmen found some armor and helmets and fashioned themselves as knights as they rode—two or three to a horse—into Jerusalem during the first Crusade. They looked for sponsorship by declaring themselves a secret society on a mission to protect Christian pilgrims. Eventually, they were given a place to establish camp near the ruins of Solomon's Temple, and changed their name from Poor Fellow Christian Soldiers to the Knights Templars. They came to be highly regarded for military success, with a command of ten thousand soldiers, though the leaders were noted to be shrewd in acquiring valuable relics and subsequently vast wealth. The knights had retrieved the Shroud of Turin and kept it hidden for more than a hundred years. It was said to be part of their initiation ceremonies, including a ritual that required knights to kiss the image's feet three times. When the Templars became more powerful than the king of France, Pope Clement V disbanded the society, citing rumors of sexual immorality, sodomy, and worshiping false idols. After conviction of "grave sins," every Templar house was raided simultaneously on Friday, October 13, 1307. Two thousand knights were imprisoned and tortured, with sixty known executions. The Grand Master, Jacques de Molay, was torched at the stake in 1314, though through the flames he called out a curse against the pope and King Philip. Both died within the year. Thereafter the shroud and other Templar relics were made public, in addition to forever marking Friday the 13th as an unholy and unlucky day.

UNARIUS SCIENCE

THE INHABITANTS OF PLANET VENUS

The Unarius Academy of Science has a library of more than one hundred books, all of which were supposedly written by intelligent extraterrestrials. In the 1950s, the author-scribe, Ernest Norman, didn't have to go white-eyed or into a trance to absorb this channeled information, but was able to have his alien source speak clearly into a tape recorder. The goal or "mission" of this effort was to enlighten the world about fourth-dimensional physics. Accordingly, the extraterrestrials told Norman that the Chinese are descendants from an outer-space civilization, part of a thirty-three-planet confederation inhabited by advanced beings. The Chinese once had a flourishing empire on Mars but migrated to earth when the red planet's environment changed. One of Norman's guides, Mal-Var, lived in the capital of Venus, in the city of Azure, and explained how mental illnesses are treated there with positive energy. His words were developed into a course of study available to Unarian followers. The inhabitants of Venus are a type of energy being that cannot be seen by the naked eye. However, spiritually developed humans can visit the Venusian metropolis during sleep, and are reportedly given a warm welcome upon arrival. In addition, Unarians learn of meditative techniques to travel freely to other worlds.

Thirty-three spaceships, with a teacher from each planet, were scheduled to arrive on earth in 2001. They were not intending to be revered as gods, but instead would arrive to conduct seminars. Either due to dyslexia or miscalculation, the date for the arrival of alien teachers was modified to 2010. During the 1950s and through the '60s, more than fifty thousand people devoutly followed Unarian theology. Currently, less than three thousand continue to decipher the original text channeled through Norman and believe it the only superior route to salvation.

MAYAN SPACESHIP CALENDAR; THE END IN 2012

Complex calendars were invented by Mayans as early as the third century B.C. There are a number of systems for counting days and seasonal changes that overlap among Mayan calendars, but one called the Venus Cycle used an annual measurement of 585 days. Some Unarians believe this was devised by

Mal-Var's people, and that the Mayans were a culture the alien guides visited often. That Mayans had accurately understood the length of a Venus solar year before the invention of telescopes proves to Unarians that the Mayan culture was assisted by aliens. Interestingly, the Mayan calendar stopped counting time on December 21, 2012. Either this date marks the end of the world, or for the more optimistic, it is the day when extraterrestrials will give us a new calendar to follow. (*See also:* Bede)

UNICORNS

MYSTICAL BEASTS

Early Greek naturalists believed unicorns existed somewhere in India. They were said to have a spiraled black and red horn, with a snow-white hide. Aristotle referred to the creature in his writings as an "Indian ass." The notion of the unicorn as a mystical living thing took hold during medieval times, when legends of its powers grew. By then, the unicorns' horn was a valuable commodity and thought capable of detecting poisons. It was used as a prized drinking mug by royalty. It was also commonly known that powder found in the broken-off horn of a newly killed unicorn was potent enough to cure any illness and even

resurrect the dead. The beast came to be associated with purity; thus many believed only virgins could find the few remaining. It was also thought to impregnate pious virgins who happened to merely spot them in the wild, and the beast was ultimately held as a symbol of fertility. (*See also*: Immaculate Conception)

In 2008, a roe deer with a single horn was born in a wildlife sanctuary in Italy. The creature is currently under study to determine this anomaly, and its genetic relation to the legendary unicorn. (See also: *St. Guy*)

Even if there are no fossils of single-horned, horselike species, the hunting of beasts with horns was good business in ancient times. Horned animals were admired for their strength and used by many early religions to symbolize hallowed authority. Both male and female royals wore headpieces and turbans that featured a horn protruding from the top, until the fashion demanded having larger and larger ones to balance on the head to signify greater wealth and prestige. Scientists believe that this historically noted fashion rage may have killed off any version of the real unicorn that may have existed during the fifth century B.C. Others believe that unicorns are still alive, but most serious devotees insist the mystical creatures now reside in peace in the astral plane, even if they do indeed appear occasionally to the lucky and spiritually pure. A number of groups conduct unicorn studies, most notably the International Society of Cryptozoology, a scholarly organization dedicated to "documenting and evaluating evidence of unverified animals of unexpected form or size, or unexpected occurrence in time or space."

UNICORNS IN RELIGION

In the book of Genesis, Adam was instructed to name the animals. The first to get a name was the unicorn—and from that time on, unicorns had the highest place among beasts. When Noah loaded all the animals onto the ark, the unicorn declined, since it was noted as a powerful swimmer. When the rains stopped, Noah's search birds initially could find no dry land and took rest by perching on the swimming unicorn's horn. Subsequently, the weight of these birds eventually drowned all previously surviving unicorns. With so many unicorn legends remaining alive for eons, told through oral history and in writings, the early Christian Church saw unicorn mythology as similar to the conception and suffering of Jesus Christ, and thus modified unicorn imagery and its symbolism. In art, the unicorn was depicted asleep on the lap of a virgin (see illustration above), and there was even a cult in the twelfth century worshiping the Virgin Mary, calling her Maria Unicornis, or Mary of the Unicorn.

UNITARIAN UNIVERSALISM

THE LIBERAL RELIGION

It might seem fitting that Unitarian Universalism is listed near the end of alphabetically cataloged religions, since it's a sort of catchall congregation, welcoming members with all theologies and beliefs. As for conversion, you become a Unitarian Universalist when you say you are, without the need for baptism or any specific ritual, and it's merely suggested that one sign a membership book. Nearly 50 percent consider themselves atheists and agnostics, while others are polytheists, such as Wiccans or pagans, while some lean toward Christianity, Buddhism, or Hinduism. Theoretically, a Satanist can also join, though he or she must have an "open mind" and try to live by seven principles, summed up by the last tenet: "Respect for the interdependent web of all existence." There's no specific holy text, and the Bible is thought to be "mythical or legendary," though recommended to be read with a "critical eye." The central belief is similar to what Superman said before taking to the air: The comic hero fought for "truth, justice, and the American way." UUs fight for (or at least rally for and support) "peace, liberty, and justice for all." Originally, Unitarian churches developed from the Protestant reforms in the 1600s and primarily discounted the Trinity, advocating belief in only one Supreme Being. The Universalist Church, before merging with this synergic religion, also stemmed from Protestant reformists and believed that God would not have made any person for the purpose of sending them to hell. It is believed all are intrinsically good, and early Universalists became the first to actively form groups to protest slavery and then later work as suffragettes. The two churches combined in the 1960s and claim about eight hundred thousand members. There are still Sunday services, with sermons and hymns, and the basic symbol is a chalice with a candle in it, rendered in an abstract way that somewhat resembles a martini glass with a flaming stirrer. Marriage rites and funeral services are individual and, along with suggestions made by ministers or deacons, can contain passages from any religious or spiritual tenet. The vigorous pursuit of righting what are considered political or social injustices is considered the best way to put UU beliefs into action. (*See also*: Deism)

URANTIA BROTHERHOOD

RELIGION REVEALED TO SLEEPERS

In the 1930s, fifty-five supernatural beings supposedly formed a committee in order to channel the true history of the earth and Jesus directly into the heads of a few former Seventh-Day Adventist members. The information was transmitted to these few chosen conduits while they slept. The wisdom revealed became a two-thousand-page handwritten tome, *The Urantia*. It was originally kept hidden and studied in secret by a handful of devotees, then finally published in 1955. Afterward, followers of the sect swelled to the tens of thousands. However, the information needed to be effective in this religion was anything but straightforward. *The Urantia* (*Urantia* meaning "earth") described a complicated galaxy of other beings and spirits that have been eternally affecting humanity. Accordingly, the universe is managed by an angelic hierarchy, which includes many Supreme Beings, Ascending Beings, Sons of God, and Universe Power Directors. The goal of life is to evolve spiritually and to reach the "Isle of Paradise," an afterlife location somewhere in the heavens that only an enlightened soul can find. Much of the channeled text is dedicated to rectifying biblical history, for instance, and it claims that Jesus was actually once named Michael of Nebadon. However, the Creator-Father didn't just send Michael, but had made over six hundred thousand versions of Jesus throughout history; it's supposed that as many as ten or more duplicates of Jesus are living at this very moment. The religion is centered on study of *The Urantia*. Less than 1,500 full-time members remain faithful to the task.

ST. URSICINUS

PATRON SAINT OF STIFF NECKS

The sixth-century abbot Ursicinus became particularly obsessed with the pain Jesus suffered on the way to Calvary Hill. He'd get a stiff neck himself just thinking about how difficult it must have been to carry a heavy cross. During his mission to convert pagans in Switzerland, he took special interest relieving the stiff-neck sufferer. He was apparently skilled at massage and miracles involving the neck, and was subsequently named the patron of all related maladies. Those with chronic neck pains are urged to pray on his feast day, December 20, for extra relief.

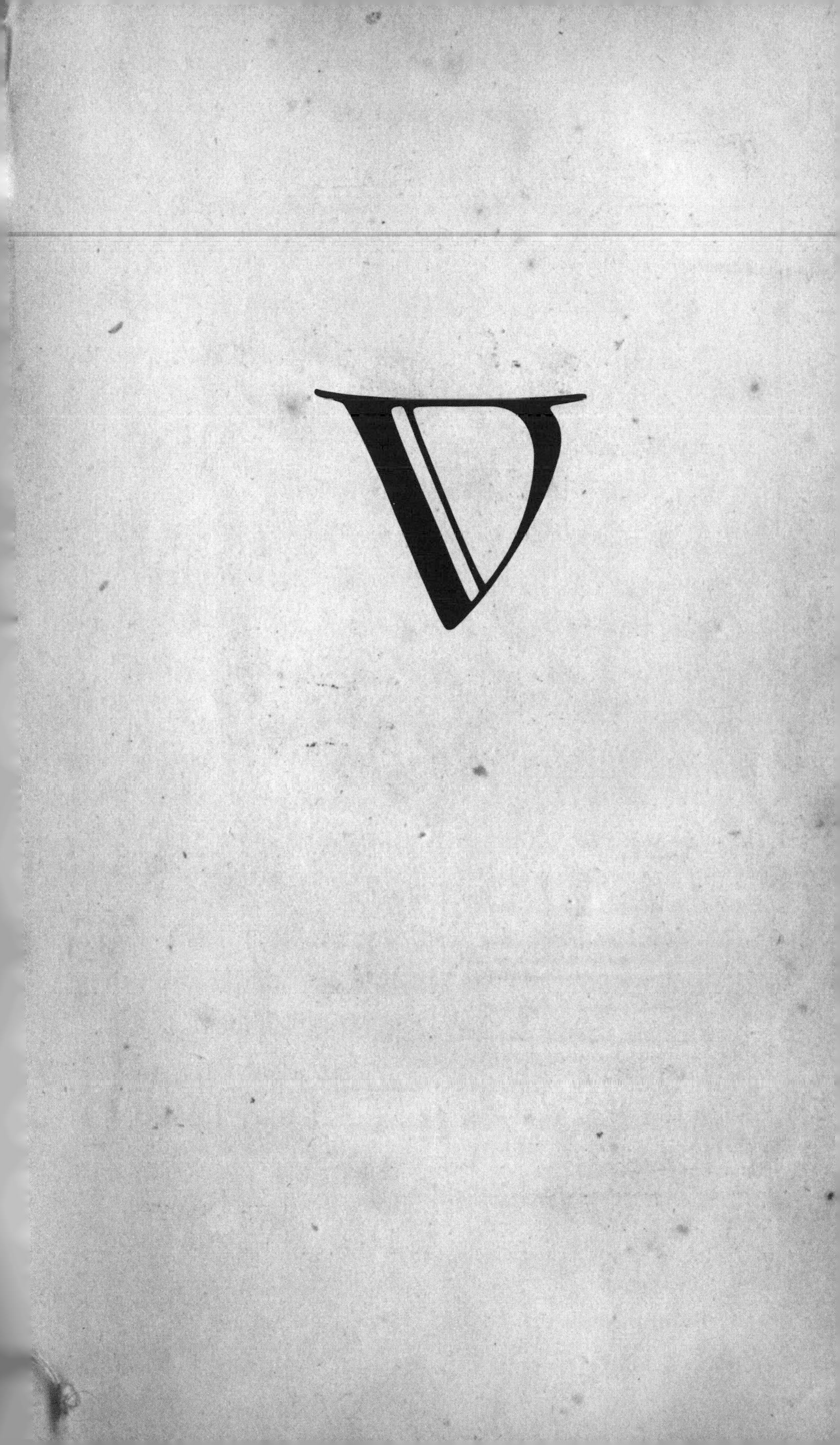

VAMPIRE, TEMPLE OF THE

"WE ARE DEADLY SERIOUS"

Since 1989, the Temple of the Vampire has been incorporated as an official religion, upholding a belief that vampires are not fantasy and should be afforded worship. The religion traces its lineage to ancient Sumeria. Its theology is contained in an official text, *The Vampire Bible*, which "enables you to acquire authentic power over others, build real wealth, achieve vibrant health, and even live beyond the usual human lifespan." It does not condone drinking blood, at least publicly, but it does offer techniques on how to consume others' "life force." The goal of the religion is to become one of the Undead, or achieve an immortal "Vampiric condition." Initiates must learn how to become a predator of humans by following study guides and moving up from the ranks of priest or priestess, sorcerer/sorceress, and finally to become an enlightened "Adept." This is the only afterlife belief, and if you do not "promote the spread of Vampirism," when you die your existence ends as dust. In addition, you will also be unprepared for the impending Vampire Apocalypse, when humanity will become the rightfully subjugated minions to the advanced vampire race. The moral code for a temple member is simple: "Test everything—believe nothing." You must acquire the superior attitude of a vampire, or act as such until you become one. One sign that vampire transformation is progressing, for example, would be to quit wasting time working at some menial job and instead through your powers make others give you money willingly. As the temple Web site claims, "This is not a game. We are all deadly serious." It is also inferred that buying the sacred fifty-two-page starter book from any source other than the temple is risky, bringing the possibility of malevolent curses. The society is so secret it will not divulge the names of its members or the number of followers.

VEDANTA SOCIETY

MEDITATION GURUS

Vedantas strive to find the divine within and achieve "cosmic consciousness" through the study of Sanskrit literature, especially the Upanishads, a Hindu holy book written throughout the ages beginning in the first century B.C. It was made popular in the 1890s by Swami Vivekananda, who is often credited with bringing knowledge of Hinduism and yoga to the West. There are more than sixty-five thousand members of the various Vedanta groups dedicated to "liberation for oneself and service to mankind." It requires finding the best swami for your present reincarnated state and spending considerable time in yoga meditation, usually repeating the classic mantra—"OM"—to trigger enlightenment. One dedicated meditator, Ashrita Furman of Queens, New York, attributes his holding of 237 Guinness world records to his practice with mantras. His achievements include eating the most M&Ms with chopsticks, and skipping a mile in record time.

YOUR OWN "OM"

OM is considered a sacred sound in Hinduism and many Eastern religions. It was originally the prayer said before reading the Vedas, and at closing meant "shout praise." It's thought most effective when an OM rises from the lungs and is exhaled through the nasal cavity, keeping the mouth closed. Also, it has come to be regarded as the most succinct of all universal sounds a human can make to express divine harmony and wisdom, particularly when directed to the Upanishads. It's also believed that when a sun or galaxy gets sucked into a black hole, all the noise of those lost worlds leaves only one sound echoing through the universe—OM.

The first study concerning the physical affects of meditation was conducted at Yale University in 1931 and concluded that yoga breathing exercises, called *pranayama*, improved cellular and blood oxygenation by 24.5 percent. In the 1950s, field scientists M. A. Wenger and B. K. Bagchi spent months traveling through India checking heart rates and pulses of meditating gurus. They discovered that although they appeared to be in a coma-like state, a meditator's heart rate increased during moments of ecstasy. They also found practitioners who could make their foreheads sweat at will, and one who was able to regurgitate on command to cleanse his body. More recent studies focus on heart rhythm and brain wave activity, with some reporting that during deep meditation, beta waves, associated with heightened thought activity, intensified even when the subject appeared asleep. Many gurus have been tested with monitors and rectal thermometers during meditations to prove the ability of the mind to control vital signs. Neuroscientists believe insertion of electrodes in the brain can stimulate the same effect. A battery-operated remote control will switch on an implant to allow a user to schedule transcendental experiences, and even gain enlightenment as required.

In the 1960s, Maharishi Mahesh Yogi, founder of Transcendental Meditation, personally gave the Beatles a mantra that was turned into lyrics in their song "Across the Universe." The chorus, "Jai guru deva om," roughly translates as "I give thanks to the heavenly teacher."

VESTAL VIRGINS

TEMPTING TENDERS OF THE FLAME

The goddess Vesta, noted as the tender of the hearth fire in mythology, eventually became the most revered deity in Rome, signifying its very well-being and good fortune. Eighteen priestesses were assigned to keep a perpetual symbolic flame burning in the goddess's temple, and they were

virtually the only nonroyal females in Rome afforded such a high position and status. Potential Vestal Virgins were chosen between ages six to ten and judged on lineage, looks, and good health. Both parents had to be examined as well, in order to see how a Vestal Virgin might look in thirty years. That was the length of commitment required of this holy position. Their main function was to tend a continually burning fire in the circular Vesta Temple, located close to the Roman Forum. The virgins also supervised the preparation of ceremonial baked goods. In addition, a Vestal Virgin had to remain celibate for the entire length of service, though she could marry after thirty years. Treated as living goddesses, they were carried about in special coaches, had front-row seats at the coliseum, and lived in luxury. Their power was such that if by chance a prisoner on his way to execution were to bump into one during the course of her duties, an immediate pardon was granted. Romans looked to the Vestals' purity as a divine omen and even their blood was considered sanctified. For an ordinary citizen to simply flirt with one of these priestesses would warrant having an eye poked out. If anyone assaulted the women, or if a virgin herself were to break the celibate vow, the accused would meet capital punishment. Since a virgin's blood could not be spilled, any Vestal priestess who by chance let the fire dim or was discovered less than chaste was buried alive in a vault and typically given only a few days' worth of food. The sacred Roman fire was finally put out and the cult disbanded in 394 A.D., after an uninterrupted run of nearly five hundred years. The last known chief Vestal Virgin priestess was one Coelia Concordia. She sidestepped being buried alive by converting to Christianity.

VOODOO

HAITIAN ZOMBIE WORSHIPERS

Voodoo is an Americanized form of *Vodou*, which is a West African word for spirit and primarily refers to angel- and apparition-worshiping religions. Archeological evidence indicates that these beliefs and rituals are six thousand years old. The people brought as slaves from various African regions to the French colony of Haiti held to their traditional ways of Vodou worship more vehemently than in other colonies, even if it was synthesized with numerous Christian canons over time. When Haiti's slaves staged the first successful rebellion and dispelled all whites in 1804, Vodou became the official religion for the next seventy years. Now Haiti is officially nearly 100 percent Catholic, even if many still practice ancient Vodou rituals, especially after the religion was made legal once again in 2003. Currently, saints' images are often used to represent Vodou spirits. In the religion's original mythology, Bondyè was the supreme creator, though he was disinterested in mankind's daily affairs. Contact and intercession with the supreme deity was made through lesser spirits, called Loa. Houngans and Mambos, or the priests and priestesses of Vodou, knew of the rituals compensatory to do this. The magic spells and rituals they used, as is the case today, were primarily for healing. Group ceremonies are conducted around a pole, or *poteau-mitan*, with iconic symbols drawn on the floor around it; the spot is considered the portal to communicate with spirits. Animal sacrifices using the blood of goats, chickens, and dogs are dripped on participants or sipped from a chalice. Reports of orgies refer to a rarely performed fertility ritual. A women seeking fecundity will lie naked and prostrate while a Houngan slits the throat of a white chicken and drips the blood on her body. He will then have sex with her, and can call on the drum players to do so as well if the woman wishes. Voodoo sorcerers, or *bokars*, practice darker magic and are believed to know how to revive the dead and turn corpses into zombies. The will of the deceased is controlled by the *bokar*. Anthropologists investigating this phenomenon surmised that individuals considered zombies had not actually died but rather had been given various toxins, some citing fugu, the poisonous puffer fish, as the cause for the trancelike state.

PLEASE CHECK YOUR SKULLS BEFORE TAKE-OFF

In 2006, when a Vodou priestess from Miramar, Florida, returned from Haiti, she was arrested at Fort Lauderdale Airport: A still dirt-crusted skull of a man who had died thirty years ago was discovered in her carry-on luggage. More skulls come through the Miami International Airport, so many they have a "Bone Room" to collect confiscated human parts. The skulls are very powerful and contain spirits to help in life's affairs, which Vodou advocates—there are eight million worshipers—claim are no different than Christianity's saintly relics.

WAY INTERNATIONAL, THE

BIBLE-RESEARCH CULT

In the 1940s, Victor Wierwille, a radio host and Evangelical minister, heard God through the headphones during a snowstorm when the power was supposedly cut off. Amid the static, a voice informed the talk-show host that he had been selected to fix the Bible and tell the world of the true interpretation: "He spoke to me audibly, just like I am talking to you now." Some of Wierwille's new findings included evidence that Jesus had three crucified beside him instead of two, and that the Christian savior was born through a natural union of Mary and Joseph; instead of divine incarnation, Jesus was conceived from Joseph's "sinless sperm." Wierwille denied the concept of the Trinity, though he gave lessons on how to communicate with the Holy Ghost. He eventually incorporated his religious views into the Way International in the mid-1950s.

The sect was accused of becoming a mind-control cult after Wierwille's various indoctrination methods were learned of, particularly the sect's policy of "Mark and Avoid," which demanded that followers abandon their family and friends. In addition to scripture studies, orgies and firearm instruction were incorporated into the training program. The combination seemed odd, though Wierwille also preached about the coming apocalypse and clarified that skill in handgun use might be required for his chosen ones at doomsday. Wierwille also claimed that Adam and Eve's original sin was really masturbation, surmising that the "forbidden fruit" were their own genitals. For Wierwille, sex with another was divinely better and less sinful than when practiced alone. Unsurprisingly, sex abuse allegations began to emerge against him. When Wierwille died of carcinoma at the age of sixty, in 1985, the organization claimed more than one hundred thousand members.

He had supported the organization with a minimum 10 percent tithe of all devotees' personal income. Today the group has less than five thousand followers and considers itself a nondenominational biblical research center.

WESTBORO BAPTIST CHURCH

PREACHING "GOD HATES AMERICA"

Although the Westboro Church is not linked with other Baptist associations, the Topeka, Kansas, congregation holds the distinction in recent times as having organized the most demonstrations and picket lines to spread their views and gain recruits. They believe in Jesus, but hate homosexuals, AIDS fund raising, Catholics, Jews, Muslims, and even American soldiers killed in Iraq, to name a few. They attend funerals of fallen military personnel, or ceremonies for deceased homosexuals, with the intent to disrupt services and make protest signs stating that "God hates America," reasoning that Jesus is displeased with America's tolerance of gay rights. This congregation believes any tragedy and subsequent suffering, from earthquakes in China to Indian tsunamis, are events that make "God happy." In 2008, the church had about seventy members, of which sixty or more were related by blood or marriage to the spiritual leader, the disbarred lawyer and self-proclaimed "pastor" Fred Phelps, Sr. (*See also*: KKK)

RELIGIOUS HATE CRIMES

According to FBI data, 16 percent of all hate crimes in the United States—more than ten thousand per year—are motivated by a religious bias. A study of data from 2004 found that 1,003 Jews suffered attacks; there were 57 incidents against Catholics, 43 against Protestants, and 143 against Muslims.

WHIRLING DERVISHES

SPINNING INTO ECSTASY

In Sufism, a mystical offshoot of Islam, one achieves a state of ecstasy and connection with the divine by spinning around and around for hours on end while reciting the word *Allah*. Practitioners belong to the Mevlevi Order of the Sufi religion, following the example of a thir-

teenth-century Persian poet, Rumi, who advocated music, poetry, and dance as the way to reconnect to the root of God inherent in everyone. Through whirling one dies and then becomes symbolically resurrected to a more "perfect" state. Sufi—their name is a derivative of the Persian word for "coarse wool"—attribute the origin of the whirling dance to followers of Rumi. The master was mesmerized while observing a cloak maker in the thirteenth century plying his trade. This man stood under a bale of yarn and with one hand pulled down a strand, while with the other hand he kept the wool from tangling. In a flawlessly choreographed motion, the yarn seemed to effortlessly feed to the weaver, though in turn transforming the cloak maker into a rotating human spindle. If the man broke concentration for a second, the wool knotted, requiring the process to begin again from scratch. To stay in the present, the man supposedly repeated, "la illaha illa'llah," or "There is no God but God." One Afghan writer, Idries Shah, is credited with broadening Sufi understanding in modern times and was praised by Nobel laureate Doris Lessing for writing books that changed her life. In one, *The Wisdom of the Idiots*, Shah explained that in Sufism one realizes that the more you learn, the more you actually know nothing—or in effect, one becomes wise, or a spiritually enlightened idiot. Medical experts studying the effects of spinning cite fluctuation of inner-ear fluids as one cause of rapture; it overrides brain signals to help keep the whirling body upright and balanced. Dancing dervishes do report some dizziness but attest that focus, and God, keep them from falling. (*See also*: Marijuana in Religion)

WICCA

MODERN WITCHES

Wicca is the modernized version of witchcraft, defined as a revival in accessing hidden truths through magic and "natural sciences." Today, Wicca is a loosely formed organization of living witches, mostly female, dispersed among the population. There are many ways to practice what believers call "earth-based spirituality," but most are generally encouraged to use incantations, spells, and hexes that cause no harm to others or themselves. A synthesis of pre-Christian pagan religions, Wiccans adhere primarily to pantheistic views, defined as gods or deities being evident in all things, living and nonliving, equally distributed from a person, to a tree, to a stone. The supreme male deity is thought to be a hunter god, a human figure with a stag head with antlers. Such images were discovered in cave paintings dating back sixteen thousand years. The female deity consists of a trinity, or Triple Goddess, depicted as a maid, mother, or crone. There are no strict commandments, no churches, no leaders of papal magnitude, though there are some accepted documents and practices considered universal to be deemed a full-fledged Wiccan. The most important is obedience to the Three-Fold Law, or the Law of Return, explaining how all actions, positive or negative, will come back and affect the spell caster with three times its intensity. (*See also*: Karma) There is no satanic worship or animal sacrifice ceremonies, and devotees insist rituals are benign, referred to as "white magic." Seasonal events, such as the winter solstice and other natural occurrences, including various phases of the moon, are focal points for worship. Modern witches usually form small groups called covens, which ideally number no more than thirteen, but as with all aspects of this Supreme Court–recognized religion, it's left up to individuals to find a spiritual connection through their self-styled path.

Training to become a Wiccan normally includes the study of astrology, channeling, necromancy, out-of-body experiences, and means to induce trancelike states through "herbs" and even "sex magic." For example, the sex ritual will have consenting adults participating in passionate and preferably intense intercourse to tap into the Mother Goddess's powers. An orgasm is a religious experience and for the female it is referred to as "drawing down the moon" of the Mother Goddess, and for the male "drawing down the sun," or tapping into the Horned God's divinity. Many Wiccans also prefer to worship in the nude, called Skycladding and believed to be the best way to intensify one's innate divine-sensing antenna. Due to Wiccans' tendency toward secrecy, it's hard to count participants, though there were believed to be nearly one million in the United States in 2008, with more than fifty thousand women per year abandoning Christian congregations to join one that is considered more in harmony with nature. Benefits for the most devout include learning to fly and shape-shifting.

WITCH TRIALS

WHEN JUSTICE BURNED

In 1401, the English Parliament made its first law specially designed to weed out sorcery and witchcraft, considered heresy, and called for capital punishment, administered by burning at the stake. Christian church doctrine thought this execution best since it was dogmatically opposed to "shedding blood." In total, it's conservatively estimated that more than fifty thousand people had been executed for suspicion of witchcraft since the passing of the first English law. The last sentenced to death for witchcraft in England was Mary Hicks and her nine-year-old daughter Elizabeth, both hanged in 1716. In America, the Puritans who came to escape religious intolerance had a sixteen-month period of equal anti-witch hysteria. In January 1692, two girls, ages nine and eleven, began to act crazy, and were observed on their knees barking like dogs. When physicians could find no cure, it was diagnosed that Satan had possessed them.

Puritans believed in angels and devils and the powers of the invisible world. It happened that the young girls deemed affected by witchcraft were the daughter and niece of a disliked clergyman, who had a slave, Tituba, brought with him from Barbados. Tituba had knowledge of Obeah practices and was known to entertain with stories about black magic. When Tituba was interrogated and confessed to indeed conversing with the devil—some say she was either coerced through a beating or did it as a way to gain influence under her clergy master's direction—she kicked off a frenzied hunt for witches, targeted toward the nonconformists among the population. When Tituba accused two neighbors by name, and alluded that many men and women were secretly practicing witchcraft, arrests and trials began in earnest, totaling more than 150. By June of that year, the first, sixty-year-old Elizabeth Bishop, was hanged, primarily because she had once made a scarlet-colored dress for herself, and was known to act a tad eccentric. Convictions were obtained by allowing "astral" evidence, such as tests made by "witch-cakes" prepared with the possessed victim's urine. These pastries were given to dogs to reveal signs. Other methods to determine guilt of sorcery included dunking the suspect in water, to the brink of drowning, to see whether a witch floated, or by pressing under heavy stones until admitting guilt. Before it ended, thirty had been hanged or died in prison for the alleged crime of witchery.

In America, where hanging was considered less brutal than burning, sixty-six people (sixty-five of them escaped slaves) were executed by burning at the stake, ten in New York in the summer of 1741, and one (for witchcraft) in Illinois in 1779. Tituba, of Salem, Massachusetts, infamy was not executed, and although put in prison, she was released when bought by another as a slave. Her fate remains unknown. Recently, witch hunts have occurred in Africa, and there were reports of nearly a dozen elderly women with "red eyes" who were killed for witchcraft in the Republic of Congo in 1999. In 2008, a woman in Saudi Arabia was convicted of sorcery and awaits beheading.

WITCH OF ENDOR

OLD TESTAMENT NECROMANCER

Saul, the first king of Israel, found himself in deep trouble. The mighty Philistines assembled a huge army around his fortress. He tried to find answers on how to handle the impending massacre by interpreting his dreams, but his fitful sleep offered no clues. There seemed to be no reliable rabbis, prophets, or sanctioned holy men around to help decipher what to do in this crisis. When Saul took power, the first thing he did was run all witches, magicians, and soothsayers from town, believing them evil. With limited options, and the Philistines' battle trumpets rattling his nerves, he summoned the famed Witch of Endor, a woman by the name of Zephaniah. Saul wanted to contact the ghost of the prophet Samuel, dead only a few months, for opinions. The Witch of

Endor was famous for her ability to conjure up dead spirits. At the time, even rabbis believed a person's soul lingered around a body for up to a year, though summoning the dead or necromancy was considered by the more devout as a sin. Zephaniah thought it was a trap and refused to go to Saul, but he instead went to her in the small town of Endor, meeting in secret the night before his troops clashed at the Battle of Gilboa in 1000 B.C. The Witch of Endor did indeed call back Samuel's ghost, but she either refused to tell Saul what he said or Samuel was too aggravated to offer coherent advice. Nevertheless, Saul lost the battle disastrously and begged his servants to kill him before the conquering force found him. In the end, Saul was nailed spread-eagle to the wall of Beth-shan by the Philistines, and he has since been held up as an example of what can happen to a person who wants to talk to the deceased. Necromancers take a different view, saying that Saul, with his previous persecution of their ilk, got what he deserved.

X- Y- Z

X-POWER

THE SPIRITUAL LETTER

Before X became the twenty-fourth letter of the English alphabet, it was the symbol for the Roman number 10. It was confusing to those who read Latin when encountered in the middle of English words; the X was almost eliminated from the written language. X's earliest origins are traced to Egyptian hieroglyphics and it seems to have been used as the symbol to represent a fish. By itself, it has become an acceptable and legal signature mark, though in math it came to signify the unknown. One sees it everywhere from the word *X-ray* to *The X-Files* television series ("the truth is out there"), and X marks the spot on a treasure map. X held religious significance to early religions and eventually the occult. The X figured significantly in a pre-Egyptian culture, known as the UrRean people of the Nile Delta. Their King Assur who turned god but was resurrected, and in detail it mirrors the story of Jesus four thousand years before Jesus was born. The UrRean afterlife rituals developed to follow Assur's resurrection and journey through the underworld, from which the notions of Egyptian mummification and pyramids arose. Among the UrReans there was a Hawk-God and a Sun-God, but X was the sacred symbol that showed how Assur's blood passed on to humans. Multiple tattoos of X's on an UrRean was a sign of connection to the divine creator. In the Middles Ages the lowercase *x* came to mean death, and "damned," since the dead who were baptized got a small cross next to their name, instead of an x. The x-corpses were usually dumped in a ditch rather than buried in a sanctified graveyard. (*See also*: Charles Manson)

XENOPHANES

THE PHILOSOPHER WHO NEVER SLEPT IN THE SAME PLACE TWICE

The ancient Greek philosopher Xenophanes is credited as the first Western thinker to conceive of one supreme god, as opposed to the list of anthropomorphic deities that abounded in Greek religion. He envisioned an intelligent god, though as nonhuman energy and as a creature that always was and will be. He described a god as without organs or senses—as man understands them—yet knowing everything; it "toiled without moving." Xenophanes wondered why many religions envision gods that look like people, noting: "If horses or oxen or lions had hands and could produce works of art, they too would represent the gods after their own fashion." He was a wanderer and known to never sleep in the same spot more than once in sixty-seven years. He died in 480 B.C., at the age of ninety-two.

ANTIDOTE FOR DIVINE BOREDOM

Like many religions, the ancient Greek faith was conceived to explain natural phenomena. Every city had a protecting spirit deity, and over time a mythology about their birth, lives, and characteristics became widely accepted. There was no division of religion and state in ancient Greece. Athens, for example, protected as it was by the goddess Athena, required a daily public ritual of some sort be offered to the gods. The Olympic Games and most theatrical productions were religious events, honoring one god or many. Incidentally, Athena was also the goddess of wisdom and was born out of the head of Zeus, the chief of all gods who lived in the heavens, at a place called Olympus. Humans were created because these gods were bored, and when still jaded, they invented love to really stir things up.

XENOGLOSSY

SPEAKING IN COHERENT TONGUES

If a Brooklynite were to awake one morning and begin to speak in a perfect British accent and be unable to stop, that individual might be suffering from Foreign Accent Syndrome, though one so afflicted will usually not speak in a completely unknown language. Xenoglossy occurs when a person can speak fluently in a language that would be impossible for them to know; the same Brooklyn resident might suddenly speak in perfect Swahili for instance. In the New Testament, the description of the Pentecost explained how men from Jerusalem could suddenly speak to any foreigner encountered and somehow fully know the other's language, with the purpose of spreading the word. That was truly "speaking in tongues," but it is not the same phenomenon that occurs during Pentecostal or charismatic ceremonies. Those possessed by the Holy Sprit in modern times use no known languages, not even archaic ones, and mostly speak in what are considered to be untranslatable sounds. New Age thought believes actual cases of xenoglossy can take place under hypnotism. The few cases that have been studied support proof of past life and reincarnation, since that remains the only "reasonable" explanation why an entirely different language can come to mind.

PROVING REINCARNATION

Throughout history, and in modern times, more people have believed in reincarnation than in other afterlife theories. Reincarnation is a growing field of academic study, with most investigators focusing on the memories of young children to prove it. It's been concluded that by age three a person can verbalize details of their past life, but most lose nearly all memories by seven. University of Virginia professor Ian Stevenson spent more than forty years trying to verify reincarnation. He found one boy who claimed he was once a mechanic in Beirut, and that he had died in a car accident. The child named his previous brothers and sisters, and other details that matched such an incident, which had occurred three years before the child was born. Another child said he had owned a medical supply company in Delhi, India, and had been shot by his brother. Such a circumstance took place eight years before that baby was born. When the youngster was taken to his supposed past-life family, he knew many of his former relatives by their private nicknames.

鳥居氏清倍圖
清倍

YELLOW BAMBOO

MARTIAL-ARTS MAGIC

Chi is the word for life force in Eastern theologies. To harness the power of chi for the purpose of doing good is called White Magic. Yellow Bamboo is a martial arts group that strives to achieve the magical level of a chi master by following certain physical exercises and meditations, and to practice "good sorcery." For example, once at master level you can knock down an opponent without touching him. An invisible beam of chi emanating from the palm of the hand can even deflect the path of a fired bullet. Certain chi masters can also levitate. The quest to control chi is considered an ancient technique, as seen in movies that depict slow-motion, roof-hopping martial arts experts sailing over long distances, then the next second able to dodge nunchuks at lightning speed. Other than in cinema, no living chi masters have yet to verify these powers under laboratory conditions, though this is a fact regarded as an irrelevant issue to the more than thirty thousand Yellow Bamboo devotees.

RIP VAN WINKLE SYNDROME

A number of chi masters have demonstrated their powers over the "life force" by going into a deep sleep and then being buried alive. The record belongs to the 1830s Indian master Sadhu Haridas. After getting into deep thought, he was wrapped in linens and placed in a padlocked chest that was buried in a garden. Six weeks later, he was exhumed and revived back to life by pouring warm water over his head. In Hebrew scriptures, chi might be used to explain how Honi HaM'agel was noted to go in a trance for seventy years while he waited for a tree to grow. In Christian tradition, seven young men fleeing religious persecution during the reign of Roman emperor Decius hid out in a cave at Ephesus. Instead of dragging them out, the Romans merely sealed the cave with a boulder, and the group was soon forgotten. But 155 years later the stone was rolled away and the men were revived, believing they had been asleep for only one day. In the short story "Rip Van Winkle," by American author Washington Irving, Rip was a likable though lazy guy. After he headed to the mountains to find reprieve from his nagging wife, he sat down for a nap that lasted twenty years. It's unlikely Van Winkle was a chi master, though he did encounter ghosts of Henry Hudson's lost crew before falling into his slumber.

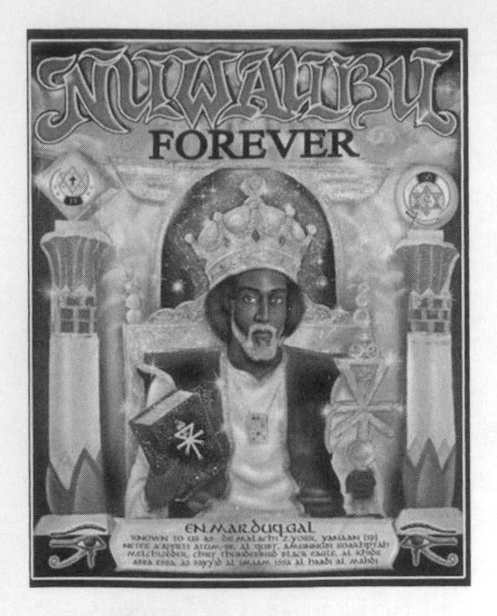

MALACHI Z. YORK

UNITED NUWAUBIAN NATION OF MOORS

Malachi York was a Black Panther, then a soul singer before turning to religion. In the 1990s he built a commune, Tama-Re, with pyramid buildings and Moorish architecture on four hundred acres in Georgia. There his followers came to live under his rule and utopian vision. He considered himself as a descendant of Egyptian kings and other ancient deities or people of holy lineage. Although he demanded chastity of his followers, and separated sexes into different dormitories, he said these regulations did not apply to him. York was subsequently convicted of molesting minors on the compound—more than one hundred children—and sentenced to 135 years in 2004. From prison he sends updated doctrines on all issues of Nuwaubian life, though he is noted to change course and ideology frequently. York's ideas range from Islamic thought to belief in cosmic beings, cosmology, and the search for mystical beasts. (*See also*: Unicorns) There are still five hundred followers (down from twenty-five thousand) dedicated to following York's principles. They believe he was framed and are primarily assigned to work for his release.

BEN YAHWEH

CHURCH FUNDED BY VICE

Ben Yahweh (Hulton Mitchell, Jr.) was the charismatic black leader of the Church of Love. He preached that he was a prophet "Son of the Son of God" from a black Israelite sect, a lost tribe of Israel and the only true Jew. He also stressed that white people were devils. He sent his disciples out to build a multimillion-dollar empire through prostitution and drugs, extortion, arson, and murder in the Miami area. He and sixteen of his inner circle were convicted in 1992 of killing at least fourteen white people and resistant blacks.

YIN AND YANG

OPPOSITE ATTRACTION

Yin-Yang is a Chinese philosophical concept that explains how opposite forces occurring in the natural world are intertwined and interact. The circular icon for Yin and Yang of two sperm-shaped markings, one white and the other black, with opposite-colored eyes, derives from I Ching symbols, with some claiming it signifies creation arising from the void. Others have postulated it was part of an astrological map indicating the arrival of winter and summer solstice, representative of the natural overlapping of seasons. Yin, the white symbol, has a female aspect, ranging in qualities from soft to cold, wet, or tranquil. The black Yang has properties considered masculine, varying from hard, solid, or aggressive. If one portion is too weakly manifested, for example, or the other more prevalent, then the universal principle of harmony is offset, causing all kinds of grief and suffering. In Chinese medicine, martial arts, and meditative practices, followers of the Yin-Yang belief seek to keep both aspects in balance.

ZEN

WHY EXPLAINING ZEN IS UN-ZEN

In the fifth century A.D., a Buddhist monk from Persia, or present-day Iran, traveled barefoot to India. He kept on going, deep into China, drawing followers wherever he went. This monk had just spent nineteen years staring at a cave wall and afterward encouraged a simpler version for attaining enlightenment based on the scholarly writings of a third-century Buddhist scholar, Toa-Sheng. The wandering founder of Zen is often referred to as a Bodhidharma, a name meaning "Law of Wisdom," and serves as a composite of the various mystical monks considered by numerous Zen sects as the religion's founder. Bodhidharma was not an apostate of Buddhism, but urged a downplay in sig-

nificance of dogmas, instead recommending focus on tapping into the divinity, or the "Buddha within," through meditation and intuition. Zen was called Chan Buddhism in China and pronounced "Zen" when transported to Japan in the twelfth century. In theory there should be no words describing Zen, because explanations are limited and add to further delusion.

Zen teachers prefer riddle-like questions designed to expand consciousness, such as "What is the sound of one hand clapping?" Sitting in a lotus position and meditating to a point where the mind is empty of all conflicting thought is the primary practice advocated by Zen. Awakening, defined as experiencing the Buddha (God), is attained in part through a meditation practice called zazen. Teachers are responsible to guide Zen devotees toward that higher path, rather than spending time memorizing ancient texts, and are noted to use statements, or fragmentary answers to address questions. For example, a student might ask how to stop cramps in his legs, but will be given another thing to ponder: "The sun is warm. The grass is green." Zen instructors might also inflict a bamboo stick to an initiate's back during meditative sessions if deemed necessary. The same stick is used when monks in training gather at communal tables for meals, and knuckles will be cracked if the teacher deems an action not "Zen-like"—even if defining what is "Zen-like" is immediately not Zen. Nevertheless, it takes many years to master this alternate thought processing and be deemed a Zen master, or attain the state defined as the "eye turned inward." The goal is to be so absolutely present in each moment, until for example, if when washing dishes one actually sees, feels, and smells the soap, and the warm water rinsing the plate, and one is devoid of all other thought; then progress has been made. Awareness of each of life's linked segmented moments becomes a nonstop meditation. Zen believes that being free from time and worry, or suffering in general, allows the soul to achieve enlightenment, and thereby stop the cycle of karma to reside in a perfect state in nirvana.

DEEPLY DETACHED

The Zen tradition of overcoming inflicted pain, or rising above suffering, such as learning to sit unmoving with legs folded for great lengths, has its origin in legends about the founder. One story has a disciple imploring his teacher to guide him to the enlightened level. The teacher made the student sit lotus-style an entire winter in the snow outside a monastery. When spring arrived and the student again asked to learn the secrets, the teacher told the student to first cut off his own hand, a lesson seen by Zen advocates as the need to deflate ego and abandon attachment. When Bodhidharma supposedly died in China, they buried him. But shortly after, they saw the two-hundred-year-old master walking way off in the distance, carrying one shoe in his hand, heading back barefoot toward Iran, from whence he came. When they dug up the grave, they found an empty coffin, except for one shoe.

ST. ZITA

PATRON SAINT OF MAIDS AND SERVANTS

In the thirteenth century, twelve-year-old Zita was placed as a servant in the home of wealthy Italian wool merchants. From the start, the other servants told her to slow down and not take to her chores so diligently, or otherwise they would all have to work harder. They constantly tried to make it seem that Zita had messed up and so she was beaten regularly by the homeowners, though Zita was noted to persist with a cheery attitude. Eventually the other servants thought they had caught Zita red-handed; they discovered that the humble servant girl was giving away the household food and even her employers' clothing to the poor. Although it was a serious crime, she was allowed to continue, especially after two angels were seen at her side helping with chores. She remained a servant in the same house for forty-eight years. Zita believed she was given this job as her penance and assigned by God to spend her life as a maid. Despite stealing, and since it was done for the poor, Zita was canonized in 1696.

ZOROASTRIANISM

THE ORIGINAL MAGI

Before Zoroaster was born, in 630 B.C., his mother was visited by angels. A beam of light provided the sperm, which is nearly identical to the story of the Annunciation that Christians believe only happened to Jesus. Also known as Zarathustra (a name meaning "camel handler"), Zoroaster reportedly came into the world laughing and was said to be already wise enough to argue with scholarly elders by age twelve. By thirty he had chosen hermitage, even if some records indicate he was married three times and had numerous children. Zoroaster became known as a Persian prophet who established one of the earliest mono-

theistic religions, and whose ideas became synthesized into many aspects of Judaism and Christian beliefs. Zoroaster claimed to communicate with angels that controlled fire, water, and plants, to name a few, all operating under Ahura Mazda, the uncreated creator of all who was in constant battle with a dark lord strikingly similar to Satan. His ideas of heaven, hell, and Judgment Day also found their way into Christian doctrines, including the notion that man was given free will to decide between good and evil. Zoroastrianism spread throughout the area and into China, becoming the world's most popular religion until the rise of Islam in sixth century A.D. Most records about Zoroaster's later years were destroyed in a fire when Alexander the Great conquered Persia. Most believed he was betrayed by a loyal follower for thirty pieces of gold and murdered while praying in a sanctuary. Others say he meditated so deeply that his body simply vanished from the physical world.

When a Zoroastrian died it was considered an impure event. No one wanted to handle the corpse, and so the body was placed on a platform until pecked clean by scavenger birds. At that point the soul would leave the body and attempt to walk across a bridge to paradise. However, if your actions while alive didn't bring joy into the world you'd likely be tossed over the side and become a tortured plaything of the source of all evil, Angra Mainyu.

In this religion a secret sect of priests were called Magi, considered as sorcerers and from which the word *magic* derives. They specialized in fire rituals, part of the religion's purification ceremonies, and were considered learned in astrology. In the New Testament, the three kings depicted in the nativity scene were Zoroastrian priests following a star that was believed to signify the birth of the awaited "King of Enlightenment." They unfortunately checked in with King Herod upon arrival into Jerusalem. With this knowledge, Herod then ordered the "Massacre of Innocents," which set out to kill every male under two years old who could one day possibly vie for Herod's throne—as many as sixty-four thousand toddlers were murdered in the purge.

Today, the Parsi religion most resembles old-time Zoroastrianism, though it has dwindled in numbers to less than seventy thousand worldwide, since they do not believe in missionaries or conversion practices, and membership traditionally requires that at least one parent have been born into it.

SOURCES

Abell, Troy D. *Better Felt Than Said: Holiness-Pentecostal Experience in Southern Appalachia*. Baylor University Press, 1983.

Abulafia, Anna Sapir. *Christians and Jews in Dispute*. Ashgate, 1998.

Alcock, James E. *Science and Supernature: A Critical Appraisal of Parapsychology*. Prometheus, 1990.

Allis, Oswald T. *Prophecy and the Church*. Baker, 1969.

Andrews, Edward Deming. *The People Called Shakers*. Oxford University Press, 1953.

Ankerberg, John, et al. *Behind the Mask of Mormonism*. Harvest House, 1996.

Appel, Willa. *Cults in America: Programmed for Paradise*. Holt, Rinehart & Winston, 1983.

Arens, W. *The Man-Eating Myth*. Oxford University Press, 1979.

Armstrong, Karen. *Great Transformation: The Beginning of Our Religious Tradition*. Vintage, 2007.

———. *Islam: A Short History*. Modern Library, 2002.

Asad, Talal. *Genealogies of Religion: Discipline and Reasons of Power in Christianity and Islam*. Johns Hopkins University Press, 1993.

Aune, D. E. *Prophecy in Early Christianity and the Ancient Mediterranean World*. Eerdmans, 1983.

Aurand, Ammon. *The Amish and the Mennonites*. Aurand, 1939.

Aurobindo, Sri, ed. *The Bhagavad Gita*. All India Press, 1992.

Ayoub, Mahmoud. *The Qur'an and Its Interpreters*. State University of New York Press, 1984.

Bachrach, Bernard. *Early Medieval Jewish Policy in Western Europe*. University of Minnesota Press, 1977.

Bainbridge, William S. *Satan's Power: A Deviant Psychotherapy Cult*. University of California Press, 1978.

———. *The Sociology of Religious Movements*. Routledge, 1997.

Baker, Robert A. *They Call It Hypnosis*. Prometheus, 1990.

Bapat, P. V., ed. *2500 Years of Buddhism*. Government of India, Publications Division, 1956.

Barber, T. X. *LSD, Marijuana, Yoga, and Hypnosis*. Aldine, 1970.

Barclay, William. *Epilogues and Prayers*. SCM, 1972.

Barker, Eileen, ed. *Of Gods and Men: New Religious Movements in the West*. Mercer University Press, 1983.

Barr, James. *The Semantics of Biblical Language*. Oxford University Press, 1961.

Barrett, David B. *Schism and Renewal in Africa: An Analysis of Six Thousand Contemporary Religious Movements*. Oxford University Press, 1969.

Barrett, David V. *The New Believers: Sects, "Cults," and Alternative Religions*. Cassell, 1991.

Barrett, Leonard E. *The Rastafarians*. Beacon, 1977.

Bartlett, John R. *Jews in the Hellenistic World: Josephus, Aristeas, the Sibylline Oracles, Eupolemus*. Cambridge University Press, 1985.

Barton, J. *Oracles of God: Perceptions of Ancient Prophecy in Israel after the Exile*. Longman & Todd, 1986.

Barton, Tamsyn S. *Power and Knowledge: Astrology, Physiognomics, and Medicine under the Roman Empire*. University of Michigan Press, 2002.

Bascomb, W. *Ifa Divination: Communication between Gods and Man in West Africa*. Indiana University Press, 1969.

Basham, A. L. *The Wonder That Was India*. Grove Press, 1959.

Basso, Keith H. *Western Apache Witchcraft*. University of Arizona Press, 1960.

Bastien, Remy. *Vodoun and Politics in Haiti*. Random House, 1971.

Bauer, J. B., ed. *An Encyclopedia of Biblical Theology*. Crossroad, 1981.

Baumgarten, Albert I. *The Phoenician History of Philo of Byblos*. Brill, 1981.

Beard, Mary, et al. *Religions of Rome*. Cambridge University Press, 1998.

Beckerlegge, Gwilym, ed. *World Religions Reader*. Routledge, 2001.

Beecher, Maureen Ursenbach, and Lavina Fielding Anderson, eds. *Sisters in Spirit: Mormon Women in Historical and Cultural Perspective*. University of Illinois Press, 1987.

Beit-Hallahmi, Benjamin. *The Illustrated Encyclopedia of Active New Religions, Sects and Cults*. Rosen, 1998.

Bell, H. Idris. *Cults and Creeds in Graeco-Roman Egypt*. Liverpool University Press, 1953.

Ben-Gurion, David. *Rebirth and Destiny of Israel*. Philosophical Library, 1954.

———. *The Jews in Their Land*. Doubleday, 1974.

Benoit, Pierre. *Aspects of Biblical Inspiration*. Priory, 1965.

Berger, David. *From Crusades to Blood Libels to Expulsions: Some New Approaches to Medieval Antisemitism*. Touro College Press, 1997.

Bettenson, H. *Documents of the Christian Church*. Oxford University Press, 1967.

Beversluis, Joel. *A Sourcebook for Earth's Community of Religions*. CoNexus, 1995.

Bialik, Hayyim Nahman, and Joshua C. Ravnitzky. *The Book of Legends*. Schocken, 1992.

Bin Gorion, Micha Joseph. *Mimekor Yisrael: Classical Jewish Folktales*. Indiana University Press, 1976.

Blackburn, Simon. *Oxford Dictionary of Philosophy*. Oxford University Press, 2005.

Blackmore, Susan J. *In Search of the Light: The Adventures of a Parapsychologist*. Prometheus, 1986.

Blofeld, John. *The Tantric Mysticism of Tibet*. Dutton, 1970.

Blumenthal, H. J. *Soul and Intellect: Studies in Plotinus and later Neoplatonism*. Variorum, 1993.

Boadt, Lawrence. *Reading the Old Testament: An Introduction*. Paulist, 1984.

Borghouts, J. F. *Ancient Egyptian Magical Texts*. Brill, 1978.

Borst, Arno. *Medieval Worlds: Barbarians, Heretics, and Artists*. University of Chicago Press, 1992.

Bossy, John. *Christianity in the West, 1400–1700*. Oxford University Press, 1985.

Bourguignon, Erika. *Possession*. Chandler & Sharp, 1976.

Bowen, John R., ed. *Religion in Culture and Society*. Allyn & Bacon, 1998.

Bowker, John. *The Oxford Dictionary of World Religions*. Oxford University Press, 1997.

———. *What Muslims Believe*. Oneworld, 1999.

Boyarin, Daniel. *Dying for God: Martyrdom and the Making of Christianity and Judaism*. Stanford University Press, 1999.

Boyer, Pascal. *Religion Explained*. Basic Books, 2002.

Bradley, Robert I. *The Roman Catechism in the Catechetical Tradition of the Church*. University Press of America, 1990.

Brandon, George. *Santeria: From Africa to the New World*. Indiana University Press. 1993.

Brandon, S. G. F. *A Dictionary of Comparative Religion*. Scribner's, 1970.

Braude, Ann. *Radical Spirits: Spiritualism and Women's Rights in Nineteenth-Century America*. Macmillan, 1989.

Bray, Gerald. *Biblical Interpretation Past and Present*. Intervarsity, 1996.

Bromley, David G. *Strange Gods: The Great American Cult Scare*. Beacon, 1981.

———. *The Politics of Religious Apostacy: The Role of Apostates in the Transformation of Religious Movements*. Praeger, 1998.

Brown, Patricia Leigh, and Carol Pogash. "The Pleasure Principle." *New York Times*, March 15, 2009.

Brown, Peter. *The Cult of the Saints: Its Rise and Function in Latin Christianity*. University of Chicago Press, 1982.

Brown, Slater. *The Heyday of Spiritualism*. Hawthorn, 1970.

Buckland, Raymond. *Witchcraft from the Inside*. Llewellyn, 1971.

Budge, E. A. Wallis. *Amulets and Talismans*. Dover, 1978.

Bugliosi, Vincent, and Curt Gentry. *Helter Skelter*. Norton, 1974.

Burkert, Walter. *Lore and Science in Ancient Pythagoreanism*. Harvard University Press, 1972.

Burland, Cottie A. *The Arts of the Alchemist*. Macmillan, 1968.

Burnham, Kenneth E. *God Comes to America: Father Divine and the Peace Mission Movement*. Lambeth, 1979.

Burns, Paul, et al., eds. *Butler's Lives of the Saints*. Liturgical Press, 1999.

Burtchaell, James T. *Catholic Theories of Biblical Inspiration Since 1810*. Cambridge University Press, 1969.

Buxton, R. *Imaginary Greece: The Contexts of Mythology*. Cambridge University Press, 1994.

Cabezon, Jose Ignacia, ed. *Buddhism, Sexuality, and Gender*. State University of New York Press, 1992.

Calvin, Jean, and Henry Beveridge. *Institutes of the Christian Religion*. Logos Research Systems, 1997.

Casciaro, J. M., and J. M. Monforte. *God, the World and Man in the Message of the Bible*. Four Courts, 1996.

Catholic Encyclopedia Online. http://www.catholic.org.

Chadwick, H. *The Early Church*. Dorset, 1986.

Chan, Wing-tsit. *Religious Trends in Modern China*. Columbia University Press, 1953.

Chisholm, Robert B., Jr. *Interpreting the Minor Prophets*. Zondervan, 1990.

Christoper, A., et al., eds. *Magika Hiera: Ancient Greek Magic & Religion*. Oxford University Press, 1991.

Clift, Wallace. *Jung and Christianity*. Crossroad, 1986.

Cohen, Daniel. *Prophets of Doom*. Millbrook, 1999.

Cohen, Jeremy, ed. *Essential Papers on Judaism and Christianity in Conflict: From Late Antiquity to the Reformation*. New York University Press, 1991.

Cohn, Norman. *The Pursuit of the Millennium*. Harper Torchbooks, 1957.

Coil, Henry Wilson. *A Comprehensive View of Freemasonry*. Macoy, 1977.

Cook, Michael. *The Koran: A Very Short Introduction*. Oxford University Press, 2000.

Cook, S. L. *Prophecy and Apocalypticism: The Post-exilic Social Setting*. Fortress, 1995.

Corydon, Bent. *L. Ron Hubbard: Messiah or Madman*. Lyle Stuart, 1987.

Crapanzano, Vincent, and Vivian Garrison. *Case Studies in Spirit Possession*. Wiley, 1977.

Crawford, J. R. *Witchcraft and Sorcery in Rhodesia*. Oxford University Press, 1968.

Critchley, Simon. *The Book of Dead Philosophers.* Random House, 2008.

Daner, Francine J. *The American Children of Krishna: A Study of the Hare Krishna Movement.* Holt, Rinehart, & Winston, 1976.

Davies, R. Trevor. *Four Centuries of Witch-Beliefs.* Methuen, 1947.

Davis, John, ed. *Religious Organization and Religious Experience.* Academic Press, 1982.

Dawkins, Richard. *The Blind Watchmaker: Why the Evidence of Evolution Reveals a Universe Without Design.* Norton, 1988.

De Vaux, Roland. *Ancient Israel: Its Life and Institutions.* McGraw-Hill, 1961.

de Groot, J. J. M. *The Religion of the Chinese.* Macmillan, 1910.

Delehaye, Hippolyte. *The Legends of the Saints: An Introduction to Hagiography.* University of Notre Dame Press, 1961.

Douglas, Mary. *Purity and Danger: An Analysis of the Concepts of Taboo.* Routledge, 1991.

Doyle, Arthur. *The History of Spiritualism.* Arno, 1975.

Drane, John William. *Introducing the Old Testament.* Lion, 2000.

Draper, Theodore. *The Rediscovery of Black Nationalism.* Viking, 1969.

Drury, Nevill. *Dictionary of Mysticism and the Esoteric Traditions.* Prism, 1992.

Earhart, H. Byron, ed. *Religious Traditions of the World.* HarperCollins, 1993.

Edersheim, Alfred. *The Temple, Its Ministry and Services as They Were at the Time of Jesus Christ.* Logos Research Systems, 2003.

Eliade, Mircea, ed. *The Encyclopedia of Religion.* Macmillan, 1987.

———. *Shamanism: Archaic Technique of Ecstasy.* Pantheon, 1964.

Ellis, E. Earle. *The Old Testament in Early Christianity: Canon and Interpretation in the Light of Modern Research.* Baker, 1992.

Elwood, Robert S. *Many People, Many Faiths.* Prentice-Hall, 1992.

Endress, Gerhard. *Islam: An Historical Introduction.* Columbia University Press, 2002.

Enns, Paul P. *The Moody Handbook of Theology.* Moody, 1997.

Enroth, Ronald. *Youth Brainwashing and the Extremist Cults.* Zondervan, 1977.

Enroth, Ronald, et al. *A Guide to Cults and New Religions.* Intervarsity, 1983.

Erasmus, Charles J. *In Search of the Common Good*. Free Press, 1985.

Esack, Farid. *The Qur'an: A Short Introduction*. Oneworld, 2002.

Evans, G. R. *The Language and Logic of the Bible: The Road to Reformation*. Cambridge University Press, 1985.

Evelyn, Brooks. *Righteous Discontent: The Women's Movement in the Black Baptist Church, 1880–1920*. Harvard University Press, 1993.

Faivre, Antoine. *The Eternal Hermes: From Greek God to Alchemical Magus*. Phanes, 1995.

Farmer, William R., and D. Farkasfalvy. *The Formation of the New Testament Canon*. Paulist, 1983.

Fee, Gordon D., and Douglas Stuart. *How to Read the Bible for All Its Worth: A Guide to Understanding the Bible*. Zondervan, 1982.

Festinger, Lionel. *When Prophecy Fails*. Harper & Row, 1966.

Firth, Raymond. *The Fate of the Soul*. Cambridge University Press, 1955.

Fischer-Schreiber, Ingrid, et al., eds. *The Encyclopedia of Eastern Philosophy and Religion: Buddhism, Hinduism, Taoism, Zen*. Shambhala, 1994.

Flood, Gavin. *An Introduction to Hinduism*. Cambridge University Press, 1996.

Forster, Genevieve. *The World Was Flooded with Light*. University of Pittsburg Press, 1967.

Fournier, K. *Evangelical Catholics*. Thomas Nelson, 1990.

Fowden, Garth. *Pagan Philosophers in Late Antiquity*. Princeton University Press, 1973.

Foxe, John. *Fox's Book of Martyrs*. Zondevan, 1926.

Frend, W. H. C. *The Rise of Christianity*. Fortress, 1984.

Froelich, Karlfried, ed. *Biblical Interpretation in the Early Church*. Fortress, 1984.

Frye. Northrop. *The Great Code: The Bible and Literature*. Harcourt Brace Jovanovich, 1982.

Fuchs, Stephen. *Rebellious Prophets: A Study of Messianic Movements in Indian Religions*. Asia Publishing House, 1965.

Fuller, R. C. *Alternative Medicine and American Religious Life*. Oxford University Press, 1990.

Gardner, Gerald B. *Witchcraft Today*. Citadel, 1970.

Gardner, Martin. *Fads and Fallacies in the Name of Science: The Curious Theories of Modern Pseudoscientists and the Strange, Amusing and Alarming Cults that Surround Them. A Study in Human Gullibility.* Dover, 1957.

———. *Urantia:The Great Cult Mystery.* Prometheus, 1995.

Garland, R. *Religion and the Greeks.* Classical, 1994.

Geary, Patrick J. *Living With the Dead in the Middle Ages.* Cornell University Press, 1995.

Gentz, William H. *The Dictionary of Bible and Religion.* Parthenon, 1986.

George, Timothy. *Theology of the Reformers.* Broadman, 1988.

Gerstel, David. *Paradise Incorporated: Synanon.* Presidio, 1982.

Gethin, R. M. L. *The Buddhist Path to Awakening: A Study of the Bodhi-Pakkhiya Dhamma.* Brill, 1992.

Gleason, L. Archer, and G. C. Chirichigno. *Old Testament Quotations in the New Testament: A Complete Survey.* Moody, 1983.

Glick, Leonard. *Abraham's Heirs: Jews and Christians in Medieval Europe.* Syracuse University Press, 1999.

Glock, Charles Y., and Robert N. Bellah, eds. *The New Religious Consciousness.* University of California Press, 1976.

Godwin, David. *Light in Extension: Greek Magic from Homer to Modern Times.* Llewellyn, 1992.

Goll, Jim W. *The Seer.* Destiny Image, 2004.

Gonda, Jan. *Vedic Literature: Samhitas and Brahmanas.* Wiesbaden, 1975.

Goodwin, John. *Occult America.* Doubleday, 1972.

Gordon, Henry. *Extrasensory Deception: Esp, Psychics, Shirley MacLaine, Ghosts, UFOs.* Prometheus, 1987.

Gottlieb, R. S., ed. *This Sacred Earth: Religion, Nature and the Environment.* Routledge, 1996.

Grabbe, L. L. *Priests, Prophets, Diviners, Sages: A Socio-Historical Study of Religious Specialists in Ancient Israel.* Trinity Press International, 1995.

Green, Joel B., and Scott McKnight, eds. *Dictionary of Jesus and the Gospels.* Intervarsity, 1992.

Greenberg, Moshe, ed. *The Torah: The Five Books of Moses.* Jewish Publication Society of America, 2000.

Grey, Sir George. *Polynesian Mythology and Ancient Traditional History of the Maori as Told by Their Priests and Chiefs*. Taplinger, 1970.

Grim, Patrick, ed. *Philosophy of Science and the Occult*. State University of New York Press, 1990.

Guiley, Rosemary. *The Encyclopedia of Witches and Witchcraft*. Facts on File, 1999.

Hadden, Jeffrey K. *Prophetic Religion and Politics*. Paragon House, 1986.

Hagerty, Cornelius. *The Authenticity of Sacred Scripture*. Lumen Christi, 1969.

Harmless, William, S. J. *Mystics*. Oxford University Press, 2008.

Harner, Michael J. *The Jivaro. People of the Sacred Waterfalls*. University of California Press, 1974.

Harrison, Jane E. *Prolegomena to the Study of Greek Religion*. Princeton University Press, 1991.

Hawkins, Gerald S. *Beyond Stonehenge*. Harper & Row, 1973.

Hawley, John Stratton, and Donna Marie Wull. *Devi: Goddess of India*. University of California Press, 1996.

Hawting, Gerald R. *The Idea of Idolatry and the Emergence of Islam: From Polemic to History*. Cambridge University Press, 1999.

Heelas, Paul. *The New Age Movement: Celebrating the Self and the Sacralization of Modernity*. Basil Blackwell, 1996.

Heffernan, Thomas. *Sacred Biography: Saints and Their Biographers in the Middle Ages*. Oxford University Press, 1992.

Heideking, Jurgen. *The Federal Processions of 1788 and the Origins of American Civil Religion*. Citation, 1994.

Heidt, William G. *A General Introduction to Sacred Scripture: Inspiration, Canonicity, Texts, Versions, and Hermeneutics*. Liturgical Press, 1970.

Henderson, Ebenezer. *The Twelve Minor Prophets*. Baker, 1980.

Heschel, Abraham J. *The Prophets: An Introduction*. Harper & Row, 1962.

Hillerbrand, Hans, ed. *The Reformation: A Narrative History Related by Contemporary Observers and Participants*. Baker Book House, 1978.

Hinnells, John R., ed. *A New Dictionary of Religions*. Penguin, 1995.

———, ed. *A Handbook of Living Religions*. Penguin, 1985.

Hoekema, Anthony A. *The Bible and the Future*. Eerdmans, 1979.

Hogue, John. *Nostradamus: The New Revelations.* Barnes & Noble, 1995.

The Holy Bible: Holman Christian Standard Version. Holman Bible Publishers, 2003.

The Holy Bible: King James Version. Electronic 1769 edition of the 1611 Authorized Version. Logos Research Systems, 1995.

Hopfe, Lewis M., ed. *Religions of the World.* Macmillan, 1987.

Hopkins, Martin. *God's Kingdom in the Old Testament.* Henry Regnery, 1964.

Hopkins, Thomas J. *The Hindu Religious Tradition.* Wadsworth, 1982.

Hori, Ichiro. *Folk Religion in Japan.* University of Chicago Press, 1968.

Hornung, Erik. *Conceptions of God in Ancient Egypt.* Cornell University Press, 1982.

Houk, James T. *Spirits, Blood, and Drums: The Orisha Religion in Trinidad.* Temple University Press, 1995.

Hubbard, L. Ron. *Dianetics: The Modern Science of Mental Health.* Hermitage House, 1950.

Hugh-Jones, Stephen. *The Palm and the Pleiades: Initiation and Cosmology in Northwest Amazonia.* Cambridge University Press, 1979.

Hultkrantz, Ake. *The Religions of the American Indians.* University of California Press, 1979.

Huxley, Aldous. *The Devils of Loudon.* Harper & Brothers, 1952.

Huxley, Francis. *The Invisibles: Voodoo Gods in Haiti.* McGraw-Hill, 1966.

Jabotinsky, Z'ev. *The War and the Jew.* Altalena, 1987.

Jaini, Padmanabh S. *The Jaina Path of Purification.* University of California Press, 1979.

James, E. O. *Prehistoric Religion.* Barnes & Noble, 2002.

James, William. *The Varieties of Religious Experience.* Mentor, 1961.

Jamieson, Robert, et al. *A Commentary, Critical and Explanatory, on the Old and New Testaments.* Logos Research Systems, 1997.

Javers, Ron, and M. Kildruff. *The Suicide Cult.* Bantam, 1978.

Jenkins, P. *Mystics and Messiahs: Cults and New Religions in American History.* Oxford University Press, 2000.

Jensen, Joseph. *God's Word to Israel.* Glazier, 1982.

Joachim, W. *The Comparative Study of Religions*. Columbia University Press, 1958.

Johnson, K. Paul. *The Masters Revealed: Madame Blavatsky and the Myth of the Great White Lodge*. State University of New York Press, 1994.

Johnson, Paul. *A History of Christianity*. Penguin, 1984.

Jorgensen, Danny L. *The Esoteric Scene, Cultic Milieu, and Occult Tarot*. Garland, 1992.

Juergensmeyer, Mark. *Global Religions: An Introduction*. Oxford University Press, 2003.

Jungmann, Joseph A. *The Mass of the Roman Rite: Its Origins and Development*. Benziger Brothers, 1961.

Kaltner, John, et al. *The Uncensored Bible*. HarperCollins, 2008.

Kamen, Henry. *The Spanish Inquisition*. Yale University Press, 1997.

Kanter, Rosabeth. *Commitment and Community: Communes and Utopias in Sociological Perspective*. Harvard University Press, 1972.

Kapleau, Phillip. *Three Pillars of Zen*. Beacon, 1967.

Karleen, Paul S. *The Handbook to Bible Study: With a Guide to the Scofield Study System*. Oxford University Press, 1987.

Keddiee, Nikki R., ed. *Scholars, Saints, and Sufis: Muslim Religious Institutions in the Middle East since 1500*. University of California Press, 1976.

Keene, M. Lamar. *The Psychic Mafia*. Prometheus, 1997.

Keiser, Jacqueline L. *The Anatomy of Illusion: Religious Cults and Destructive Persuasion*. Thomas, 1987.

Kelly, J. N. D. *The Oxford Dictionary of the Popes*. Oxford University Press, 1989.

Kendrick, T. D. *The Druids: A Study of Keltic Prehistory*. Barnes & Noble, 2003.

Kennelly, Karen, ed. *American Catholic Women: A Historical Exploration*. Macmillan, 1989.

Kenrick, Donald, and Gratton Puxon. *Destiny of Europe's Gypsies*. Basic Books, 1972.

Kibinge, Maturi. *The Christian Police*. Centurion, 2002.

King, Francis. *Ritual Magic in England*. Neville Spearman, 1970.

King, Peter J. *One Hundred Philosophers*. Barron's Educational, 2004.

Kinsley, David. *Tantric Visions of the Divine Feminine: The Ten Mahavidyas*. University of California Press, 1997.

Kitagawa, Joseph M. *Religion in Japanese History*. Columbia University Press, 1966.

———, ed. *Buddhism and Asian History*. Macmillan, 1987.

Klein, Julie Thompson. *Interdisciplinarity: History, Theory, and Practice*. Wayne State University Press, 1990.

Koch, Klaus. *The Prophets: The Assyrian Period*. Fortress, 1982.

Kohn, Michael H. *The Shambhala Dictionary of Buddhism and Zen*. Shambhala, 1991.

Kramer, Joel, and Diana Alstad. *The Guru Papers: Masks of Authoritarian Power*. North Atlantic, 1993.

La Barre, Weston. *The Peyote Cult*. Yale University Publications, 1938.

———. *They Shall Take Up Serpents*. University of Minnesota Press, 1962.

La Vey, Anton Szandor. *The Satanic Bible*. Hearst Corporation, 1969.

Larue, Gerald A. *The Supernatural, the Occult, and the Bible*. Prometheus, 1990.

Leary, Timothy, Ralph Metzner, and Richard Alpert. *The Psychedelic Experience: A Manual Based on the Tibetan Book of the Dead*. University Books, 1964.

Leslie, Charles M., ed. *Anthropology of Folk Religion*. Vintage, 1960.

Levine, Lee I. *The Ancient Synagogue: The First Thousand Years*. Yale University Press, 2000.

Leví-Strauss, Claude. *The Savage Mind*. University of Chicago Press, 1962.

Levy, G. Rachel. *Religious Conceptions of the Stone Age*. Harper Torchbooks, 1963.

Lewis, James R., ed. *The Gods Have Landed: New Religions from Other Worlds*. State University of New York Press, 1995.

Liebeschuetz, J.H.W.G. *Continuity and Change in Roman Religion*. Clarendon, 1979.

Lowenkopf, Anne N. *The Hassidim*. Sherbourne, 1971.

MacArthur, John. *Alone With God*. Victor, 1995.

Maccoby, Hyam. *Judas Iscariot and the Myth of Jewish Evil*. Free Press, 1992.

MacNulty, W. Kirk. *Freemasonry: A Journey Through Ritual and Symbol*. Thames & Hudson, 1999.

Manseau, Peter. *Rag and Bone: A Journey Among the World's Holy Dead*. Henry Holt, 2009.

Marcus, Ivan G. *Rituals of Childhood: Jewish Acculturation in Medieval Europe*. Yale University Press, 1996.

Maringer, J. *The Gods of Prehistoric Man*. Weidenfeld & Nicolson, 1960.

Marty, Martin E., ed. *Civil Religion, Church, and State*. K. G. Saur, 1992.

Masuzawa, Tomoko. *The Invention of World Religions, or How the Idea of European Hegemony Came to be Expressed in the Language of Pluralism and Diversity*. University of Chicago Press, 2005.

Mathews, Warren. *World Religions*. West, 1991.

Matt, Daniel C. *The Essential Kabbalah: The Heart of Jewish Mysticism*. Castle, 1995.

Matthews, Caitlin. *The Elements of the Celtic Tradition*. Element, 1989.

Mbiti, John. *African Religions and Philosophy*. Heinemann, 1990.

McBrien, Richard P. *The Saints*. HarperCollins, 2006.

McCutcheon, Russell T. *Manufacturing Religion: The Discourse on Sui Generis Religion and the Politics of Nostalgia*. Oxford University Press, 1997.

McDonald, H. D. *Theories of Revelation: An Historical Study 1700–1960*. Baker, 1979.

McKenna, Christina, and David M. Kiely. *The Dark Sacrament*. HarperOne, 2007.

McNamara, Jo Ann, et al. *Sainted Women of the Dark Ages*. Duke University Press, 1992.

Melton, J. G. *Encyclopedic Handbook of Cults in America*. Garland, 1992.

Meyer, Marvin, ed. *Ancient Christian Magic: Coptic Texts of Ritual Power*. Harper, 1994.

Miller, Timothy, ed. *America's Alternative Religions*. State University of New York Press, 1995.

Mills, Jeannie. *Six Years with God: Life Inside Jim Jones' People's Temple.* A&W, 1979.

Mir, Mustansir. *Dictionary of Qur'anic Terms and Concepts.* Garland, 1987.

Mooney, James. *The Ghost-Dance Religion and the Sioux Outbreak of 1890.* University of Chicago Press, 1969.

Moorman, J. R. H. *A History of the Church in England.* Black, 1973.

Morenz, Sigfried. *Egyptian Religion.* Cornell University Press, 1973.

Morris, Brian. *Anthropological Studies of Religion: An Introductory Text.* Cambridge University Press, 1988.

Most, William. *Free from All Error: Authorship, Inerrancy, Historicity of Scripture, and Modern Scripture Scholars.* Franciscan Marytown Press, 1985.

Munn, Nancy D. *Walbiri Iconography: Graphic Representation and Cultural Symbolism in a Central Australian Society.* University of Chicago Press, 1986.

Narby, Jeremy. *Shamans Through Time.* Penguin, 2001.

Negev, Avraham, ed. *The Archaeological Encyclopedia of the Holy Land.* Thomas Nelson, 1986.

Nelsen, Hart M. *The Black Church in America.* Basic Books, 1971.

Newby, Gordon. *The Making of the Last Prophet: A Reconstruction of the Earliest Biography of Muhammad.* University of South Carolina Press, 1989.

New York Times. "Skoptsy Members on Trial; Russia Trying Hard to Suppress an Extraordinary Sect." October 13, 1910.

Nicholson, Shirley, ed. *The Goddess Re-Awakening: The Feminine Principle Today.* Quest, 1989.

Nickell, Joe. *Looking for A Miracle: Weeping Icons, Relics, Stigmata, Visions and Healing Cures.* Prometheus, 1993.

Nigosian, A. S. *World Faiths.* St. Martin's, 1994.

Nirenberg, David. *Communities of Violence: Persecution of Minorities in the Middle Ages.* Princeton University Press, 1996.

Nock, A. D. *Conversion: The Old and New in Religion from Alexander the Great to Augustine.* Johns Hopkins University Press, 1998.

———. *Essays on Religion and the Ancient World.* Harvard University Press, 1972.

Northrup, Bernard E. *True Evangelism: Paul's Presentation of the First Five Steps of the Soul-Winner in Romans*. Oxford University Press, 1997.

Numbers, Ronald L. *Prophetess of Health: Ellen G. White and the Origins of Seventh-Day Adventist Health Reform*. University of Tennessee Press, 1992.

Oberman, Heiko A., ed. *Forerunners of the Reformation: The Shape of Late Medieval Thought*. Holt, Rinehart, & Winston, 1966.

Oden, Thomas. *Ancient Christian Commentary on Scripture*. Intervarsity, 1998.

Packer, J. I. *Concise Theology: A Guide to Historic Christian Beliefs*. Tyndale House, 1995.

Paden, William E. *Religious Worlds: The Comparative Study of Religion*. Beacon, 1988.

Palmer, Helen. *The Enneagram*. Harper & Row, 1988.

Palmer, Susan. *Moon Sisters, Krishna Mothers, Rajneesh Lovers: Women's Roles in New Religions*. Syracuse University Press, 1994.

Paris, Charles W. *Biblical Catechetics After Vatican II*. Liturgical Press, 1971.

Parker, William, et al., eds. *Celtic Christianity: Ecology and Holiness*. Lindisfarne, 1982.

Parrinder, Geoffrey. *Mysticism in the World's Religions*. Oneworld, 1995.

———. *Religion in Africa*. Penguin, 1969.

Pearson, Birger A. *Gnosticism, Judaism and Egyptian Christianity*. Fortress, 1990.

Peck, M. Scott. *The Road Less Traveled*. Simon & Schuster, 1978.

Peckham, B. *History and Prophecy: The Development of Late Judean Literary Traditions*. Doubleday, 1993.

Peloubet, F. F. *Perloubet's Bible Dictionary*. John C. Winston, 1947.

Pennington, Brian K. *Was Hinduism Invented? Britons, Indians, and the Colonial Construction of Religion*. Oxford University Press, 2007.

Peters, Francis E. *Muhammad and the Origins of Islam*. State University of New York Press, 1994.

Pettegrew, Andrew, ed. *The Early Reformation in Europe*. Cambridge University Press, 1997.

Pink, Arthur Walkington. *Why Four Gospels?* Logos Research Systems, 1999.

Porgoff, Ira. *The Cloud of Unknowing*. Doubleday, 1989.

Posy, Arnold. *Mystic Trends in Judaism*. Jonathan David, 1966.

Prawer, Joshua. *The History of the Jews in the Latin Kingdom of Jerusalem*. Oxford University Press, 1988.

Rahner, Karl. *Visions and Prophecies*. Herder & Herder, 1963.

Randi, James. *The Faith Healers*. Prometheus, 1987.

Redford, Donald B., ed. *Oxford Encyclopedia of Ancient Egypt*. Oxford University Press, 2002.

Reuther, Rosemary, and Rosemary Keller, eds. *Women and Religion in America*. Harper & Row, 1981.

Richards, Jeffrey. *Sex, Dissidents and Damnation: Minority Groups in the Middle Ages*. Routledge, 1990.

Robbins, Rossell H. *The Encyclopedia of Witchcraft and Demonology*. Crown, 1966.

Roberts, Alexander, and Sir James Donaldson. *The Ante-Nicene Fathers: The Writings of the Fathers Down to A.D. 325*. Eerdmans, 1975.

Robinson, Haddon W. *Biblical Preaching: The Development and Delivery of Expository Messages*. Baker, 1980.

Robinson, James B. *Buddha's Lions: The Lives of the Eighty-Four Siddhas*. Dharma, 1980.

Rodwell, John Medows, ed. *The Koran*. Bantam, 2004.

Rudin, A. James, and Marcia Rudin. *Prison or Paradise: The New Religious Cults*. Fortress, 1980.

Sagan, Carl. *The Demon-Haunted World: Science as a Candle in the Dark*. Random House, 1995.

Sasson, H. H. *A History of the Jewish People*. Harvard University Press, 1976.

Sawyer, John F. A. *Prophecy and the Biblical Prophets*. Oxford University Press, 1993.

Schiffman, Lawrence H. *Reclaiming the Dead Sea Scrolls: The History of Judaism, the Background of Christianity, and the Lost Library of Qumran*. Jewish Publication Society, 1994.

Scholem, Gershom. *Kabbalah*. Times Books, 1976.

Schreck, Alan. *Catholic and Christian: An Explanation of Commonly Misunderstood Catholic Beliefs*. Servant, 1984.

———. *The Compact History of the Catholic Church*. Servant, 1987.

Schulze, Reinhard. *A Modern History of the Islamic World*. New York University Press, 2000.

Sell, Henry Thorne. *Studies in Early Church History*. Woodlawn Electronic Publishing, 1998.

Sells, Michael. *Approaching the Qur'an: The Early Revelations*. White Cloud, 1999.

Shah, Idries. *Wisdom of the Idiots*. Octagon, 1991.

Sharma, Arvind, and Katherine K. Young, eds. *Feminism and World Religions*. State University of New York Press, 1999.

Shermer, Michael. *Why People Believe Weird Things: Pseudoscience, Superstition, and Other Confusions of Our Time*. Freeman, 1997.

Shipps, Jan. *Mormonism: The Story of a New Religious Tradition*. University of Illinois Press, 1985.

Silver, Eric. *Begin: The Haunted Prophet*. Random House, 1984.

Simoons, Frederick J. *Eat Not This Flesh: Food Avoidances in the Old World*. University of Wisconsin Press, 1971.

Smalley, Beryl. *The Study of the Bible in the Middle Ages*. University of Notre Dame Press, 1973.

Smart, Ninian. *Dimensions of the Sacred: An Anatomy of the World's Beliefs*. University of California Press, 1996.

———. *The Religious Experience of Mankind*. Scribner's, 1984.

Smith, D. Howard. *Chinese Religions*. Holt, Rinehart, & Winston, 1971.

Smith, Huston. *Buddhism*. HarperCollins, 2004.

———. *The World's Religions*. HarperCollins, 1991.

Smith, James E. *The Major Prophets*. College Press, 1992.

———. *The Minor Prophets*. College Press, 1992.

Smith, Jonathan Z. *Guide to the Study of Religion*. Continuum, 2000.

———, ed. *The HarperCollins Dictionary of Religion*. HarperCollins, 1995.

Southern, R. W. *Western Society and the Church in the Middle Ages*. Penguin, 1970.

Spitz, Lewis W. *The Protestant Reformation, 1517–1547*. Harper & Row, 1985.

Stark, Rodney, ed. *Religious Movements: Genesis, Exodus, and Numbers*. Paragon House, 1987.

Stein, Gordon, ed. *The Encyclopedia of the Paranormal*. Prometheus, 1996.

Stein, Stephen J. *Communities of Dissent*. Oxford University Press, 2003.

Steinmueller, John E. *Some Problems of the Old Testament*. Bruce, 1936.

Steinsaltz, Adin. *The Essential Talmud*. Jason Aronson, 1994.

Stent, W. R. *An Interpretation of a Cargo Cult*. Oceania, 1976.

Stewart, Gary. *Basic Questions on Suicide and Euthanasia: Are They Ever Right?* Kregel, 1998.

Stewart, Gary P., et al. *Basic Questions on End of Life Decisions*. Kregel, 1998.

Stone, Merlin. *When God Was a Woman*. Harcourt Brace, 1976.

Story, Dan. *Christianity on the Offense: Responding to the Beliefs and Assumptions of Spiritual Seekers*. Kregel, 1998.

Strong, James. *The Exhaustive Concordance of the Bible*. Woodside Bible Fellowship, 1996.

Stutley, Margaret. *Shamanism*. Routledge, 2003.

Sumption, Jonathan. *The Age of Pilgrimage: The Medieval Journey to God*. HiddenSpring, 2003.

Swanson, James. *Dictionary of Biblical Languages With Semantic Domains: Greek (New Testament)*. Logos Research Systems, 1997.

Taguchi, Paul Cardinal. *The Study of Sacred Scripture*. Daughters of St. Paul, 1974.

Telushkin, Joseph. *Jewish Literacy: The Most Important Things to Know About the Jewish Religion, Its People and Its History*. William Morrow, 1991.

Terry, Milton S. *Biblical Hermeneutics: A Treatise on the Interpretation of the Old and New Testaments*. Phillips & Hunt, 1883.

Terwiel, B. J. *Monks and Magic*. Craftsman, 1976.

Tester, S. J. *A History of Western Astrology*. Boydell, 1987.

Tooker, Elizabeth. *The Iroquois Ceremonial of Midwinter*. Syracuse University Press, 1971.

Torrey, R. A. *Difficulties in the Bible: Alleged Errors and Contradictions*. Woodlawn Electronic, 1998.

Trachtenberg, Joshua. *The Devil and the Jews: The Medieval Conception of the Jew and Its Relation to Modern Anti-Semitism*. Jewish Publication Society of America, 1983.

Treadway, Scott, and Linda Treadway. *Ayurveda and Immortality*. Celestial Arts, 1977.

Tripōlitis, Antonia. *Religions of the Hellenistic-Roman Age*. Eerdmans, 2002.

Valantasis, Richard. *Religions of Late Antiquity in Practice*. Princeton University Press, 2000.

Vauchez, André. *Sainthood in the Later Middle Ages*. Cambridge University Press, 1987.

Vogt, E. Z., and R. Hyman. *Water-Witching*. University of Chicago Press, 1978.

Vollmar, Edward R. *The Catholic Church in America: An Historical Bibliography*. Scarecrow, 1963.

Waldram, James B. *The Way of the Pipe: Aboriginal Spirituality and Symbolic Healing*. Broadview, 1997.

Walton, Robert C. *Chronological and Background Charts of Church History*. Academie, 1986.

Watts, Alan W. *The Way of Zen*. Random House, 1957.

Webb, Diana. *Medieval European Pilgrimage, C.700–C.1500*. Macmillan, 2002.

Werblowsky, R. J. Zwi. *Man, Myth, and Magic: An Illustrative Encyclopedia of the Supernatural*. Marshall Cavendish, 1970.

West, Martin L. *Early Greek Philosophy and the Orient*. Clarendon, 1971.

Whitaker, Richard E. *The Eerdmans Analytical Concordance to the Revised Standard Version of the Bible (with the Deuterocanonical/Apocryphal Books)*. Eerdmans, 1988.

Whiteford, John. *Sola Scriptura: An Orthodox Analysis of the Cornerstone of Reformation Theology*. Conciliar, 1995.

Wild, Robert. *The Post Charismatic Experience*. Living Flame Press, 1984.

Wilson, Bryan. *Religious Sects: A Sociological Study*. McGraw-Hill, 1970.

———. *Sects and Society*. University of California Press, 1961.

Wilson, Colin. *The Occult: A History*. Random House, 1971.

Wilson, Liz. *Charming Cadavers: Horrific Figurations of the Feminine in Indian Buddhist Hagiographic Literature*. University of Chicago Press, 1996.

Wink, Walter. *Jesus and Non-violence*. Fortress, 2005.

Winstead, Karen A. *Chaste Passions: Medieval English Virgin Martyr Legends*. Cornell University Press, 2000.

Wolf-Salin, Mary. *No Other Light: Points of Convergence in Psychology and Spirituality*. Crossroad, 1986.

Wood, D. R. W. *New Bible Dictionary*. Intervarsity, 1996.

ACKNOWLEDGMENTS

Much thanks to HarperCollins, namely Carrie Kania, Cal Morgan, Kolt Beringer, Justin Dodd, Jennifer Hart, and Amanda Kain. Sincere and special thanks to my editor, Peter Hubbard. I also thank Frank Wiemann of the Literary Group International.

PHOTOGRAPHIC CREDITS

Absalom: Julius Schnorr Von Karolsfeld; Adamites: Exhibit Supply Co.; Adamites: Gardiner Greene Hubbard Collection; St. Agatha: Francisco de Zurbarán; St. Agatha: National Library of Medicine; Ahab: Gustave Doré; Afterlife: Currier & Ives; Afterlife: W. J. Morgan & Co.; Sister Aimee: Library of Congress; Megachurch: Albert Levering; Akashic Record: Theodor Horydczak Collection; Nadi Leaves: Courtesy www.indiadestinationtours.com; Palm Reader: Library of Congress Prints and Photographs Division; St. Albans: Matthew Paris; Martyr Factory: Pieter Bruegel; Amen: American Stereoscopic Company; Alhamdulillah: The Courier Company; Angels: Thomas Kelly; Cherubs: L. Prang & Co.; Angels: American Lithographic Co.; Antichrist: U.S. Lithograph Co.; John Parsons: NASA; Anticlericalism: Library of Congress Prints and Photographs Division; Merry Martyr: Courtesy Society of Jesuits; Aphrodite: Detroit Publishing Co.; Apocalypse: Office of War Information, Overseas Picture Division; Four Horsemen of the Apocalypse: Albrecht Durer; Apostate: State Historical Society of Colorado; Apotheosis: Library of Congress Prints and Photographs Division; Arius: Courtesy of the Arian Church; Thomas Aquinas: Fra Angelico; Shoko Asahara: Associated Press; Ash Wednesday: Caldwell & Co.; Astrology: Sidney Hall; Jeanne Dixon: Jeanne Dixon Museum and Library; Astrology: Theodor Horydczak Collection; Sri Aurobindo: Library of Congress Prints and Photographs Division; Atlantis: Ignatius Donnelly; Atlantis: Bernard Partridge; Bab:

Theodor Horydczak Collection; Bab: *New York World*; Babel: Charles Mavrand; Noah: Gibson & Co.; Roger Bacon: Michael Maier; Bullard: Theo Buerbaum; Moses: Julius Schnorr von Carolsfeld; Modern Weather: WPA Federal Art Project; St. Barbara: Library of Congress Prints and Photographs Division; St. Barbara: Frederick Heppenheimer; Bede: Lamb Design Collection; Beguines: Currier & Ives; Beguines: National Photo Company Collection; Conrad Beissel: Works Projects Administration Poster Collection; David Berg: xfamily.org; Bible: George Grantham Bain Collection; Bitter Herbs: Library of Congress Prints and Photographs Division; Bitter Herbs: Currier & Ives; Black Magic: Carl Guttenberg; Black Mass: Henry Chaprout; Madame Blavatsky: George Grantham Bain Collection; Bloody Mary: August F. Jaccaci Company; Bullard: Shober Carqueville; John Calvin: John Sartain; Calvin: Yorkshire and Humber; Count Cagliostro: Wikimedia Foundation; Ouroboros: Theodor Johann; Cain: Gustave Doré; Canonization: R. J. Stock; Catherine of Siena: Giovanni Battista Tiepolo; Padre Pio: Courtesy The Padre Pio Foundation of America; Cayce: Courtesy the Edgar Cayce Institute for Intuitive Studies; Cayce: G. Eric and Edith Matson Photograph Collection; Church of the Universe: Courtesy the Church of the Universe, Reverend Brother Michael J. Baldasaro; Circuit Rider: Alfred Waud; Ark of Covenant: American Colony Jerusalem Photo Dept.; Ark of Covenant: Julius Schnorr von Carolsfeld; Creationism: Verdict Publishing Company; Creationism: Beck & Lawton; Cristo Redentor: Richard Lynch (www.richard-lynch.co.uk); Aleister Crowley: Wide World Photos; Crusades: Romeyn de Hooghe; Crystal Ball: Library of Congress Prints and Photographs Division; Crystal Ball: Augustino Patheo; Cult of Personality: Library of Congress; C.U.T.: Office of War Information; C.U.T.: Tancrede Dumas; Santo Daime: www.santodaime.org; King David: Courtesy Bizzell Bible Collection, University of Oklahoma Libraries; King David: Underwood & Underwood; Day of the Dead: Antonio Vanegas Arroyo; Deism: Kimmel & Forster; Devils of Loudon: Fritz W. Guerin; Dionysos: Matson Photo Service; Divine Hair: Library of Congress Prints and Photographs Division; Divining Rod: George Grantham Bain Collection; Order of the Doble-Cruz: Francisco de Paula Marti; Chalice:

Charles R. Lamb; St. Dominic: Detroit Publishing Co.; Dreamtime: Wm. H. Jackson; Drogo: American Red Cross Collection; Druids: Currier & Ives; Druids: Julius Gipkens; Dunkards: Alan Lomax; Eastern Orthodox Churches: Underwood & Underwood; Eastern Orthodox Churches: Beck & Pauli; Eckankar: Harris & Ewing Collection; Ectonic Forces: Strobridge Lith. Co.; Ectoplasm: John Massy Wright; Mary Baker Eddy: Library of Congress Prints and Photographs Division; Ecstasy: John William Gibson; Eden: Currier & Ives; Eden: Oliver Wendell Holmes Collection; Efi and Efo: Frank and Frances Carpenter Collection; Eight-Fold Path: Theodor Horydczak Collection; Thich Quang Duc: *New York World-Telegram*; Eighth Sphere: William Blake; Elan Vital: Frank Leslie; Elijah, Muhammad: *New York World-Telegram* and the Sun Newspaper Photograph Collection; Malcolm X: Marion Trikosko; Enneagram: Courtesy www.wishfulthinking.co.uk; Epicurus: Carroll University; Episcopalian: Cornelis Vermeulen; Essenes: Matson Photo Service; Essenes: Frank and Frances Carpenter Collection; EST: Mayer, Merkel & Ottmann; Church of Euthanasia: Horydczak Collection; Evangelical: Office of War Information; Eve: Currier & Ives; Eve: Walter Shrirlaw; Evil Eye: Lewis Marks; St. Expedite: Donald H. Cresswell; St. Fiacre: William Dent; Faustus: Antonio Vanegas Arroyo; Flagellants:Wm. Leach; Mortification of the Flesh: John June; Nicolas Flamel: M. S. Manservant; George Fox: Lehman & Duval; Fox Oatmeal: British Cartoon Prints Collection; Fox, Margaret and Kate: *New York World Telegram;* Freemasonry: Strobridge Lith. Co; Gandhi: George Mason University; Bindi: Frank and Frances Carpenter Collection; St. Genesius: Detroit Publishing Co.; Genevieve: Puvis de Chavannes; Count of St. Germain: Donald H. Cresswell; Gideon: Harris & Ewing Collection; Gnosticism: Camille Flammarion; Falun Gong: A. B. Frost; Sweet Daddy Grace: Cape Verdian Society; Gurdjieff: Janet Flanner-Solita Solano; St. Guy: WPA Federal Art Project; Hamasta: Edward S. Curtis; Serpent Handler: Joe Geruero; Haruspicy: Office of War Information; HHH: W. J. Morgan & Co.; Heaven's Gate: Rick A. Ross Institute; Hermes: Theodor Horydczak Collection; L. Ron Hubbard: A. Z. Baker; I AM: Henry Fueseli; Idolatry: Daniel Beard; Stoning: *Daily Mail*; Ifa: Coutinho & Sons; Immaculate Conception:

Donaldson Brothers; Imam: Joseph Hallworth; Imp: Russell Lee; Incorruptible Saints: G. Eric and Edith Matson Photograph Collection; Indulgences: H. Humphrey; Way International: Alex Spade; Inquisition: Pieter Buregal; William Irvine: George Grantham Bain Collection; Isaiah: Benjamin West; St. Issa: American Colony; Jainism: Wm. H. Jackson; Jívaro: Frank and Frances Carpenter Collection; Jesus: Thomas B. Noonan; Job: William Blake; John Frum: *Harper's Weekly*; Frum: F. W. Taylor; Jonah: J. H. Parker; John the Baptist: Gustave Doré; Jim Jones: Library of Congress Prints; Judas: American Colony; KKK: National Photo Company Collection; KKK: *Harper's Weekly*; Kaaba: Matson Collection; Kabbalah: G. Eric and Edith Matson Photograph Collection; Kama Sutra: Matson Collection; Karma: Wm. H. Jackson; Mrs. Keech: Library of Congress Prints; Kemetic: Detroit Publishing Co.; David Koresh: Sheriff's Office, Waco, Texas; Kosher Chicken: Marjory Collins; Hare Krishna: Keystone View Company; Lalibela: Giustino; Dalai Lama: George Grantham Bain Collection; Holy Laughter: Federal Art Project; Eliphas Levi: Dartmouth University; Levitating Saints: Ann Rosener; St. Lawrence: Free Artists Software; Left-Handed Path: J.M.W. Jones S. & P. Co.; St. Lidwina: Edwin Osscam; Limbo: *Harper's Weekly*; Lot's Wife: Library of Congress Prints; Lady of Lourdes: Detroit Publishing Co.; St. Lucy: Dominico Venezianio; Jesús Malverde: World Politics Review; Mandaeanism: James Fuller Queen; Charles Manson: AP; Margaret of Antioch: Catholic Forum; Margaret of Hungary: Michelangelo Caravaggio; Marian Apparitions: eBay; Conrad of Marburg: G. Barrie & Son; Masai: Frank and Frances Carpenter Collection; Karni Mata: Alfred T. Palmer; Mesmerism: What Cheer Show Print; Millerites: Thomas S. Sinclair; Merlin: Julia Margaret Cameron; Miracles: Gottscho-Schleisner Collection; Sun Myung Moon: Underwood & Underwood; Mother Ann: Library of Congress; Shaker: Samuel Kravitt; Mother Teresa: Curtis & Cameron; Mozi: Yoshiume Utagawa; Mummy: National Photo Company; Native American Church: John Collier; Nebuchadnezzar: William Blake; New Aeon: Charles Wills; Chaos: M. Leone Bracker; Isaac Newton: James McArdell; Council of Nicaea: A. Bryer and J. Herrin; St. Nick: Horydczak Collection; Nostradamus: Claude Alexander; John

Humphrey Noyes: Syracuse University; Onan: John Boydell; Oomoto: Wikipedia; Orenda: Wm. A. Drennan; Oxford Group: Library of Congress Prints and Photographs Division; St. Paul: University of St. Thomas; William Pelley: Harris & Ewing Collection; Pentecostalism: Russell Lee; Phallic Cults: Wiesman Collection; Pharmacologism: Library of Congress; Pilgrimage: Hodge & Sampson; Pieter Plockhoy: British Cartoon Prints Collection; St. Pius X Society: Library of Congress Prints and Photographs Division; Pythagoras: Library of Congress Prints and Photographs Division; Ouija: National Photo Company Collection; Phineas Parkhurst Quimby: Belfast Historical Society & Museum; Qur'an: Library of Congress Prints and Photographs Division; Raëlism: Benda Wladyslaw; Paschal B. Randolph: Library of Congress Prints and Photographs Division; Rasputin: McManus-Young Collection; Rastafarian: American Colony; Ravens: French Political Cartoon Collection; Ritual Killings: A. Welker; Roma: First National Pictures; Rosicrucians: Gravf von Lohrbach; Ruth: Matson Photo Service; Maria Sabina: Alvaro Estrada–J. D. Cress; Sadhus: Wm. Henry Jackson; Salvation Army: Falk Co.; Samson: Gerard Hoet; Santería: Frank and Frances Carpenter Collection; Satan: John L. Magee; Science of the Mind: National Library of Medicine; Seventh-Day Adventists: Dorothea Lange; Tungus Shamans: Wm. H. Jackson; Shakti: Samuel Erhart; Holy Sepulcher: W. Hammerschmidt; True Cross: Currier & Ives; Mother Shipton: British Cartoon Prints Collection; Six-Six-Six: *Harper's Weekly*; Simeon of Stylites: National Photo Company Collection; Joseph Smith, Jr.: Library of Congress Prints and Photographs Division; Skoptsy: Francisco de Quevedo; Soul's Weight: Russell Lee; Spafford: American Colony; Sun gazing: National Photo Company Collection ; Swedenberg; G. M. Woodruff; Tantra: National Library of Medicine; Tarot: B. Robinson; St. Teresa of Ávila: Giovanni Lorenzo Bernini; Technopaganism; Harris & Ewing Collection; Telekinesis: Jack Delano;Televangelists: John T. McCutcheon; Thuggee Cult: Underwood & Underwood; Johannes Trithemius: Johannes Busaeus; Shroud of Turin: Sergei Mikhailovich Prokudin–Gorskii Collection; Unicorns: British Cartoon Prints Collection; Unarius Science:Wm. H. Jackson; Unitarian Universalism: Unitarian Universalist Association;

Universal Life Church: *Modesto Bee*; Edward S. Curtis; Vestal Virgins: H. A. Thomas & Wylie Lith. Co.; Whirling Dervishes: Bain News Service; Pascal Sebah; Wicca: Lindsay and Blakiston; Witch Trials: Library of Congress Prints and Photographs Division; Jehovah's Witnesses: George Grantham Bain Collection; X-Power: Library of Congress Prints; Xenophanes: Brooklyn College; Yellow Bamboo: Kobayashi Kiyochika; Malachi Z. York: Courtesy Support The Master Teacher Dr. Malachi Z. York Fund; Yin and Yang: Wikipedia; Zen: David Murray Collection; St. Zita: Library of Congress Prints and Photographs Division; Zoroastrianism: Conference on Zoroastrism.